T0270954

THE ECONOMICS OF
ELECTRICITY MARKETS

THE ECONOMICS OF ELECTRICITY MARKETS

Darryl R. Biggar
Australian Competition and Consumer Commission, Melbourne, Australia

Mohammad Reza Hesamzadeh
KTH Royal Institute of Technology, Stockholm, Sweden

IEEE PRESS

WILEY

A co-publication of IEEE Press and John Wiley & Sons Ltd

This edition first published 2014
© 2014 John Wiley & Sons Ltd

Registered office
John Wiley & Sons Ltd, The Atrium, Southern Gate, Chichester, West Sussex, PO19 8SQ, United Kingdom

For details of our global editorial offices, for customer services and for information about how to apply for permission to reuse the copyright material in this book please see our website at www.wiley.com.

Library of Congress Cataloging-in-Publication Data

Biggar, Darryl R. (Darryl Ross)
  The economics of electricity markets / Darryl R Biggar, Mohammad Reza Hesamzadeh.
    pages cm
  ISBN 978-1-118-77575-2 (hardback)
1. Electric power consumption. 2. Electric power–Economic aspects. 3. Electric utilities. I.
Hesamzadeh, Mohammad Reza. II. Title.
  HD9685.A2B54 2014
  333.793′2–dc23

                                                                                           2014014039

A catalogue record for this book is available from the British Library.

ISBN 9781118775752

Set in 10/12pt TimesLTStd-Roman by Thomson Digital, Noida, India.

1 2014

Contents

Preface

Around the world, the electricity industry is in the process of undergoing a fundamental transition. Twenty years ago, electricity was primarily generated at large, industrial-scale generating plants, and transported in one direction to consumers via the transmission and distribution networks. The large generators were typically closely integrated into the operation of the transmission and distribution networks. Electricity consumers, on the other hand, were treated as essentially passive.

This paradigm has changed and will change further. Around the world, a number of regions have chosen to introduce competition and competitive markets into the generation of electricity. In most of these regions, the operation of generation and transmission is coordinated through market mechanisms. This required a substantial change in the way the electricity industry is organised and operated.

However, further transformations are underway. With increasing pressure for decarbonisation of the energy sector, there is increasing penetration of renewable generation and increasing take-up of electric vehicles. Changes in battery technology threatens to substantially change the way electricity is stored and consumed. Just as importantly, the IT and communication revolutions have opened up the scope for a host of new devices and appliances, allowing small-scale consumers for the first time to respond to local electricity market conditions.

The full benefits of these developments will only be achieved if the electricity industry completes its transition. From a paradigm of one-way managed electricity supply, the electricity industry is transforming to a new service model. In this new paradigm the industry exists to provide a platform for the two-way trade of electricity, with all customers large and small, integrated with and responding to local market conditions. This is an exciting time to be studying the electricity industry.

In preparing this text, we found three themes that were important and that shaped the material. The first of these was *symmetry between generators and loads*. In the future, the historic distinction between electricity producers and electricity consumers will diminish. It seems likely to us that an increasing number of participants in the electricity industry will be able to produce and consume electricity, switching between net injection and net withdrawals from the system according to local market conditions. In such a world, it seems to us essential that there be symmetry in the treatment of generators and loads. Rather than distinguishing generators and loads, we prefer to view them all as electricity market participants. The key distinction that will remain is not between generators and loads but between large and small market participants.

In a similar manner, we have also actively avoided any distinction between transmission and distribution networks. Although there are real differences in their construction and operation, these differences do not seem to us as fundamentally important for the economic analysis of electricity markets. There are only networks, and the physical limits that those network impose on power flows.

The second major theme was the importance of *understanding over sophisticated modelling*. We have sought to highlight key principles and to develop understanding. For this purpose, we have used the simplest possible models and examples wherever possible. We have avoided complex sets of equations or complex models wherever possible, even at the sacrifice of realism. In our view, a small amount of understanding is worth a large amount of sophisticated black-box modelling.

Consistent with this approach, we have not always sought to accurately capture every element of real-world electricity networks. For example, for much of this text, electricity losses have been ignored. Some engineers may be troubled by this. Nevertheless, we consider that the benefits of clarity and simplicity of presentation outweigh the additional complexity of modelling losses in every model. In our view, when it comes to learning about the electricity industry, understanding is more important than sophisticated modelling.

The third major theme was *consistency of economic approach*. We have sought to set out a consistent, coherent, economic approach to electricity markets, with reliance as far as possible on price signals and market-based incentives. In practice, every real electricity market of which we are aware is some distance away from this theoretically ideal model. Real-world electricity markets tend to be a patchwork of compromises, approximations and *ad hoc* interventions. Some of those interventions may be justified. However, we are concerned that often compromises are made due to a lack of understanding or fear of the theoretically ideal approach. We consider there is considerable value in setting out a thorough-going economic approach.

There will be debates about the extent to which this approach can be implemented in practice. Also, many of the departures from the theoretical framework (such as zonal pricing) are worth studying in their own right. Nevertheless, we consider that students of electricity markets should be exposed to the simplicity and elegance of the theoretically pure approach. One of the important implications of this approach is that reliability – which is traditionally a primary concern of the power system engineer – diminishes in importance. In a thorough-going market approach, reliability disappears as an issue entirely. Prices always adjust to balance supply and demand.

We hope you find your study of the electricity market as fascinating and challenging as we do.

Nomenclature

The following terminology is used in this book:

Symbol	Name	Units	Meaning
Q_i	Rate of production or consumption	units/time interval, kW, or MW	Rate of production of output by producer i or rate of consumption by consumer i
$C_i(Q_i)$	Cost function	\$/time interval	Rate at which costs are incurred by producer i when producing at the rate Q_i (units/time interval)
$MC_i(Q_i) = C_i'(Q_i)$ $= \dfrac{dC_i}{dQ_i}$	Marginal cost function	\$/unit, \$/kWh or \$/MWh	Rate at which costs increase with an increase in the rate of production
$\pi_i(Q_i)$	Profit function	\$/time interval	Rate at which profits are received by producer i when producing at the rate Q_i (units/time interval)
$U(Q)$	Utility function	\$/time interval	Rate at which utility is received when consuming at rate Q (units/time interval)
$\varphi_i(Q_i)$	Net utility function	\$/time interval	Rate at which net utility is received by customer i when consuming at the rate Q_i (units/time interval)
$\pi_i(Q_i)$	Profit function	\$/time interval	Rate at which profit is received by generator i when producing at the rate Q_i (units/time interval)
P	Price	\$/unit, \$/kWh, \$/MWh	Amount of additional expenditure required to acquire and additional unit

(*Continued*)

(*Continued*)

Symbol	Name	Units	Meaning
$Q^D(P)$	Demand curve	units/time interval, kW, MW	Rate of consumption as a function of the linear market price
$P^D(Q)$	Inverse demand curve	$/unit, $/kWh, $/MWh	Price consistent with a given rate of consumption
$P^S(Q)$	Supply curve	$/unit	Price consistent with a given rate of production
$P^{RD}(Q)$	(Inverse) residual demand curve	$/unit, $/kWh, $/MWh	Market price consistent with a given rate of production of a dominant firm. Residual demand curve is equal to market demand curve less supply of other firms.
ε	Elasticity	dimensionless	Elasticity is a measure of the sensitivity of the demand function to the price. For the demand function $Q(P)$ elasticity is defined as: $$\varepsilon = \frac{\mathrm{d}\ln Q}{\mathrm{d}\ln P} = \frac{P}{Q}\frac{\mathrm{d}Q}{\mathrm{d}P}$$
c	Variable cost	$/unit, $/MWh	Variable cost of production (typically assumed to be constant)
K	Productive capacity	units/time interval, kW, MW	Maximum rate of production of a producer
K_{it}	Productive capacity	MW	Maximum rate of production of a generator of type t at node i
f	Cost per unit of capacity	$/unit/time interval, $/MW	Marginal cost of adding an extra unit of capacity
F	Fixed cost of generation	$/time interval	Fixed cost of generation (often equal to fK).
p_i	Probability	dimensionless	Probability of state i occurring
V, I, P	Voltage, current, power	Volts, Amps, Watts	
W, Q, P	Real power, reactive power, apparent power	Watts, VARs and VAs (also kVA and MVA)	
λ	Lagrange multiplier on the energy balance constraint	$/MWh	Often interpreted as the price. Equal to System Marginal Cost (SMC)
μ_i, ν_i	Lagrange multiplier on generator production constraints	$/MW	ν_i is the Lagrange multiplier on the lower bound on production (usually taken as zero); μ_i is the Lagrange multiplier on the upper bound on production (equal to the generator's capacity).

(*Continued*)

(Continued)

Symbol	Name	Units	Meaning
γ	Lagrange multiplier on energy-limit constraint	\$/MWh	
N_j	Partition of the set of producers and consumers	Set	This is a partition, so $N_i \cap N_j = \varnothing$ and $\cup N_j = N$
$Q^{LD}(z)$	Load duration curve	MW	For any fraction z, the level of demand q for which $\Pr(Q \geq q) = z$
σ_t	Duration of the tth interval	Hours	
P^{DA}, Q^{DA}, P^{RT} and Q^{RT}	Day ahead and real-time prices and quantities	\$/MWh, MW	
$f(q), F(q) = \Pr(Q \leq q)$	Probability density function for demand	Probability	When demand is treated as a random variable, the probability density function determines the shape of the load–duration curve
$org(l),\ term(l)$	Originating and terminating nodes for link l	Node	Network link l joins node $org(l)$ to node $term(l)$
Z_i	Net injection to node i	MW	Net injection is the local production less the local consumption
F_l	Flow on link l	MW	
K_l	Maximum flow on link l	MW	
μ_l	Constraint marginal value for link l	\$/MW/time interval	
MS	Merchandising surplus	\$/time interval	$MS = -\sum_i P_i Z_i$
CR_l	Congestion rents associated with link l	\$/time interval	$CR_l = \mu_l K_l$
SR_l	Settlement residues associated with link l	\$/time interval	$SR_l = (P_j - P_i) F_l$
$W(K_l)$	Economic welfare	\$/time interval	The overall economic welfare for a network of a given configuration and network flow limits K_l
$H(P, \varepsilon)$	Hedge contract	\$/time interval	
$V(P, \varepsilon)$	Volume associated with a hedge contract	MW	$V(P, \varepsilon) = \frac{\partial H}{\partial P}(P, \varepsilon)$

(Continued)

(*Continued*)

Symbol	Name	Units	Meaning
$\text{FTR}(P_i, P_j, V)$	FTR hedge contract	\$/time interval	Payout on a FTR hedge contract from a node with price P_i to a node with price P_j and a volume V. $\text{FTR}(P_i, P_j, V) = (P_j - P_i)V$
$\text{CapFTR}(P_i, P_N, S, V)$	Cap FTR hedge contract	\$/time interval	Payout on a CapFTR hedge product from a node with price P_i to node with price P_N, strike price S and volume V. $\text{CapFTR}(P_i, P_N, S, V) = (P_N - P_i)VI(P_i \geq S)$

Part I

Introduction to Economic Concepts

Construction of a pole-mounted transformer substation, Waitaki Electric Power Board, New Zealand, ca 1925 (Source: Neil Rennie, Power to The People: 100 Years of Public Electricity Supply in New Zealand, Electricity Supply Association of New Zealand)

1

Introduction to Micro-economics

This book is about the economics of electricity markets. It is therefore essential that the reader understands a number of basic concepts in economics. Much of the material here can be found in introductory textbooks in economics. However, we hope that setting out this material at the start of this textbook will assist readers who do not have a background in economics.

Readers who have a background in economics may choose to skip this part. However, this presentation probably contains some new material, even for readers familiar with economics. In addition, we introduce notation and a few key ideas which are used throughout the rest of the book. We recommend at least a review of this material.

1.1 Economic Objectives

Economics is the study of the production, consumption, and exchange of goods and services in an economy – including how production, consumption, and exchange are organised, how information flows and how participants are rewarded and incentivised for playing their part. Economics seeks to both create theories which explain the patterns of behaviour and organisation that we see in the real world (so-called positive theories), and to develop policies and proposals for changing the arrangements that exist in the real world (normative theories).

But, if we are to recommend changes to existing arrangements, we need a commonly agreed set of objectives that we are trying to achieve. In our view, this common set of objectives must relate, in some way, to a common vision of the overall *economic welfare* of the society or economy as a whole.

There may be many different ways of articulating the overall economic welfare of a society or economy, if such a thing exists at all. It may never be possible to get consensus over whether or not some alternative state of the world, B, is preferred to the status quo, A. Economists tend to focus on areas where, in principle, there could be consensus – that is, situations where, in principle, every member of society could agree that B is preferred over A. These tend to be situations where there is what might be described as waste or inefficiency – where we could reorganise things so that we could achieve the same outcomes with fewer resources, or achieve better outcomes with the same resources. If these situations exist we could, in principle, leave everyone better off.

Although there is some variation in economic theory, in practice most public policy economists make certain assumptions which simplify the task of determining the total

The Economics of Electricity Markets, First Edition. Darryl R. Biggar and Mohammad Reza Hesamzadeh.
© 2014 John Wiley & Sons, Ltd. Published 2014 by John Wiley & Sons, Ltd.

economic welfare. Chief amongst these is the assumption that we can ignore income effects. In effect, this means that the benefit of an additional dollar to me is about the same as the benefit to any other member of the society. This assumption rules out the possibility of deriving any benefit from income redistribution alone. Alternatively, we can imagine that such income redistribution has already occurred through some other mechanism.

If we make this assumption, for any public policy change we can envisage, we can value the benefits and the costs imposed using a simple monetary metric. The change in economic welfare brought about by the public policy change is the simple sum of the monetised benefits and costs. An arrangement which maximises the economic benefits less the costs is said to be *efficient*. This is the usual meaning of the term economic efficiency in public policy analysis.

This notion of economic efficiency does not incorporate everything which the broader public might consider important. In particular, it does not usually directly deal with controversial questions about how income should be distributed in the economy. Neither does it normally directly address questions of fairness or equity.[1] Nevertheless, this notion of economic efficiency captures important, and broadly acceptable, notions of social welfare, and for most economists represents a legitimate goal for economic policymakers.

It is valuable to break down this notion of efficiency further. It is useful and helpful to distinguish between *short-run* and *long-run* concepts of efficiency. In the short-run we have an existing stock of assets in place. Short-run efficiency relates to getting the most out of the existing stock of assets: producing as much as possible and allocating those goods and services to those customers which value them most highly. In the longer run we can change the stock of assets, creating new assets or removing old assets. Longer-run efficiency includes the notion of efficiency in changing the stock of assets over time. We can distinguish between both *production assets* (used to make other goods or services, including electricity) and *consumption assets* (which are used to directly provide services to customers, such as electrical appliances or electrical machinery).

Specifically, we can distinguish between the following:

a. Efficiency in the *use* or *operation* of an existing stock of assets. This includes efficiency in the allocation of goods and services (ensuring that goods and services are consumed by those who value them most highly) and efficiency in the production of goods and services (ensuring that goods and services are produced at the lowest possible cost, given the existing stock of assets).
b. Efficiency in investment in the *creation* of new assets (or the disposal of old assets), including investment by seller(s) in new production assets (of the right size, in the right location, in the right amount, of the right type, and so on) and investment in developing new goods and services, and investment by buyer(s) in assets which increase their value for the goods or services produced (new consumption assets).

Many textbooks distinguish between allocative, productive and dynamic efficiency. These terms are defined in different ways by different economists. We prefer the following definition:

Allocative and productive efficiency are short-run efficiency concepts, relating to the efficient use of the existing stock of assets. *Allocative efficiency* refers to ensuring that the goods and

---

[1] This does not mean to imply that economic policies will not accord with principles of fairness or equity – but rather that those terms are interpreted in an economic way where they are addressed at all.

services produced are allocated to those who value them most highly. *Productive efficiency* refers to ensuring that goods and services are produced at the lowest possible cost. In contrast, dynamic efficiency is a longer-run concept, relating to changes in the existing stock of assets. *Dynamic efficiency* refers to efficient decisions regarding investment in new assets (what, where, when, and what type of investment), including investment in developing new products and services over time.

Result: Economic efficiency has both a short-run and long-run dimension. In the short run, economic efficiency is about the efficient use of a given set of production and consumption assets (productive and allocative efficiency). In the longer run, economic efficiency is about efficient decisions in the creation of new assets or the disposal of old assets (dynamic efficiency).

This book is about the design of arrangements to achieve these economic efficiency objectives in the electricity industry. Any particular arrangement is only desirable to the extent it achieves these objectives. In particular, this book will explore the extent to which competitive markets in the electricity sector can achieve the objectives above. We will see that in many situations, competitive markets deliver economically efficient outcomes. In other situations, competitive markets will not achieve these outcomes and we must substitute alternative arrangements, such as direct price controls. Particular institutional arrangements, such as competitive markets, are not an end in themselves. They are only the *means to an end* – a means to the achievement of the objectives set out above.

In many parts of the text that follow we will first seek to determine the efficient outcome (that is, the efficient use/operation of existing assets and/or the efficient investment outcomes) and then seek to determine whether particular market arrangements can achieve those outcomes, and under what conditions the market arrangements might achieve those outcomes.

When it comes to achieving efficient outcomes using a given stock of assets, economists typically focus separately on the buying (or demand) side of the market and the selling (or supply) side of the market. The next two sections focus in turn on the buying side of a market and the supply side of a market. We will then bring these ideas together to look at what it means to achieve efficiency in the use of a given stock of assets. In the subsequent section we will explore whether or not short-run efficient outcome can be achieved using a competitive market. First, however, we review the principles of constrained optimisation.

1.2 Introduction to Constrained Optimisation

Optimisation lies at the heart of economics. Under conventional economic theory all economic actors are assumed to be maximisers of an objective function. This underlying assumption is made even more explicit in the smart markets introduced in Section 1.6. It is therefore essential for all students of electricity markets to have some understanding of the theory of constrained optimisation.

In this text we will often see a constrained optimisation problem expressed in the following form:

$$\max f(x)$$

Subject to the following conditions:

a. For $i = 1, \ldots, n, \quad g_i(x) = c_i \leftrightarrow \lambda_i$
b. For $j = 1, \ldots, m, \quad h_j(x) \leq d_j \leftrightarrow \mu_j$

Here x is a vector of k variables, and the functions $f(\cdot), g_i(\cdot), h_j(\cdot)$ are all functions from $\mathbb{R}^k$ to $\mathbb{R}$. The equations $g_i(x) = c_i$ and $h_j(x) = d_j$ are known as constraint equations.

The variables λ_i and μ_j are known as *Lagrange multipliers* and their value will become apparent shortly. Each constraint equation has its own Lagrange multiplier. Here we are following the convention which uses the symbol $\leftrightarrow$ to show the Lagrange multiplier which is associated with each constraint equation.

Let us suppose that we have a set of values x, λ, μ which satisfy the following conditions, known as the *Karush–Kuhn–Tucker* (or KKT) conditions. Then x is a solution to the constrained optimisation problem above.

The KKT conditions are as follows:

1. For $l = 1, \ldots, k,$

$$\frac{\partial f}{\partial x_l} - \sum_i \lambda_i \frac{\partial g_i}{\partial x_l} - \sum_j \mu_j \frac{\partial h_j}{\partial x_l} = 0$$

This condition is known as the First Order Condition

2. For $i = 1, \ldots, n, \quad g_i(x) = c_i$
3. For $j = 1, \ldots, m, \quad \mu_j \geq 0, \quad$ and $\quad h_j(x) \leq d_j$ and $\mu_j(h_j(x) - d_j) = 0$

In other words, the problem of finding a solution to the constrained optimisation problem above reduces to the problem of finding a solution to the KKT conditions (1)–(3).

It is worth noting that the Lagrange multipliers have a particular interpretation. The Lagrange multipliers measure the extent to which the objective function can be improved following a small change in the constraints. For example, let us define $\mathcal{L}$ to be the value of the objective function at the solution of the constrained optimisation above. Then the Lagrange multiplier λ_i is the change in the objective function with respect to a small change in the parameter c_i. Similarly, the Lagrange multiplier μ_j is the change in the objective function with respect to a small change in the parameter d_j.

$$\lambda_i = \frac{\partial \mathcal{L}}{\partial c_i} \quad \text{and} \quad \mu_j = \frac{\partial \mathcal{L}}{\partial d_j}$$

1.3 Demand and Consumers' Surplus

Let us focus more closely on the buying (or demand) side of a market for a particular good or service, such as electricity. We will focus on an abstract buyer or customer of this service. Although we will use the word customer, we do not intend to limit ourselves to small customers or consumers. Rather this customer could be a large business, such as an aluminium smelter, a small business, such as an office or restaurant, or a residential household.

To model the behaviour of customers in a market, in principle we need to specify two things: (a) the range of actions or choices that the customer faces; and (b) some form of objective which the customer is seeking to pursue.

1.3.1 The Short-Run Decision of the Customer

In principle, customers can take a range of actions which affect the value they receive from a good or service. This is particularly true in the case of electricity. Customers do not consume electricity directly; instead they consume the services of a range of machinery, pumps, heaters, devices and appliances which consume electricity. The demand for electricity at any one point in time depends on the stock of past investments made by the customer. More generally, the demand for a particular good or service in the economy will depend on the past actions taken by customers.

For the moment we will focus on short-run decisions of the customer. Let us assume that the customer has made a set of decisions in the past regarding devices which consume electricity. The key remaining decision of the customer is how much electricity to consume at a given point in time – or more precisely, the rate at which electricity is consumed.[2]

1.3.2 The Value or Utility Function

In order to complete the model of customer behaviour, we need to specify the customer's objective. Many introductory textbooks in economics start by introducing the demand curve. However, we will follow a slightly unconventional path and start with the notion of a *value or utility function*. This approach is straightforward and allows us to draw simple parallels between the demand and the supply sides of each market.

Let us suppose we have a customer which is consuming a particular good or service at the rate Q (units per interval of time). Let us assume that we can express the utility or value (also known as surplus) that this customer receives from consuming this particular good or service in the form of a function, known as a utility function $U(Q)$ ($ per interval of time). (This customer could itself be a firm, in which case the utility is equal to the profit of the firm from the activity).

This utility will depend on a number of factors, such as the investments the customer has made in equipment which uses the good or service in question, or, if the customer is itself a firm, the demand for the final product produced by the firm and the substitutes available for the good or service in question. Typically, the customer is assumed to obtain higher utility from consuming at a higher rate. In other words, $U'(Q) > 0$ (here the prime symbol signifies the first derivative of the utility function with respect to the rate of consumption). Also, by assumption the rate at which value increases with consumption decreases the higher the rate of consumption (i.e. $U''(Q) < 0$).

In practice, a customer will typically not consume just a single good or service, but several different goods and services at the same time. The utility function can be a function of the rate of consumption of each of these goods and services. For example, if a customer consumes two goods, at the rates Q_1 and Q_2, the rate at which the customer receives value or utility might be denoted by $U(Q_1, Q_2)$ ($/interval of time).

1.3.3 The Demand Curve for a Price-Taking Customer Facing a Simple Price

Let us suppose that the customer obtains a particular good or service through arm's length transactions in a market. The simplest assumption we can make is that the customer pays a

---

[2] Strictly speaking, the customer does not directly choose the rate of consumption of electricity – instead he/she chooses the rate at which to enjoy the services for which electricity is used (such as the rate of manufacture of aluminium), and the rate of consumption of electricity follows.

simple constant price P ($ per unit) for this good or service independent of the rate that he/she consumes. Many goods and services have more complicated pricing schemes, but for the moment, it is convenient to assume that each customer pays a simple constant price. As a consequence, if the customer consumes the good or service at the rate Q (units/interval of time), the customer must make a payment equal to PQ ($/interval of time) each time period.

Let us assume that the customer is a *price-taker* – that is, the customer has no influence over the market price, regardless of how much he/she consumes. The customer is assumed to choose a rate of consumption which maximises his/her net surplus or net utility – that is the utility from consumption less the revenue paid to obtain the good or service. In other words, the customer is assumed to maximise the following expression:

$$\varphi(Q) = U(Q) - PQ$$

What rate of consumption maximises the net utility? The first-order condition for the maximum is as follows:

$$\frac{d\varphi}{dQ} = U'(Q) - P = 0$$

which implies that the optimal rate of consumption is where the marginal utility is equal to the price:

$$U'(Q) = P$$

The first derivative of the utility function $U'(Q)$ ($ per unit) is known as the *inverse demand curve* and will be denoted by $P^D(Q)$. The inverse demand curve shows, for each rate of consumption, the corresponding price that a price-taking customer is prepared to pay to sustain that rate of consumption. The inverse demand curve is downward sloping: An increase in the market price corresponds to a lower rate of consumption, and vice versa.

The result above shows that a price-taking customer (that is a customer who cannot influence the market price) will choose to consume at a rate where the inverse demand curve is equal to the market price. This is illustrated in Figure 1.1.

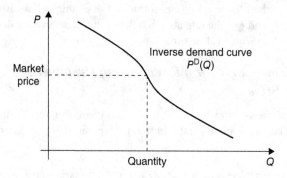

Figure 1.1 A typical inverse demand curve

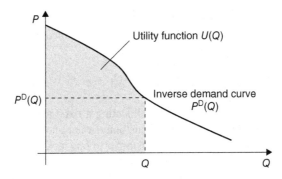

Figure 1.2 The utility function is the area under the inverse demand curve

> *Result*: For a price-taking consumer facing a simple linear price, the consumer's demand function is given by (the downward sloping part of) the marginal utility curve.

Because the inverse demand function is downward sloping, it has an inverse, which is known as the demand curve and will be denoted here by $Q^D(P)$. The demand curve shows, for a given level of the market price, the rate at which this customer is willing to consume.

Given the inverse demand curve for a good or service, we can work out the corresponding utility function, and vice versa. Since the inverse demand curve is the first derivative of the utility function, the utility function is found by integrating the inverse demand curve. This gives the level of the utility function up to some constant (here denoted by c):

$$U(Q) = \int^Q P^D(Q)dQ + c$$

This has been illustrated diagrammatically in Figure 1.2. The utility or value function (up to a constant) is equal to the area under the (inverse) demand curve:

Where there are many different customers, all consuming the same good or service at different rates $Q_1, Q_2, \ldots$, the total value function for the market as a whole from consuming this particular good or service is just the sum of the value functions of each customer. This is also known as the total or gross *consumers' surplus*.

$$CS(Q_1, Q_2, \ldots) = U(Q_1, Q_2, \ldots) = \sum_i U_i(Q_i)$$

If we have a fixed total rate of consumption of the good or service available, Q, say, how should that rate of consumption be allocated between customers to achieve the highest possible overall value or surplus? This problem can be written mathematically as follows:

$$\max_{Q_1, Q_2, \ldots} U(Q_1, Q_2, \ldots) \text{ subject to } \sum_i Q_i = Q$$

This is a constrained optimisation problem. We can solve this problem by setting out the Lagrangian and computing the KKT conditions as explained in the previous section. The first-order condition for this problem is

$$U'_i(Q_i) = P^D_i(Q_i) = \lambda$$

where λ is a Lagrange multiplier.

In other words, the problem of efficiently allocating a fixed total rate of consumption of a given good or service between any number of customers can be solved by setting a market price λ for the good or service and allowing each customer to buy at whatever rate he/she desires at that market price. The market price should be chosen in such a way that the total demand (i.e. the total rate of consumption) at that price is equal to the total rate of consumption required:

$$\sum_i Q^D_i(\lambda) = Q$$

We are starting to see how competitive markets can solve allocation problems. As we will discuss in more detail later, if all participants are price-takers, the market solves the problem of efficiently allocating a given rate of consumption of a good or service between customers with different needs and preferences.

This analysis has focused on the case where the customer consumes a single good or service. In practice, of course, a customer will typically consume several goods or services at one time. The value that a customer places on any one good will typically depend on the rate at which he/she is consuming another good. This complicates the analysis above a little bit. We can distinguish two extreme cases, where the utility function depends only on the total rate of consumption of the two goods:

$$U(Q_A, Q_B) = U(Q_A + Q_B)$$

In this case the two goods are said to be perfect substitutes and can be treated as though they are really one good in practice. The other extreme case is where the utility function can be separated into two separate functions:

$$U(Q_A, Q_B) = U_A(Q_A) + U_B(Q_B)$$

In this case, the two goods are independent of each other.

1.4 Supply and Producers' Surplus

Now let us focus on the selling or supply side of a market for a particular good or service. Without loss of generality we can assume that goods and services are produced by an economic entity which we will refer to as a *firm*. A firm purchases certain goods and services, known as *inputs*, and converts them into different goods and services, known as *outputs*.

As with the customer side of the market, in order to model the behaviour of a firm we need to know something about its possible range of actions, and something about its objective. In the longer run, firms can take a wide range of actions, such as investing in new production capacity, marketing their goods and services, and investing in research and development for new products. But, for the moment, let us follow the approach set out in the previous section: let us take the stock of existing investments of a firm as given, and focus on the short-run decisions of the firm – primarily the decision as to the rate of production.

In the course of production a firm will incur expenditure on inputs. Some of the expenditure of a firm will take the form of investments to increase the productive capacity, or the demand for the services of the firm. Such expenditures, once made, do not vary with the rate of production of the firm at any given point in time. Let us focus for the moment on those expenditures which are directly related to the rate of production at a point in time, known as the *variable costs*. This might include the labour costs of staff involved in production, or the cost of purchasing inputs which vary with production.

More generally, when discussing the costs or expenditure of a firm we must be clear about the time frame we have in mind. In the short run, when a firm has already made sunk investments in buildings, equipment and so on, the managers of the firm need only be concerned with the expenditure which they can alter in the short run by altering the rate of production of the firm. In the longer run, the firm may be able to alter its size or scale, or change its location.

1.4.1 The Cost Function

Let us assume that the firm produces just a single good or service. The rate of production of this good or service is denoted by Q (units/interval of time). The rate at which expenditure is incurred to produce at a given rate is known as the *cost function* and will be denoted by $C(Q)$ ($/interval of time). Typically we will assume that higher rates of production correspond to higher costs (i.e. $C'(Q) > 0$). In addition, for most firms the rate at which cost increases with the rate of production itself increases as the rate of production increases (i.e. $C''(Q) > 0$).

We will often be interested in how costs change with a small increase in the rate at which output is produced, which is known as the *marginal cost*. The marginal cost function is the slope of the cost function and will be denoted by $MC(Q)$.

$$MC(Q) = \frac{dC}{dQ} = C'(Q)$$

By the assumptions above, the marginal cost function is positive and upward sloping.

Most firms produce not just a single good or service, but many hundreds or thousands of different goods and services. The costs of the firm can be a function of the rate of production of each of these goods and services. For example, if the firm produces two goods, at the rates Q_1 and Q_2, the rate at which costs are incurred can be denoted by $C(Q_1, Q_2)$ ($/interval of time). There is a separate marginal cost for each good or service, which is the partial derivative of the cost function with respect to the corresponding rate of production:

$$MC_1(Q_1, Q_2) = \frac{\partial C}{\partial Q_1}$$

1.4.2 The Supply Curve for a Price-Taking Firm Facing a Simple Price

Let us suppose the firm sells its output in a conventional market. Let us take the simplest case and suppose that the firm obtains the same price P ($ per unit) for this good or service for each unit, independent of the rate at which it produces. As a consequence, if the firm produces the good or service at the rate Q(units/interval of time), the producer receives a flow of funds at the rate PQ ($/interval of time).

Let us consider the case where the firm is a price-taker – that is where the firm cannot influence the market price by varying his/her output. In economics it is conventional to assume that the short-run objective of a firm is to maximise its *profits*. Profits are conventionally defined as the revenue the firm receives from sales of outputs in a period less the expenditure incurred by the firm in the purchase of inputs. In other words, the firm is assumed to maximise:

$$\pi(Q) = PQ - C(Q)$$

This is an adequate statement of the short-run objective of most firms. In the case of longer-run decisions (such as decisions regarding investment in assets which last multiple periods), the investment decision will change the cash flow of the firm not just in a single period but over many periods into the future. In this case it makes more sense to assume that the firm maximises the *present value* of the stream of profits (the determination of the value of a stream of cash flows, especially under conditions of uncertainty, takes us beyond the scope of this text).

Even if we limit ourselves to short-run decisions, not all economic firms will choose to maximise profits. In particular, some government-owned firms pursue a range of broader objectives. However, for many purposes, especially for short-run decisions, it is reasonable to make the assumption that firms seek to maximise profits.

The rate of production which maximises the rate at which the firm receives profits satisfies the following first-order condition:

$$\frac{d\pi}{dQ} = P - C'(Q) = 0$$

We find that, for each price, a profit-maximising price-taking firm chooses the rate of production where the marginal cost curve is equal to that price.

$$C'(Q) = P$$

The *supply curve* for a firm (which we will denote as $P^S(Q)$) shows, for each level of market price, the rate of production which the firm will choose. We have shown that a profit-maximising price-taking firm will choose to produce where its marginal cost is just equal to the market price. In other words, the (upward sloping) part of the marginal cost curve is the supply curve for the firm.

Result: For a price-taking producer facing a simple linear price the supply function is given by (the upward sloping part of) the marginal cost curve.

Under the assumptions set out above, the marginal cost curve (and therefore the supply curve for a firm) is upward sloping (Figure 1.3).

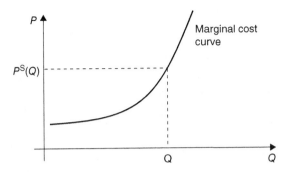

Figure 1.3   The marginal cost curve is the supply curve for a competitive firm

Given the supply curve for a price-taking firm, we can work out the corresponding cost function, and vice versa (in exactly the same way we showed earlier for the value function and the demand curve). Since the supply curve is the first derivative of the marginal cost curve, the cost function is found by integrating the marginal cost curve. This gives the level of the value function up to some constant (here denoted by c):

$$C(Q) = \int^{Q} P^{S}(Q)\mathrm{d}Q + c$$

This has been diagrammatically shown in Figure 1.4. The cost function (up to a constant) is equal to the area under the supply curve:

Where there are many different firms, all producing the same good or service at different rates $Q_1, Q_2, \ldots$, the total cost of production (\$/interval of time) for the market as a whole is just the sum of the cost functions of each producer:

$$C(Q_1, Q_2, \ldots) = \sum_{i} C_i(Q_i)$$

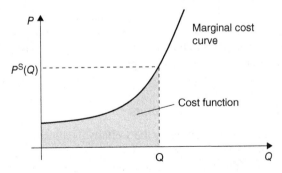

Figure 1.4   The cost function is the area under the marginal cost curve

Let us define the gross producers' surplus to be the negative of the total cost of production for the market as a whole:

$$PS(Q_1, Q_2, \ldots) = -C(Q_1, Q_2, \ldots)$$

If we want to produce a given good or service at a given total rate, Q, say, how should we allocate this total rate of production across different producers to end up with the overall lowest-cost rate of production? This problem can be written mathematically as follows:

$$\max_{Q_1, Q_2, \ldots} PS(Q_1, Q_2, \ldots) = \min_{Q_1, Q_2, \ldots} C(Q_1, Q_2, \ldots) \text{ subject to } \sum_i Q_i = Q$$

The first-order condition for this problem is

$$C_i'(Q_i) = P_i^S(Q_i) = \lambda$$

where λ is the Lagrange multiplier. In other words, the problem of efficiently allocating a given rate of production between a number of different producers with different costs can be solved by setting a market price λ for the goods or services and allowing each producer to sell as much as he/she desires at that market price. The market price should be chosen in such a way that the total supply (i.e. the total rate of production) at that price is equal to the desired rate of production.

Here we see how competitive markets, in which all the participants are price-takers, help solve the problem of producing a given total rate of production at the lowest possible ongoing costs.

The analysis above focuses on the case of a firm which produces a single output. Most real-world firms produce hundreds or thousands of different goods or services. As in the previous section, we can extend the analysis above to a firm which produces two or more goods or services at different rates. The cost function of the firm may now depend on the rate at which the firm produces two or more outputs.

1.5 Achieving Optimal Short-Run Outcomes Using Competitive Markets

Let us suppose we have a market with a large number of buyers (customers) and a large number of sellers (firms or producers). Let us suppose that both the sellers and the buyers have made past sunk investment in assets for producing or consuming the good or service in question. Given these investments, there is a remaining question as to the rate at which each seller or buyer should produce or consume so as to maximise overall economic welfare in the short run.

We are interested here, and throughout this book, in two questions: (i) What is the optimal (welfare-maximising) outcome? (ii) Can this optimal outcome be achieved using competitive markets?

1.5.1 The Short-Run Welfare Maximum

Let us focus on the first question: What is the welfare-maximising outcome?

Let us suppose that we have a set of producers. Producer i produces at the rate Q_i^S and has the cost function $C_i(Q_i^S)$. In addition, let us suppose that producer i faces some generic constraints on his/her rate of production which we will denote by $g_i(Q_i^S) \leq 0$.

Similarly, let us suppose we have a set of consumers. Consumer i produces at the rate Q_i^B and has the utility function $U(Q_i^B)$. In addition, consumer i faces some generic constraints on his/her rate of consumption which we will denote by $h_i(Q_i^B) \leq 0$.

The welfare-maximisation problem is to find a rate at which each producer should produce, Q_i^S and a rate at which each customer should consume, Q_i^B which maximises the total economic surplus, subject to the condition that the total rate of production must equal the total rate of consumption:

max TS

$$= \max \text{CS}\left(Q_1^B, Q_2^B, \ldots\right) + \text{PS}\left(Q_1^S, Q_2^S, \ldots\right)$$

$$= \max \sum_i U_i\left(Q_i^B\right) - \sum_i C_i\left(Q_i^S\right)$$

Subject to   (a) $\sum_i Q_i^B = \sum_i Q_i^S \leftrightarrow \lambda$

(b) $g_i\left(Q_i^S\right) \leq 0 \leftrightarrow \alpha_i$

(c) $h_i\left(Q_i^B\right) \leq 0 \leftrightarrow \beta_i$

From the KKT conditions for this problem, at the optimum, the following conditions must hold:

i. For each consumer: $U_i'\left(Q_i^B\right) - \lambda - \beta_i \dfrac{\partial h_i}{\partial Q_i^B} = 0$

ii. For each producer: $C_i'\left(Q_i^S\right) - \lambda + \alpha_i \dfrac{\partial g_i}{\partial Q_i^S} = 0$

In the case where there are no other constraints on the rate of consumption of producers and consumers, we have the result that, at the optimum, the marginal utility of all customers and the marginal cost of all producers are the same: $U_i'\left(Q_i^B\right) = \lambda = C_i'\left(Q_i^S\right)$.

1.5.2 An Autonomous Market Process

Although a few markets in a modern economy are organised around a central market operator or auctioneer, most markets in an economy operate autonomously without a central auctioneer or market-maker. There are various ways in which this market process could operate. The key question we would like to consider is whether or not a competitive market process can lead to the efficient outcome described above.

Let us suppose that, by some mechanism a single market price is determined. Let us assume that there are a large number of buyers and a large number of sellers, so that no buyer or seller has any influence over that market price. Each buyer or seller is assumed to be free to choose its own rate of production or consumption given the market price.

Let us focus first on the task of a producer. As before we will assume the producer seeks to maximise his/her profit, given the market price and any constraints on the rate of production. The task of the producer is, therefore, to solve the following problem:

$$\max \pi_i\left(Q_i^S\right) = PQ_i^S - C_i\left(Q_i^S\right)$$

Subject to : $g_i\left(Q_i^S\right) \leq 0 \leftrightarrow \alpha_i$

Similarly, the task of a consumer is to solve the following problem:

$$\min U_i\left(Q_i^B\right) - PQ_i^B$$

$$\text{Subject to}: \; h_i\left(Q_i^B\right) \le 0 \leftrightarrow \beta_i$$

The KKT conditions for these problems are as follows:

i. For each consumer, $U_i'\left(Q_i^B\right) - P - \beta_i \dfrac{\partial h_i}{\partial Q_i^B} = 0$

ii. For each producer, $P - C_i'\left(Q_i^S\right) - \alpha_i \dfrac{\partial g_i}{\partial Q_i^S} = 0$

Comparing these conditions with the conditions above, we can draw the following conclusion: Provided the market price P is equal to the constraint marginal value of the overall supply–demand balance constraint (which we have labelled λ) the decentralised profit-maximising decisions of producers and the utility-maximising decisions of consumers will achieve the welfare-maximising combination of production and consumption.

All that remains to show is that the market price can (and will) be set equal to the constraint marginal value on the overall supply–demand balance constraint.

In a typical market, this equilibration of supply and demand occurs through an adjustment process, through adjustments to the market price. If the market price is too high, the rate of production of the good or service will exceed the rate of consumption – inventories or stocks of the good or service will build up, or some producers will not be able to sell all they want at the price. In either case, some producers will cut their price, resulting in a fall in the market price. Conversely, if the market price is too low, the rate of production of the good or service will fall short of the rate of consumption. Inventories will decline and/or some customers will not be able to purchase at the rate they desire. In either case, customers will bid up the price. The equilibrium market price is where the total rate of production is equal to the total rate of consumption.

In other words, in a conventional, competitive economic market, where buyers and sellers are price-takers, actions taken by individual buyers and sellers, choosing their own short-term rate of production to maximise their own objectives, achieve an overall short-run welfare maximum. This is one of the key reasons why economists like competitive markets – they are a mechanism by which individuals, pursuing their own ends, can be coordinated 'as if by an invisible hand' to achieve outcomes which are efficient for the economy as a whole.

Result: Economic welfare-maximising outcomes can be achieved with a decentralised market process. Provided the market price is equal to the constraint marginal value of the overall supply–demand balance constraint in the welfare-maximisation problem, and provided all producers and consumers are price-takers, the decentralised profit-maximising decisions of producers and the utility-maximising decisions of consumers will achieve the welfare-maximising combination of production and consumption.

1.6 Smart Markets

Most markets in a modern economy operate entirely autonomously, without any form of central role for a market operator or auctioneer. However, for reasons which we will see later, wholesale electricity markets are not like this. Due to the complexities of the way that electric power flows out of electricity networks, it is not possible to separate the market for the transportation of electric power from the market for the production or consumption of power – these two markets must be integrated through the central role of the market operator.

So-called *smart markets* were developed to incorporate more sophisticated physical constraints into market processes. Smart markets are used in the allocation of take-off and landing slots at airports, in the allocation of radio spectrum and in natural gas markets. Most importantly, smart markets are used in the electricity industry to integrate physical network constraints with the trading of electric power.

1.6.1 Smart Markets and Generic Constraints

Mathematically, a smart market involves an extension to the problem discussed in the previous section. Let us suppose that, as before, we seek a combination of production and consumption which maximises economic welfare, subject to (a) the overall supply–demand balance and (b) constraints on individual producers and consumers. In addition, now let us introduce some generic constraints on the rate of production and consumption, which we will denote: $k_j(Q^S, Q^B) \leq 0$. The overall welfare-maximisation problem is now as follows:

$$\max \text{TS} = \max \sum_i U_i(Q_i^B) - \sum_i C_i(Q_i^S)$$

$$\text{Subject to:} \quad \text{(a)} \sum_i Q_i^B = \sum_i Q_i^S \leftrightarrow \lambda$$

$$\text{(b)} \, g_i(Q_i^S) \leq 0 \leftrightarrow \alpha_i; \quad h_i(Q_i^B) \leq 0 \leftrightarrow \beta_i$$

$$\text{(c)} \, k_j(Q^S, Q^B) \leq 0 \leftrightarrow \gamma_j$$

The KKT conditions for this problem are as follows:

$$U_i'(Q_i^B) - \lambda - \beta_i \frac{\partial h_i}{\partial Q_i^B} - \sum_j \gamma_j \frac{\partial k_j}{\partial Q_i^B} = 0$$

$$C_i'(Q_i^S) - \lambda + \alpha_i \frac{\partial g_i}{\partial Q_i^S} + \sum_j \gamma_j \frac{\partial k_j}{\partial Q_i^S} = 0$$

Let us proceed as before and ask whether or not this outcome can be achieved through a decentralised competitive market process. Comparing with the conditions for profit maximisation and utility maximisation above, we see that these outcomes can be achieved in a competitive market process provided that every producer and consumer is a price-taker and, in addition, the *i*th consumer faces the price:

$$P_i^B = \lambda + \sum_j \gamma_j \frac{\partial k_j}{\partial Q_i^B}$$

Similarly, the ith producer must face the price:

$$P_i^S = \lambda - \sum_j \gamma_j \frac{\partial k_j}{\partial Q_i^S}$$

The parameters λ and γ_j must be chosen in such a way that the overall supply–demand balance constraint is satisfied, and each of the generic constraints $k_j(Q^S, Q^B) \leq 0$ is satisfied.

1.6.2 A Smart Market Process

We have demonstrated that we can, in principle, decentralise the welfare-maximisation task using a competitive market process, provided we can set the prices correctly. But how might we achieve this objective?

Let us therefore consider a slightly different market process – one in which there is a central market operator which operates a constrained-optimisation process. Specifically, let us assume that the centralised market process operates through a series of steps as set out below:

a. Time is divided into a series of time intervals. Each interval of time, every seller submits to the market operator an offer function showing, at each price, the rate at which that producer is willing to produce. Similarly, every buyer submits to the market operator a bid function which shows, for every price, the rate at which he/she is willing to consume.
b. The market operator then carries out mathematical optimisation to find the combination of rate of production and consumption which maximises the total surplus (the sum of the utility functions of the customers and the cost functions of the buyers) under the assumption that the offer curve of each producer accurately reflects its true supply curve, and the bid curve of each buyer accurately reflects its true demand curve.
c. The market operator then sends out instructions to each buyer as to the rate at which that buyer should consume and, to each seller, the rate at which that producer should produce. The market operator also announces a market price (which may differ across producers and consumers).
d. Each producer and each customer, given the market price they face, choose a rate to produce or consume, respectively. Each producer is then paid the corresponding price for its production during an interval. Similarly, each customer must pay the corresponding price for its consumption during the interval.

Does this market process achieve an overall efficient allocation?

Let us suppose that each producer announces a cost function $\hat{C}_i(Q_i^S)$ and each consumer announces a utility function $\widehat{U}_i(Q_i^B)$. These announced functions may differ from the underlying 'true' cost and utility functions.

Given these announced functions, the market operator then carries out an optimisation task as follows:

$$\max \widehat{TS} = \max \sum_i \widehat{U}_i(Q_i^B) - \sum_i \hat{C}_i(Q_i^S)$$

$$\text{Subject to} \quad (a) \sum_i Q_i^B = \sum_i Q_i^S \leftrightarrow \lambda$$

$$(b)\, k_j(Q^S, Q^B) \leq 0 \leftrightarrow \gamma_j$$

Now, if the announced cost and utility functions accurately reflect the true cost and utility functions of the producers and consumers (as modified by any production or consumption constraints), then this maximisation yields the overall welfare maximising outcome. Furthermore, this optimisation yields the constraint marginal values λ and γ_j from which the prices can be calculated.

All that remains to show is that the producers and consumers will announce to the market operator cost and utility functions which reflect their true cost and utility functions – allowing for any production or consumption constraints.

Intuitively, this is straightforward to verify when there are no producer or consumer production or consumption constraints. Let us suppose a customer is a price-taker in the sense that he/she cannot, by changing his/her bid function, influence the market price. For each potential market price P, the customer must announce the rate at which he/she is willing to consume. But the amount that the customer is willing to consume at that price is just the amount which maximises the net value $U(Q) - PQ$, which is just the amount given by the customer's demand curve $Q^B(P)$. In other words, if the customer is a price-taker in the market, he/she can do no better than submit a bid curve which perfectly reflects his/her demand curve. The same is true, of course, for the producers in the market. If each producer is a price-taker, it can do no better than submitting an offer curve which perfectly reflects its supply curve $Q^S(P)$.

The argument is very similar when we take into account production or consumption constraints. We have seen that a welfare-maximising price-taking consumer which faces a price P will choose a quantity that satisfies

$$U_i'(Q_i^B) - \beta_i \frac{\partial h_i}{\partial Q_i^B} = P$$

In other words, the consumer has an incentive to offer to the market operator a demand curve $\widehat{U}_i'(Q_i^B)$ which satisfies

$$\widehat{U}_i'(Q_i^B) = U_i'(Q_i^B) - \beta_i \frac{\partial h_i}{\partial Q_i^B}$$

Similarly, each producer has an incentive to offer to the market operator a bid curve $\widehat{C}_i'(Q_i^S)$ which satisfies

$$\widehat{C}_i'(Q_i^S) = C_i'(Q_i^S) + \alpha_i \frac{\partial g_i}{\partial Q_i^S}$$

In summary we have proven the following result:

Result: Even when there are generic constraints on production and consumption, the welfare-maximising outcome can be achieved using a generalised market process known as a smart market, involving a central market operator. In this market process:

a. Each producer and each consumer submits a bid or offer curve which reflects his/her true supply and demand function (after taking into account private production or consumption constraints).

b. The market operator carries out a constrained optimisation task, seeking to maximise economic welfare based on the announced bid and offer curves subject to (i) an overall supply–demand balance constraint and (ii) generic production-consumption constraints. This process determines a set of prices (which may vary between producers and consumers) and a rate of production and consumption for each producer and consumer.

c. Given the announced prices, each producer and each consumer chooses to voluntarily comply with the assigned rate of production or consumption.

The market process described here appears needlessly clumsy. It involves the communication of large amounts of information (bid and offer curves and despatch instructions) to and from a centralised market operator. And it requires the central market operator to carry out a large constrained-optimisation problem. In most markets, this degree of information sharing and centralised control is unnecessary and burdensome. However, we will see later that this is not the case in wholesale electricity markets. Wholesale electricity markets require a centralised process of the kind described here.

1.7 Longer-Run Decisions by Producers and Consumers

The previous section focused on short-run actions by producers and consumers. In practice, however, producers and consumers make a number of different forms of investments which affect either the cost of production or the value in consumption of the good or service.

For example, a producer might make a decision to expand its production capacity or to lower its production cost. A customer might make a decision to install new equipment which increases its demand for a particular good or service.

The difference between a short-run decision and a longer-run decision in practice depends on the extent to which the particular action can be changed in response to a change in the market price. If the action of the producer or consumer can be changed in response to each and every movement in the market price, the action is a short-run action. If, however, the action cannot be reversed as the market price changes, the decision is a longer-run (medium-term or long-term) decision.

The range of possible longer-run investments by producers and consumers is very large. For example, producers might face a decision as to what amount of productive capacity to install, at what location, at what time and of what type. Electricity consumers might face a decision as to what type of electrical devices to install, with what capabilities and of what size. Each of these decisions will require a slightly different economic model. We cannot cover all of these decisions here in this section. Instead, let us focus on a couple of simple questions.

1.7.1 Investment in Productive Capacity

To begin with, let us consider the decision to add more productive capacity. Let us suppose that there is a single producer and a single customer (which may represent the aggregate of many smaller producers and customers). To keep things as simple as possible, let us suppose that the cost of production of the producer varies linearly with the rate of production, up to the physical maximum rate of production, which we will call the 'capacity' of the firm, and which we will label K (units/interval of time). We will suppose that each additional unit of output costs c ($/unit) and an additional unit of capacity costs f ($/unit/interval of time). In other words, let us assume that the cost function of the producer takes the form

$$C(Q, K) = cQ + fK$$

The rate of production of the firm is assumed to be able to be varied much more quickly than the capacity of the firm. In order to justify a change in the rate of production, we need to introduce some variability in the supply or demand conditions. Let us assume that the supply conditions of the firm are fixed over time, but that demand is varying. Let us suppose that there are several different states of the world, labelled s. In state of the world s, which occurs with probability p_s, the utility of customers for the good or service is given by $U_s(Q)$.

Since there is some uncertainty in this world, there is some risk. To keep things simple let us make the simplest possible assumption, which is that overall social welfare is neutral to risk. In practice this means that the overall social welfare is indifferent between receiving a fixed, certain pay-off and any uncertain pay-off which has the same expected or average value.

Let us assume that there is always sufficient demand relative to costs that it is always efficient to produce something.

To find the socially-efficient level of capacity K we must solve the following optimisation problem:

$$\max \sum_s \left[p_s(U_s(Q_s) - C(Q_s, K)) \right]$$

$$\text{Subject to } Q_s \leq K \leftrightarrow \mu_s$$

The KKT conditions for a maximum of this problem are

$$\mu_s = U_s'(Q_s) - c \quad \text{and} \quad \mu_s(K - Q_s) = 0 \quad \text{and} \quad \sum_s p_s \mu_s = f$$

In other words, the optimal level of capacity is the level where the average distance between the marginal value of consumption and the marginal cost of production at times when the capacity of the firm is exhausted is equal to the cost of additional capacity (Figure 1.5). This is a simple example of a classic *peak load pricing* problem.

We will see later that this conclusion generalises to the case where there are different types of producers, with different costs of production and different costs of expanding capacity. As in the example above, we find that for each different type of producer, the capacity of that type of producer is at the optimal level when the average gap between the marginal value of the customer and the marginal cost of production at times when the capacity of that type of producer is exhausted is equal to the cost of adding capacity of that type.

Now let us ask the question: will this level of capacity be chosen in a competitive market? Let us assume that we have a competitive market. In the short run, as we have seen above, the market price will be equal to the common marginal value of all the customers and the marginal cost of production. Let us suppose the market price in state of the world s is P_s. Let us consider

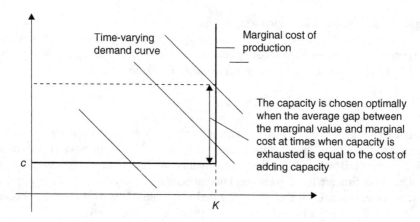

Figure 1.5   Determining the optimal level of capacity

the level of capacity chosen by a producer which seeks to maximise its expected profit. In other words, the producer solves the following problem:

$$\max \sum_s p_s[P_s Q_s - C(Q_s, K)]$$

Subject to $Q_s \leq K$

The first-order conditions for this problem show that (a) the firm produces at capacity when the market price exceeds the marginal cost of production and (b) the level of capacity is optimal if the average gap between the market price and the marginal cost of production at times when the market price exceeds the cost of production is equal to the cost of adding capacity.

But from the earlier analysis we know that the market price is equal to the marginal value of consumption. It therefore follows that in a competitive market a firm which can adjust capacity continuously will choose a level of capacity which is socially optimal.

It is straightforward to check that this result generalises to the case where there are several different types of production technologies. We carry out this analysis in Part IV. In other words, under the simple assumptions set out in this example, competitive markets will deliver socially efficient outcomes in both the short run and the long run.

There are many extensions to these simple results which we will explore further throughout this text.

1.8   Monopoly

The previous sections used the simplifying assumption that producers and customers could not influence the market price – in other words, that they were price-takers. This is a reasonable assumption where there are a large number of producers and customers, none of which is very large relative to the size of the market. However, this is not always the case. In some markets there is a single dominant firm, or a small group of firms. In either case it is likely that these firms will have some impact on the price or prices at which the good or service is transacted in the market.

In fact, a firm which faces little or no competition not only will be able to influence the price or prices at which the good or service is transacted in the market but will often also have some control over the way that goods and services are charged for in the market. For example, a firm with little or no competition may be able to charge different prices for different types of customers. Alternatively, such a firm may be able to charge its customers a separate fee for the right to consume at all, and then another fee per unit consumed known as a two-part tariff. These different pricing practices are known generically as *price discrimination*. Some forms of price discrimination arise in competitive markets. As a general rule, however, price discrimination is much more common in markets which lack competition.

Let us put aside this question of how the prices are structured, and simply focus on one component – the price for additional units of the good or service per interval of time.

As discussed in more detail in Part VII, a firm which is not a price-taker has some influence over this price and is said to have *market power*. In particular, if the firm produces at a very slow rate, the market price, which allocates this slow rate of production amongst customers, is likely to be very high. Conversely, if the firm produces at a very high rate, customers are likely to require a low price to absorb that high rate of production.

Let us suppose that for some group of customers, other things equal (in particular, holding constant the rate of production of all other firms), for each rate of production Q of the firm, there is some corresponding price P. The fact that the firm can influence the market price implies that the market price is a function of the rate of production of the firm. We can represent this by saying that the firm faces a downward sloping *residual demand curve*, which we will label $P^{RD}(Q)$.

As before we will consider first the short-run decisions of such a firm – in particular, the decision as to the rate of production. Let us suppose that a firm facing a downward-sloping residual demand curve seeks to choose a level of output which maximises its short-run profits. Will it choose a level of output which maximises total economic surplus? The problem this firm faces can be expressed as follows:

$$\max P^{RD}(Q)Q - C(Q)$$

The first-order condition for this problem is

$$P^{RD}(Q) + Q\frac{dP^{RD}}{dQ} = C'(Q)$$

The left-hand side of this expression is the *marginal revenue curve*. This expression says that a firm with market power will produce at a quantity where the marginal revenue (using the residual demand curve) is equal to the marginal cost of production.

Since the residual demand curve is downward sloping, the second term on the left-hand side is negative, so a profit-maximising firm with market power, selling to a group of customers at a fixed price independent of the rate of consumption, chooses to produce at a rate where the marginal cost of producing at that rate is less than the price at which it sells the output.

Earlier we saw that if customers are small, they will behave as price-takers and will consume at a rate where their marginal utility is equal to the price they are charged. But, we have just seen above: this price is higher than the marginal cost of production. So, a profit-maximising firm with market power, selling at a fixed price per rate of consumption, chooses a volume which is too low relative to the efficient outcome. This results in a reduction in economic welfare relative to a theoretical ideal, which is known as the *deadweight loss*.

We can re-write the expression above as follows:

$$\frac{P^{RD}(Q) - C'(Q)}{P^{RD}} = -\frac{Q}{P^{RD}}\frac{dP^{RD}}{dQ} = \frac{1}{\varepsilon^{RD}}$$

Here ε^{RD} is the elasticity of the residual demand curve, which is defined as follows:

$$\varepsilon^{RD} = -\frac{P^{RD}}{Q}\frac{dQ}{dP^{RD}}$$

The expression on the left-hand side above is known as the *Lerner index* or the price-marginal-cost margin. This expression then says that a profit-maximising firm facing a downward-sloping residual demand curve $P^{RD}(Q)$ will choose a rate of production where the Lerner index is equal to the inverse of the elasticity of the residual demand curve.

1.8.1 The Dominant Firm – Competitive Fringe Structure

As an example, let us consider the case of a large firm which faces competition from a number of very small rivals. This market structure is known as a *dominant firm* with a *competitive fringe*. The good or service produced by the dominant firm is assumed to be a perfect substitute in the eyes of the customers for the good or service produced by the rivals.

First let us ask what is the short-run social welfare maximising rate of production and consumption. Since the product of the dominant firm and that of the rivals are perfect substitutes in the eyes of the customers, the total rate at which the value is received by the customers depends only on the total rate of production: $U(Q^{DF} + Q^{CF})$. The total cost of production is the sum of the cost of production of the dominant firm and the competitive fringe: $C^{DF}(Q^{DF}) + C^{CF}(Q^{CF})$. The social welfare maximising rate of production and consumption is the solution to the following problem:

$$\max U(Q^{DF} + Q^{CF}) - C^{DF}(Q^{DF}) - C^{CF}(Q^{CF})$$

Here the maximum is taken over both the production of the dominant firm Q^{DF} and the production of the competitive fringe Q^{CF}. Solving this problem we find that (as before) social welfare is maximised when the marginal cost of increasing the rate of output by the dominant firm and by the competitive fringe is the same and is equal to the marginal value for the good or service.

Let us suppose that all the rivals in the competitive fringe are sufficiently small, so that they are price-takers. As a consequence, we can construct a supply curve for the rivals which we will denote by $Q^{SCF}(P)$. Since the rivals are price-takers, this supply curve reflects the marginal cost curve of the rivals:

$$C^{CF'}(Q^{SCF}(P)) = P$$

Similarly, we will suppose that there are a large number of small customers who are price-takers. We can therefore aggregate the demand curves of each customer to form the market demand curve, which we will denote by $Q^B(P)$. For any price chosen by the dominant firm, it will sell a product at a rate equal to the demand of customers less the supply rate of the competitive fringe. This is the residual demand curve:

$$Q^{RD}(P) = Q^B(P) - Q^{SCF}(P)$$

Now suppose that this firm seeks to choose a price which maximises its profits. As noted above, the dominant firm chooses the price which maximises:

$$\max Q^{RD}(P)P - C\left(Q^{RD}(P)\right)$$

This is maximised by a price P which satisfies

$$P = C'\left(Q^{RD}\right) - Q^{RD}\frac{dP}{dQ^{RD}}$$

In other words, the dominant firm will choose to sell at a price which is above its own marginal cost of production. The extent of the gap depends on the slope of the residual demand curve. Since the customers choose their rate of consumption, and the competitive firms choose their rate of production, based on the price they face, it follows that a short-run economic inefficiency will arise: Customers will consume at a rate which is too low relative to the efficient level, and the competitive fringe will produce at a rate which is too high.

From this analysis we can identify the two components of the deadweight loss: Customers consume at a rate which is inefficiently low relative to the efficient level, and the competitive fringe produces at a rate which is too high. Overall short-term economic efficiency could be improved by increasing the rate at which the monopoly supplier produces, reducing the rate at which the competitive fringe produces and increasing the rate at which customers consume.

We can re-write the expression above in the form of the Lerner index:

$$\frac{P - C'\left(Q^{RD}\right)}{P} = -\frac{Q^{RD}}{P}\frac{dP}{dQ^{RD}} = \frac{1}{\varepsilon^{RD}}$$

where $\varepsilon^{RD} = s\varepsilon + (s/1-s)\varepsilon^{SCF}$ and s is the market share of the dominant firm, ε is the elasticity of the market demand and ε^{SCF} is the elasticity of supply for the competitive fringe.

1.8.2 Monopoly and Price Regulation

The analysis above has highlighted the short-run economic harm from deadweight loss. There are other important economic harms from market power. Specifically, firms with market power can often price discriminate between customers. At the same time, customers will often be required to make material sunk investments in reliance on the monopoly service. This is particularly the case in, say, the electricity industry where customers must invest in significant consumption assets to draw economic value from the electricity service. The customer may be concerned that the monopolist will raise its prices after the customer has made a sunk investment. Facing the threat of hold-up, the customer may be reluctant to make investments in the first place.

In the longer run, the primary economic harm from market power appears to be the threat that market power poses to investment – especially investment by customers in reliance on the monopoly service.

There are two potential solutions to this hold-up problem: vertical integration and long-term contracts. Both are common in monopoly industries. Historically, in many countries, electricity network businesses were (and still are) government owned – a form of vertical integration. In other countries, electricity network businesses are privately owned but are subject to a form of long-term contract, known as public utility regulation.

1.9 Oligopoly

In some markets it is possible to accurately model the decisions of one producer or consumer in isolation from the decisions of all the others. For example, in the case of a competitive market we made the assumption that each firm was a price-taker – that is, each firm was too small to influence the market price. This allowed us to consider the actions of each firm separately from every other firm.

In the case of the dominant firm with a competitive fringe, we could not completely ignore the impact of the competitive fringe when considering the output decisions of the dominant firm. But in that case the reaction of the competitive fringe was particularly simple and could be summarised in the residual demand curve.

In certain other markets it is not possible to consider the decisions of one firm in isolation from the decisions of the others. This occurs in markets where there are several firms each large enough to have some impact on the market price. This market structure is known as an *oligopoly*.

A situation where the actions of one player affect the pay-off or earnings of another player is said to be a situation of *strategic interaction* and is conventionally modelled using the tools of *game theory*.

Previously we noted that, in order to model the behaviour of a producer or consumer, we need to specify (a) the actions they can take and (b) their objectives. In the case of a game of strategic interaction we also need to specify an *equilibrium concept* – that is a means of selecting predicted or equilibrium outcomes.

The simplest equilibrium concept is the *Nash equilibrium*. A Nash equilibrium is a set of actions, one for each player, which has the property that for each player the action specified for that player is the best (highest pay-off) for that player given the actions specified for the other players.

In addition, in a game of strategic interaction it turns out that it is important to distinguish between a game that is played once (referred to as a *one-shot game*) and a game that is played many times (referred to as a *repeated game*). For the moment let us focus on one-shot games.

The two simplest economic models of oligopoly are the games between firms known as the Cournot and Bertrand games. These games differ in the assumptions that each firm makes about the actions of the other firms in response to its own actions. In the *Cournot game*, each firm assumes that the other firms will hold their output fixed when the firm in question adjusts its price or output. In the *Bertrand game*, in contrast, each firm assumes that the other firms hold their prices fixed when the firm in question adjusts its price or output.

1.9.1 Cournot Oligopoly

To make this clear, let us suppose that we have two firms, each producing a single good or service. Let us suppose that the two goods are perfect substitutes in the eyes of consumers, so that consumers care only about the total rate of production of the goods.

Let us suppose that firm 1 chooses a rate of production Q_1 (units/interval) and firm 2 chooses a rate of production Q_2. Let us assume that we can represent the objectives of consumers in a single utility function $U(Q_1, Q_2) = U(Q_1 + Q_2)$. Then, if we assume that the good is sold at a simple linear price and if all consumers are price-takers, we can derive a single demand curve $P(Q_1, Q_2) = P(Q_1 + Q_2)$.

Let us suppose that the cost function of firm i is given as $C_i(Q_i)$. As we saw earlier, the efficient outcome is one in which the marginal cost of production is the same for the two firms (this may imply that one firm produces nothing if the other firm has a lower marginal cost of production).

Let us suppose that each firm chooses the rate of production which maximises the rate at which it earns profits under the assumption that the other firm holds its rate of production fixed. Each firm therefore solves the following problem:

$$\max Q_i P(Q_1 + Q_2) - C_i(Q_i)$$

The first-order condition for this maximisation yields

$$C_i'(Q_i) = P(Q_1 + Q_2) - Q_i \frac{\partial P}{\partial Q_i}$$

This yields two equations in two unknowns which can be solved to find the equilibrium rate of production of the two firms. This is left as a problem in Section 1.10. It is easy to check that this outcome is inefficient: The total rate of production is less than the efficient level. Furthermore, there is some production inefficiency in producing that less-than-efficient amount – the gap between the price and the marginal cost is larger for the larger firm.

It turns out that if we extend this model by allowing for more firms, we find that the Nash equilibrium approaches the competitive level as the number of firms in the industry increases.

In this simple model each firm chose its own rate of production assuming that other firms hold their own rate of production fixed. But, in order to hold their own rate of production fixed, they will typically have to adjust their prices. It might be thought that this seems unnatural. After all, isn't it the case that firms compete by adjusting their prices rather than their quantities? As noted above, there is a related game in which firms choose their own prices assuming that other firms hold their own prices fixed. This is known as a Bertrand game.

1.9.2 Repeated Games

The analysis here uses the notion of a Nash equilibrium. A Nash equilibrium has the desirable characteristic that if the other players happen to play the Nash equilibrium actions then no player will unilaterally want to change his/her action. The game the firms were playing above was a *one-shot game*: The firms come into existence, choose their actions, receive the pay-off, and then go out of existence. But this raises deep questions. If the firms did not exist in the past how do they reach the Nash equilibrium in the first place? What if there is more than one Nash equilibrium? In that case how do the firms know which equilibrium to coordinate on?

In almost all practical circumstances, firms do not just interact once. Instead, they interact repeatedly in the market. This ongoing interaction allows for a much richer set of strategies and actions than can be represented in a one-shot game. In particular, firms can react to the choices made by other firms in the past, rewarding or punishing them. When players interact repeatedly over time, the appropriate modelling tool is a *repeated game*. The range of equilibrium possible outcomes is much larger in a repeated game than in the one-shot games discussed above.

The reason for this is that in a repeated game it is typically possible to use much more severe 'punishment strategies'. In a one-shot game, if the other player does not play the best co-operative outcome there is nothing that either side can do. But in a repeated game, if one player does not co-operate the other player can switch to playing a strategy which the first player really does not like.

If the game is repeated for long enough, the parties may be able to sustain an equilibrium which yields a much higher pay-off than the one-shot game. In fact, in some circumstances it may be possible for the firms to collude to sustain an outcome which is as though the firms are operated as a single firm – a monopoly. This is likely to involve a lower rate of production and a higher price than the Cournot game discussed above. The easiest way to achieve this collusion is simply for the two firms to merge (also known as *horizontal integration*). But even if merger is not feasible, they may be able to come to an explicit or implicit agreement as to how much to produce and how to share the proceeds.

Most countries have a competition law which places strict limits on mergers between rivals, and which prohibits anticompetitive agreements between firms. Nevertheless, in some oligopoly markets, firms may be able to sustain arrangements which are less-than-fully competitive without an explicit agreement. This is known as *tacit collusion*.

As we will see, wholesale electricity markets are prone to the exercise of market power. In some markets there may be a single dominant firm, facing a downward sloping residual demand curve, as discussed above. In other cases, however, there will be several firms each of which has a degree of market power, and the market is better modelled as an oligopoly. Since electricity generators repeatedly interact with each other in the wholesale market, they may be able to sustain various forms of collusion or co-operative outcomes. Market power in the electricity market is discussed further in Part VII.

1.10   Summary

Economists model the short-term behaviour of consumers using a utility function (also known as the consumers' surplus). Consumers are assumed to maximise their net utility (utility from consumption less the amount they must pay to consume). Faced with a simple linear price a price-taking consumer will consume to the point where the marginal utility is equal to the price. This yields a downward sloping demand curve for each consumer which (if all consumers face the same price) can be aggregated to form the market demand curve.

Similarly, the short-run behaviour of producers can be modelled using the concept of the cost function. Producers are assumed to maximise their profit (their revenue from production less the expenditure they incur). Faced with a simple linear price, a price-taking producer will produce to the point where the marginal cost is equal to the price. This yields an upward-sloping supply curve which (if all producers face the same price) can be aggregated to form the market supply curve.

The efficient level of production and consumption in a market occurs where the marginal utility of each customer is the same and is equal to the marginal cost of each producer. In a competitive market this can be achieved through a normal market process.

In a smart market the normal market process is replaced with a centralised mechanism in which producers and consumers submit bids and offer functions to a central market operator. The central market operator carries out a constrained optimisation problem which may include physical constraints. This market process also yields an efficient outcome under the assumption that all players are price-takers. Electricity markets are a form of smart market.

In the longer run producers and consumers make investments – for producers, investments in productive capacity; for consumers, investments in assets which increase the value of the good or service traded in the market. The short-run variation in the spot price in a competitive market can provide efficient signals for expansion of productive capacity.

Some firms are not price-takers, but have the ability to influence the wholesale market price. Such firms are said to have market power. When selling at a simple linear price a monopoly will not choose the efficient level of output, but will choose to produce too little. This results in a loss of welfare known as deadweight loss. By adopting more complicated price structures (known as price discrimination) the monopolist will often be able to reduce the deadweight loss.

In some markets the action of one participant affects the pay-off of another. These situations of strategic interaction are best modelled using the tools of game theory. The most common equilibrium concept is the Nash equilibrium. In economics, game theory is used to model the short-term strategic interaction of a small number of firms, also known as oligopoly. One common approach is to model a game in which each firm chooses its level of output assuming that the other firms hold their output fixed. This is known as a Cournot game. The resulting outcome lies between the monopoly and the competitive outcomes.

Questions

1.1 Show that the (inverse) demand curve is downward sloping.

1.2 Let us suppose there are two products in a market. The rate of consumption of these products is denoted by Q_A and Q_B. The utility function for consumers of this product is

$$U(Q_A, Q_B) = 100A_0 + 10Q_A + 20Q_B - 2Q_A Q_B - 3/2Q_B^2$$

Find the demand curves for products A and B. How does the demand curve for product A change in response to a change in the rate of consumption of product B?

1.3 Suppose we have a monopoly provider of a particular good or service which charges a simple linear price. Suppose that the demand curve for this service is linear and takes the form:

$$P(Q) = A - BQ$$

Suppose that the cost of production takes the form: $C(Q) = cQ$. Show that the monopoly profit-maximising rate of production is equal to half of the socially efficient rate of production. Show that deadweight loss is equal to 25% of the total potential economic welfare in this market and that the deadweight loss decreases as P decreases provided $P > c$

1.4 Two firms produce two products, A and B, which are partial substitutes for each other. Given the price for A and B, P_A and P_B respectively, the rate at which customers choose to consume in the market is $Q_A(P_A, P_B)$ and $Q_B(P_A, P_B)$ respectively. Suppose that the cost of production is given by $C_A(Q_A) = c_A Q_A$ and $C_B(Q_B) = c_B Q_B$ respectively. Write down the profit of each firm as a function of the two prices in the market. Assuming that each firm chooses its price to maximise its profit while assuming that the other firm holds its price fixed, find an expression for the Bertrand–Nash equilibrium.

Further Reading

The material in this chapter is a modified form of the material which can be found in numerous introductory economics textbooks. For even more on these subjects you might be interested to explore the subject of Industrial Organisation which deals with the behaviour of firms in markets. One useful reference is Church and Ware (2000). See also Luenberger (1995). The concept of smart markets was introduced by Rassenti, Smith and Bluffing (1982).

Part II

Introduction to Electricity Networks and Electricity Markets

South Canterbury Electric Power Board linemen framed by their earthing sticks, New Zealand, ca 1927 (Source: Neil Rennie, Power to The People: 100 Years of Public Electricity Supply in New Zealand, Electricity Supply Association of New Zealand)

Electricity is so ubiquitous we mostly take it for granted. However, the reliable delivery of electric power to homes and businesses requires a very large number of interconnected parts and a large number of people, working together as a coordinated system. But what exactly is an electric power system? How and why is it organised the way it is? How does electric power flow over the electricity network or grid? What are the concerns that need to be taken into account when the system is designed or operated? In these chapters, we start to answer these questions.

2

Introduction to Electric Power Systems

All electric power systems consist of three primary physical elements: Devices for converting energy in other forms into electrical energy (also known as generators), devices associated with transporting or delivering electrical energy to where it is consumed (wires and networks), and devices for converting that electrical energy into the myriad goods and services that make up modern life (such as lights, heaters, motors and electronic devices).

Although there are countless small stand-alone electric power systems in operation (such as those in remote locations or remote communities), for reasons which we will discuss further below, for the past century the bulk of electric power consumed has been provided through a centralised, interconnected electricity network connecting hundreds or thousands of generating units to many millions of customers. Although, as we will see, there are some forces that tend towards decentralisation and possible splintering of future electric power systems, it seems likely that for the foreseeable future most electric power will be produced and consumed through large electric power systems connecting very large numbers of generating units and electricity consumers across hundreds or thousands of kilometres. It is these centralised interconnected electric power systems that are the focus of the electricity power industry and the subject matter of this book.

To start, however, we must have some basic understanding of how electricity flows in simple circuits. From the smallest microscopic circuits to large multinational networks, the laws governing electricity power flows are exactly the same. To understand how power systems work we must introduce some basic notions of DC and AC electricity.

2.1 DC Circuit Concepts

Electricity is the movement of electrons along electrical conductors (wires) that form a closed electrical circuit. Electricity is not useful in itself, but it is useful as a source of energy. Energy is an essential input into virtually every industrial process and most personal consumption processes in a modern economy. Electricity is useful and valuable in that it allows energy to be easily and reliably transported from one location to another and to be easily transformed from one form into another.

The Economics of Electricity Markets, First Edition. Darryl R. Biggar and Mohammad Reza Hesamzadeh.
© 2014 John Wiley & Sons, Ltd. Published 2014 by John Wiley & Sons, Ltd.

Figure 2.1 Gustav Kirchoff (Source: http://en.wikipedia.org/wiki/File:Gustav_Robert_Kirchhoff.jpg)

The physical force or pressure that induces the electrons to move is known as *voltage* and is measured in Volts. The rate of flow of electrons through the conductor at any point in the circuit is known as the *current* and is measured in Amperes or Amps.

It is common to draw an analogy between the flow of electrons in a wire and the flow of water in a pipe. Voltage is analogous to the pressure on the water in the pipe, and current is analogous to the rate of flow of water (in litres per second) through the pipe. The rate at which the flow of water delivers energy depends on both the pressure and the rate of flow. Water under high pressure but at a very low rate of flow delivers very little energy. Similarly a large flow rate of water but with very low pressure also delivers very little energy. In the same way, as we will see shortly, the rate at which electrical energy is delivered depends on both the voltage and the current.

In the nineteenth century, a German physicist named Gustav Robert Kirchhoff (1824–1887, see Figure 2.1) formulated two physical laws that describe how voltage and currents vary around an electrical circuit. Kirchhoff observed that electrons are neither created nor destroyed in an electrical circuit – they only move from place to place. Therefore, the flow of electrons into any point on a circuit must equal the flow out. This is *Kirchhoff's first law*: The current flow into any point on a circuit is equal to the current flow out.

Second, as electrons move around a circuit the voltage drops across those elements of the circuit where energy is being consumed, and increases across those elements of the circuit that inject energy. Since the total energy injected must be equal to the total energy withdrawn, the total voltage drop across all the elements of around any loop in an electrical network is always zero. This is *Kirchhoff's second law*: The sum of the voltage drop around any loop of a circuit is zero.

2.1.1 Energy, Watts and Power

The conventional physical measure for energy is joules. In the wider public debate, the concept of energy is often confused with the concept of power. *Power* is the *rate at which energy is delivered* and is conventionally measured in Watts (abbreviated W). One watt is equivalent to delivering energy at the rate of one joule per second. In informal usage it is common to talk of power 'flow'. However, strictly speaking, electric power does not flow – only electrical energy flows. Power is the rate at which energy is flowing.

In an electrical circuit the rate at which a network element is consuming energy (that is the power consumed in the network element) is equal to the voltage drop across that network element times the current flow through the element. More precisely: if the voltage drop across an element

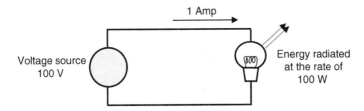

Figure 2.2   A voltage drop of 100 V and a current flow of 1 A equals a power of 100 W

in an electrical circuit is V and the current flow through the element is I, the rate at which that element is consuming energy (i.e. the power dissipated by that element) is equal to V times I.

Figure 2.2 illustrates this in a simple electrical circuit in which a lightbulb is connected to a voltage source. If the voltage drop across the lightbulb is 100 V and the current flow through the lightbulb is 1 Amp, the rate at which the lightbulb is consuming electrical energy is 100 W. If this lightbulb remains on for one second it has consumed precisely 100 joules of energy. In practice, however, electrical energy is usually measured in Watt-hours (Wh). One Watt-hour equals 3600 joules. If the lightbulb remains on for 1 h it has consumed 100 Wh of energy.

In the context of modern electric power systems, the units of Watts and Watt-hours are typically too small to be convenient. The rate of consumption of a typical household is typically measured in kilowatts (kW) and the total energy consumed in kilowatt-hours (kWh).

At the scale of an entire power system, we typically need units one thousand or one million times larger. The volume of electrical energy consumed in a power system is usually measured in megawatt-hours (MWh). Electrical energy is also sometimes measured in gigawatt-hours (GWh) (one billion watt-hours) or terawatt-hours (TWh) (one trillion watt-hours). Electrical power (the rate of flow of energy) is usually measured in megawatts (MW) and occasionally in gigawatts (GW). The capacity of a generator or the rating of a network element is also usually measured in megawatts.

The peak rate at which a generator can produce electrical energy is known as its capacity. Since this is a measure of the rate at which energy is produced, it is measured in MW.

2.1.2 Losses

In practice, no practical conductor conducts electricity perfectly. All practical conductors have some innate resistance to current flow (so-called superconductors have virtually zero resistance but these require extremely low temperatures – close to absolute zero – and therefore are not practical for other than specialised use in power systems). Because of this resistance, when a conductor carries electrical current, some energy is lost, primarily in the form of heat.

Earlier we noted that the rate at which energy is lost in a conductor is equal to the voltage drop across the conductor times the current flow. Ohm's law (Georg Ohm, 1789–1854, see Figure 2.3) tells us that the voltage drop across a conductor is itself proportional to the current flow. (The constant of proportionality is known as the resistance and is measured in Ohms). Therefore, we can conclude that the rate at which energy is lost in a conductor is proportional to the square of the current flow.

Mathematically, let P^L be the rate at which energy is lost in a conductor. We know that the rate at which energy is lost is equal to the voltage drop across the conductor multiplied by the

Figure 2.3   Georg Ohm (Source: http://en.wikipedia.org/wiki/File:Georg_Simon_Ohm3.jpg)

current through the conductor: $P^L = VI$. By Ohm's law, $V = IR$ where R is the resistance of the conductor; hence, we can conclude that the power loss is proportional to the square of the current flow $P^L = I^2R$.

Suppose we wish to design a power system that must deliver a total power flow of P MW. Since the power flow is equal to the voltage drop around the system multiplied by the current flow, if we hold the total power flow constant, the higher the voltage, the lower the current required to be flowing in the system.

However, we just noted that the energy loss in a conductor is proportional to the square of the current. Therefore, for a constant power output, *the power losses in the conductors are inversely proportional to the square of the voltage*.

Let P be the total delivered power in a system with a voltage V. The current flowing in the system is therefore the ratio of the power divided by the voltage: $I = P/V$. For a given power flow through the conductor, the power loss in the conductor is inversely proportional to the square of the voltage: $P^L = I^2R = P^2R/V^2$.

This is the primary reason why higher voltages are used for long-distance transmission. Doubling of the system voltage cuts down the rate at which energy is lost in the conductors by a factor of four. By increasing the voltage from 240 V to 500 kilovolts (kV), the rate at which energy is lost in transporting the electricity is reduced by a factor of about 4 million.

As already noted, the power that must be dissipated by a network element is proportional to the square of the current flowing through that network element. This energy is primarily dissipated in the form of heat. This heat causes transmission lines to sag or deform, and transformers to overheat and burn out. Cooling systems, fans and so on can assist in the removal of this heat. However, even with cooling systems in place, each network element has a maximum amount of current that it can safely handle. As we will see later, these so-called 'thermal' limits are one of the primary sources of technical operating limits on power systems.

2.2   AC Circuit Concepts

Very early power systems (in the 1870s in the United States) provided electricity at a constant voltage. Such systems were known as DC or direct current. Within a decade, however, these systems were superseded by power systems that were based on a voltage that varied in a constant, regular way, known as a 'sinusoidal' or 'sine wave'. This is known as alternating current or AC.

Around the world, AC voltage oscillates with a frequency of either 50 or 60 cycles per second (or Hertz, abbreviated Hz). The frequency is 60 Hz in North America, for example, and 50 Hz in Europe, the United Kingdom, Australia and New Zealand. In Australia, the peak

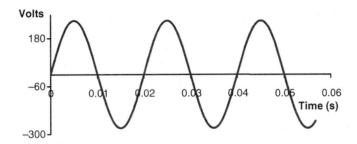

Figure 2.4 The path of voltage in a 50 Hz AC network

voltage is around 325 V, but it is conventional to express AC voltages and currents using the so-called root mean square (RMS) values which, in Australia is 230 V (the root mean square of a sine wave is found by dividing the peak amplitude by the square root of 2). Other countries have an RMS voltage of 120, 220 or 240 V. 120 V is common in North America, while 230 V is common in Europe, Australia and New Zealand.

Figure 2.4 illustrates how the voltage varies in an AC circuit with a frequency of 50 Hz. In this case the voltage varies over one complete cycle every fiftieth of a second (20 ms).

The primary advantage of alternating current is that, through a device known as a *transformer*, the voltage can easily be scaled up or down. A transformer consists of an iron core around which two coils of wire are wound. The degree to which the voltage is scaled up or down depends on the ratio of the number of loops or turns in each of the two coils of wire. If the output coil has 10 times as many turns as the input coil the voltage is scaled up by a factor of 10. An input voltage of 10 000 V is thereby converted into an output voltage of 100 kV. This is illustrated in Figure 2.5.

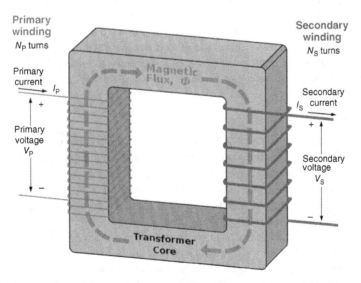

Figure 2.5 The transformation of the voltage in a transformer depends on the ratio of the number of turns (Source: http://en.wikipedia.org/wiki/File:Transformer3d_col3.svg)

The use of AC voltage and transformers makes possible the long-distance transportation of electricity. As we saw in the previous section, the higher the voltage, the lower the transmission line losses. By transforming the output of generators to very high voltage (as high as 500 or even 1000 kV), power can be transported long distances with very low losses. At the destination the electric power is transformed back down to a voltage that is safe to be distributed to industrial and residential consumers.

The use of very high voltages is, however, not without additional cost. The higher the voltage, the greater the strength of the electric fields between the wires of the network and between the wires and the ground. At some point, if these electric fields become strong enough the insulation between the wires (which is usually the air) breaks down and the current flows from one conductor to another in the form of a large spark known as an arc. The higher the voltage, the greater the distance that is required between the wires and between the wires and the ground. These large distances require larger and taller transmission towers with wider easements (see Figure 2.13). High voltages also require larger and more expensive transformers, circuit breakers and other substation equipment.

The frequency in an interconnected AC power network is *the same* at every point in the network. It is as though the network is one big synchronised machine – synchronised across thousands of miles. In Australia, the interconnected AC power system connects generators and loads in Queensland, on the one hand, through New South Wales (NSW), Victoria and South Australia, on the other. In North America, there are three separate interconnected AC power networks known as Interconnections. The Eastern Interconnection includes all US and Canadian states north and east of Texas. The Western Interconnection includes all the US and Canadian states north and west of Texas. Texas operates a separate AC power network that is not connected to these two interconnections (Figure 2.6).

Although the interconnected AC power system frequency is nominally 50 Hz (or 60 Hz in the US), it is not *exactly* this frequency all the time. Small imbalances in the supply/demand balance cause the frequency to rise or fall. The AC power system operates like a large engine. When a load is applied to the engine, the rate at which the engine is spinning tends to slow down, corresponding to a fall in the frequency. On the other hand, when a large load is suddenly removed, the engine tends to speed up. A large component of the operation of a power system in real time is making the continual adjustments necessary to ensure that the power system frequency remains within a very narrow band.

2.3 Reactive Power

As already mentioned, the rate at which energy is consumed in a network element (the power of that element) is equal to the voltage drop across that element times the current flow through the element. However, in an AC network the voltage is constantly changing, so the current flow is also changing, and following a sine wave of its own. The rate at which energy is produced or consumed by a network element is therefore more complicated than in a simple DC circuit.

In the case of a network element, which takes the form of a pure resistance, the current and voltage are perfectly 'in phase' – that is, they move up and down together or in sync, as in Figure 2.7. In this case, even though the voltage and current are both alternately positive and negative, the rate at which energy is consumed (which is the product of the voltage and the current) is always positive, as the lower graph in Figure 2.7 illustrates.

North American Electric Reliability Corporation Interconnections

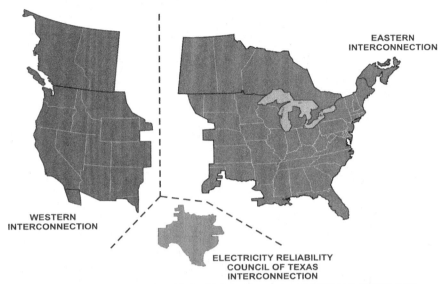

Figure 2.6 North American reliability corporation interconnections (Source: http://energy.gov/sites/prod/files/oeprod/DocumentsandMedia/NERC_Interconnection_1A.pdf)

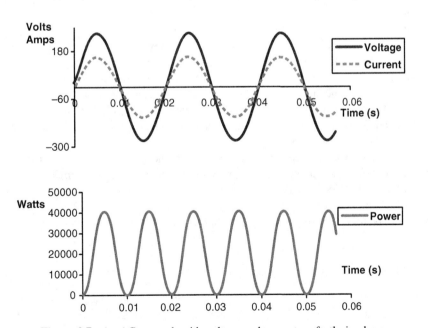

Figure 2.7 An AC network with voltage and current perfectly in phase

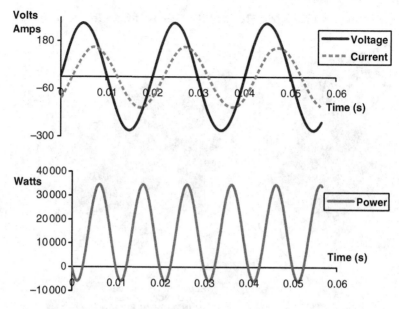

Figure 2.8 An AC network with current leading voltage by 45°

However, some network elements cause the voltage and current to be out of step with each other. In particular, *capacitors* (large metal plates separated by a very small distance) tend to cause the current to be ahead of the voltage (as in Figure 2.8) while *inductors* (coils of wire – such as those inside an electric motor) tend to cause the current to lag the voltage.

When the voltage and current are out of phase with each other, the network element is one moment absorbing a certain amount of energy from the power system and the next moment it is giving some of it back. This can be seen in Figure 2.8. The power is alternately positive and negative. The power is positive when the network element is withdrawing power from the system. The power is negative when the network element is giving some of that power back to the system. The energy that is alternatively withdrawn and then given back does no useful work on average but still causes current to flow in the conductors and therefore must be taken into account when evaluating thermal limits.

When the voltage and current are out of phase with one another, it is conventional to divide the total power flow at a point in an electrical power system into two components: The *real power*, which is always positive, and which can be used to do useful work, and the *reactive power*, which alternates between positive and negative. Since the reactive power averages to zero it canot be used to do any useful work, but it does contribute to heating and losses in the power system.

2.3.1 Mathematics of Reactive Power

Suppose that at a given point in an AC network the voltage follows a sine wave described by $V^{\max} \sin(\omega t)$ and the current follows a similar sine wave given by $I^{\max} \sin(\omega t + \theta)$ where θ is a parameter that measures the 'phase difference' between the voltage and the current – that is the degree to which the voltage and the current are out of phase. (when $\theta = 0$ the voltage and current are perfectly in phase; when $\theta = 180°$ the voltage and current are perfectly out of phase, and the power flows in the opposite direction). The instantaneous power flow is therefore

$$P = V^{max}I^{max} \sin(\omega t) \sin(\omega t + \theta) = VI \cos\theta(1 - \cos 2\omega t) + VI \sin\theta \sin 2\omega t$$

where $V = V^{max}/\sqrt{2}$ and $I = I^{max}/\sqrt{2}$. We can average this instantaneous power flow to find the average power flow over time. The first term averages to $P^{av} = VI \cos\theta$. This is the *real component* of the power flow. The second term averages to zero. This is the *reactive component* of the power flow and is equal to zero on average – reactive power does not deliver any useful work on average. It is conventional to define the quantity of reactive power as equal to $Q^{av} = VI \sin\theta$.

As already noted, even though the reactive power does no useful work on average, it still gives rise to real current flows, which can produce heating in network elements even when no real power is transported. This is clearest in the case when the phase difference between the voltage and the current is 90°. In this case, there is no real power delivered, yet there is still substantial current flowing through the system.

Since only the real power delivers useful energy, it is conventional to reserve the term 'watts' to refer to only the real component of the total power flow. The total power is referred to as the 'apparent power' and is measured in Volt–Amperes (VA, or more usually mega-Volt–Amperes – MVA). The reactive power is measured in Volt–Amperes-Reactive (VAR).

In an AC network with voltage V and current I, and with a phase difference θ between the current and the voltage, the apparent power is VI, the real power $VI \cos\theta$ and the reactive power $VI \sin\theta$. This is conventionally represented on a diagram as in Figure 2.9.

It is clear from this figure that if the real power is W, the reactive power Q and the apparent power P, then the following relationship must hold: $P^2 = W^2 + Q^2$.

Suppose we have a device which, when there is no reactive power present, can carry a maximum amount of electric power given by P. Suppose there is reactive power present in the amount Q. By the diagram above, this reduces the thermal limit on the real power that the device can handle to the amount $R = \sqrt{P^2 - Q^2} = P\sqrt{1 - Q^2/P^2}$.

In other words, when a device has a thermal limit P in the absence of reactive power, the corresponding thermal limit in the presence of reactive power is equal to $P \times PF$ where PF is known as the 'power factor', which equals one when the power flows are purely real $(PF = \sqrt{1 - Q^2/P^2})$. The larger the reactive power in the system, the smaller the power factor and therefore the lower the effective thermal limit.

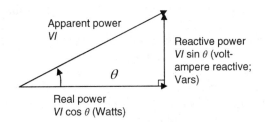

Figure 2.9   Schematic relationship between real, reactive and apparent power

Figure 2.10   High-voltage capacitor banks (Source: http://en.wikipedia.org/wiki/File:Condensor_bank_
150kV_-_75MVAR.jpg)

A power factor of 0.8 corresponds to reactive power equal in magnitude to 60% of the real
power flows. A power factor of 0.9 corresponds to reactive power equal in magnitude to 43% of
the real power flows.

2.3.2   Control of Reactive Power

Reactive power can be controlled by adding devices that increase or decrease the phase angle
between the current and the voltage. A device known as a capacitor (also known as a condenser,
which consists of two very large sheets of conductors separated by a very small distance)
induces the current to lead the voltage. Such devices are said to 'produce' reactive power.
Conversely inductors (large coils of wire, also known as reactors) induce the current to lag the
voltage and are said to 'consume' or 'absorb' reactive power (Figure 2.10).

These devices can be used to offset any reactive power flows that might exist at a certain
point on the network. For example, where the current lags the voltage capacitor banks may be
switched into the circuit in parallel with the load. The quantity of capacitors that must be
switched in will depend on the quantity of offsetting reactive power that is required.

In practice, the quantity of these inputs required will vary continuously as the level of
reactive power varies with the quantity and type of load. In principle, the system operator can
respond by switching in or out of the network different amounts of capacitance or inductance.
There are also flexible devices for controlling reactive power. These include 'synchronous
condensers' (which is essentially a type of generator without an external energy source, which
can be set to produce or consume reactive power) and so-called 'Static VAR compensators',
which are purely electronic devices that allow the level of reactive power to be controlled in
a continuous and automatic manner. Many generators are also able to produce or consume
reactive power on demand.

Reactive power arises in a power system partly as a consequence of the electrical character-istics of the network itself and partly as a consequence of the electrical characteristics of end user loads. Many loads on the power system – especially large motors – take the form of large coils or inductors and therefore consume reactive power. To offset this tendency, distribution companies often require their large customers to maintain a power factor close to one – that is, to provide their own devices (usually capacitors) to offset the reactive power induced by their electrical equipment.

Small residential and commercial customers are not currently charged or compensated for the reactive power they produce or consume. Instead, any reactive power produced or consumed by such loads must be offset by the power system itself. At certain times this can be a significant requirement. For example, on particularly hot days up to half of all the energy used may be consumed by air-conditioning motors, which tend to consume reactive power.

The power system itself can also produce or consume reactive power. A transmission line has some of the characteristics of both a capacitor and an inductor. As the voltage builds up on a transmission line, an electric field develops between the line and the ground, in the same manner as in a capacitor. Similarly, the continuously varying current flow in a transmission line induces a changing magnetic field around the line, in the same manner as in an inductor. These capacitance and inductance characteristics of a transmission line produce and consume reactive power just like other network elements.

The strength of these different effects depends on the loading on the transmission line. A lightly loaded line is likely to be net-capacitive – that is producing reactive power. As the loading on the line increases it is likely to become net-inductive – consuming reactive power. The consumption of reactive power is not linear in the loading on the line – a single transmission line carrying a given level of power consumes significantly more reactive power than two parallel lines each carrying half the same power. Transmission line outages can therefore have a significant effect on reactive power requirements.

2.3.3 Ohm's Law on AC Circuits

We saw earlier that in simple DC circuits the current flow through an element is equal to the voltage drop across the element divided by the resistance (measured in Ohms). An analogous relationship holds for AC circuits. In fact, for AC circuits the current flow through a network element is equal to the voltage drop across the element divided by the *impedance*. Impedance, like resistance, is measured in Ohms. In an AC circuit in which the voltage and current are perfectly in phase, the impedance is equal to the resistance of the network element. However, when the voltage and current is out of phase (i.e. when reactive power is present) the impedance is larger than the pure resistance of the network element. The additional component of the impedance is known as *reactance*.

Since capacitors and inductors cause the current to be out of phase, capacitors and inductors induce reactance, even when they have no resistance to the current flow. The reactance of a capacitor is proportional to the frequency whereas the reactance of an inductor is inversely proportional to the frequency.

Impedance, resistance and reactance are related in a manner that is closely analogous to the relationship between apparent power, real power and reactive power, as in Figure 2.11.

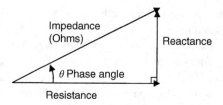

Figure 2.11 Schematic relationship between impedance, reactance and resistance in an AC circuit

2.3.4 Three-Phase Power

Only two conductors are necessary to complete an electrical circuit (in fact, in many power system applications only one conductor is required as the circuit can be completed by electrons returning through the earth). However, the flow of power in a simple AC circuit is not constant. In the simplest case of a circuit with a pure resistive load, illustrated in Figure 2.7, the power flows in 'pulses' (with two pulses for each complete cycle of the alternating current). These pulses can cause vibrations in electrical equipment and motors.

We can achieve a constant power flow over time using three conductors carrying alternating currents that are exactly 120° out of phase with one another. This is illustrated in Figure 2.12.

We can check that the total power flow is constant by adding up the power flow in each phase. In the case of a simple resistive load the power flow in each conductor is

$$P_1 = 2VI \sin^2 \theta, P_2 = 2VI \sin^2 \left(\theta - \frac{2}{3}\pi \right), P_3 = 2VI \sin^2 \left(\theta - \frac{4}{3}\pi \right)$$

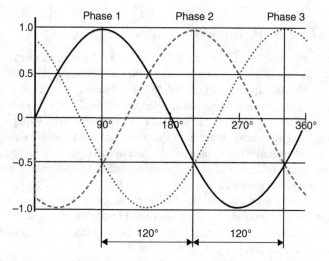

Figure 2.12 The movements of voltage and current in three-phase power (Source: http://commons.wikimedia.org/wiki/File:3_phase_AC_waveform.svg)

Summing these, we find that the total power flow is constant and equal to three times the average power flow in any one conductor: $P_1 + P_2 + P_3 = 3VI$. This is due to the fact that for any value of θ

$$\sin^2 \theta + \sin^2 \left(\theta - \frac{2}{3}\pi \right) + \sin^2 \left(\theta - \frac{4}{3}\pi \right) = \frac{3}{2}$$

Three-phase power is an efficient way of delivering large amounts of electrical energy. The fact that the power flow is constant tends to reduce vibration in electrical equipment and motors, and the three-phases can be used to create a magnetic field that rotates in a specific direction, simplifying the design of electric motors.

Three-phase power is widely used in AC electric power networks. It is typically used for all high voltage transmission lines, and most distribution lines. Nearly all large electricity consumers (particularly commercial and industrial loads) are provided with three-phase power. Some large apartment buildings are also supplied with three-phase power. Some light commercial or domestic appliances (such as large stoves) require three-phase power, but most residential customers, especially in rural areas, are only supplied single-phase power.

2.4 The Elements of an Electric Power System

Historically, all AC power systems were organised as set out in Figure 2.13. Energy is converted from some other form into an electrical form by *generators*. *Transformers* step-up the output of generators to the high voltages needed for transmission. The power is carried by high-voltage *transmission networks* closer to the major load centres, where it is stepped down to lower voltages, distributed closer to end users via the *distribution networks* before being stepped down to the voltages at which it is distributed into residential homes and businesses. The following sections examine each of these components in a little more detail.

However, it is worth emphasising that Figure 2.13 reflects a classic or traditional structure of an electric power system. Traditional power systems were constructed around relatively few large generators typically located close to major energy sources. The output of these generators was then carried often relatively large distances, in one direction, to where the electrical energy is consumed.

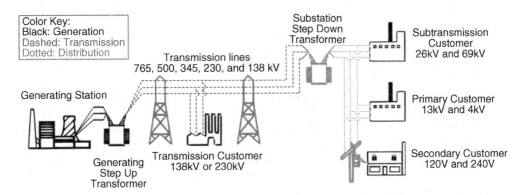

Figure 2.13 The basic elements of a power system (Source: http://en.wikipedia.org/wiki/File:Electricity_grid_simple-_North_America.svg)

However, in recent years there has been increasing take-up of small-scale generation, often located at the same physical location as energy consumers. With increasing take-up of electric vehicles, many small customers will have the ability to store reasonable amounts of electrical energy, which could be reinjected into the power system when required. It seems increasingly likely that there will be a blurring of the distinction between electricity generators and consumers. Instead, many end-customers will be able to switch from withdrawing power from the network to injecting power if local conditions change. Flow on the network will change according to changing supply and demand conditions.

At present, large electricity generators and loads are typically well integrated into the wholesale market – that is, their output is closely monitored and metered, they are typically paid (or pay) the wholesale price for the amount they produce or consume, they usually submit bids and offers directly to the wholesale market, and take instructions from the wholesale market as to how much to produce or consume. In contrast, small customers are typically almost entirely passive and unresponsive to wholesale market conditions.

However, with increasing penetration of small-scale generation, increasing scope for storage (perhaps through electric vehicles), more sophisticated metering, and greater use of IT and communications infrastructure, there is scope for integrating even smaller customers into the wholesale market. This vision of the future of the electricity industry is sometimes referred to as the *Smart Grid*, and is discussed further in Section 2.9.

The next few sections look in more detail at the primary physical components of the electric power system: electricity generation, electricity networks and electricity consumption.

2.5 Electricity Generation

Electricity generation is the transformation of other forms of energy into electrical energy. There are seven fundamental methods of transforming other forms of energy into electrical energy, but in practice only the first two or three are of any significance in existing electric power systems.

- *Electromagnetic induction* – the use of moving magnets to convert kinetic energy into electrical energy. This is by far the most common approach to generating electrical power in practice, as we will discuss further below.
- *The photoelectric effect* – the conversion of photons (light energy) into electrical energy. In the last decade the declining cost of solar generation and government subsidies for solar power has dramatically increased the volume of electric power generated in this way.
- *Electrochemistry* – the generation of electricity through chemical reactions, such as those that occur within a battery. Historically, although widely used in small-scale applications where portability is essential, batteries have not been common in the electric power industry. However, with increasing penetration of electric vehicles, this situation may soon change. Another device that falls into this category is the fuel cell. Fuel cells convert chemical energy in the fuel (which is typically hydrogen but may be a hydrocarbon such as methanol) into electrical energy. Some fuel cell reactions can be reversed allowing them to store as well as release electrical energy.
- *The thermoelectric effect* – the conversion of heat energy (across temperature differences) directly into electrical energy, via the use of thermocouples, thermopiles, or thermionic

converters. To date there has been virtually no use of this approach to generating electricity in the electrical power industry.

- *Piezoelectric effect* – the generation of electricity from the mechanical strain of electrically anisotropic crystals. To date there has been no use of this approach in the electrical power industry, although patents have been filed that make use of this effect to generate electricity from waves.
- *Nuclear energy* – directly capturing the energy contained in alpha or beta emissions. To date this approach has only been used where very long-life remote use is required, such as on spacecraft.
- *Static electricity* – the energy stored in charges separated by mechanical action. The most dramatic example is lightning. Although different approaches have been tested, to date it has not proven possible to commercially capture the electrical energy in lightning.

As noted here, of these approaches, by far the most common in practice is the first. Almost all commercial electricity generation is carried out by electromagnetic induction – the conversion of kinetic energy into electrical energy.

In most cases the kinetic energy that turns the rotor of the generator comes from a *turbine*. We can further distinguish between generators according to the fluid that moves the turbine. In the case of *hydroelectric generation* the turbine is turned by the energy of falling water; in the case of *wind generation* the turbine is turned by the energy of the wind; in the case of some gas-powered generators (known as *gas turbines*) the blades of the turbine are turned directly from the gases produced by the combustion of natural gas or oil.

In almost all other applications, the blades of the turbine are turned by steam under pressure. This steam, in turn, is manufactured from some heat source. The source of the heat varies from one generator to another. In *nuclear power generators*, the heat comes directly from the nuclear reaction itself. In *fossil fuel generators* (coal, gas, or oil), the heat comes from burning the *fossil fuel*. Other sources of heat are *biomass* (the burning of waste), solar energy (capturing the heat

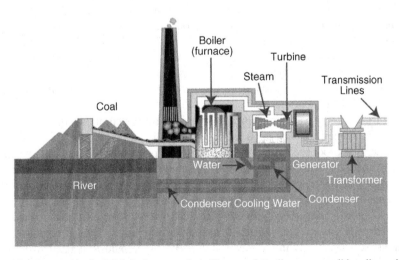

Figure 2.14 Conventional fossil-fuel generation (Source: http://commons.wikimedia.org/wiki/File: Coal_fired_power_plant_diagram.svg)

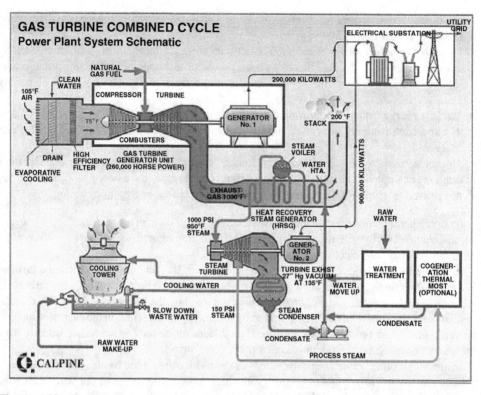

Figure 2.15 Combined-cycle gas turbine (Source: http://www.power-technology.com/projects/san_joaquin/san_joaquin3.html)

energy from the sun, through a series of concentrating mirrors, known as *solar thermal generation*), or *geothermal* (underground) energy sources (Figure 2.14).

In some cases two or more approaches are combined. In the case of combined-cycle gas generators, hot burning gas is used to directly turn a turbine. The hot exhaust that emerges from the gas turbine is used to produce steam, which is converted into electricity in a second generator, thereby increasing the overall efficiency of the process. In cogeneration facilities, the heat energy (usually in the form of steam) is directly used to provide local heating services, alongside the production of electricity (Figure 2.15).

Although turbines of various forms are by far the most common, in some applications (especially emergency or back-up power or balancing power on very high demand days) electrical energy is produced using a standard diesel or petrol engine turning a small generator.

Although the bulk of the electric power consumed comes from the conversion of kinetic energy through electromagnetic induction, increasing amounts of power come from other sources. One of the most important is *solar photovoltaic generation*. Solar generation uses solar cells to directly convert sunlight into electrical energy. Solar cells are small enough to be easily installed on the roofs of households or small businesses. Typically roof-top solar PV installations have a capacity in the range of 2–5 kW. Some industrial-sized solar farms have been established around the world, with most less than 100 MW, but some of 200 MW or up to 400 MW in size (Figure 2.16).

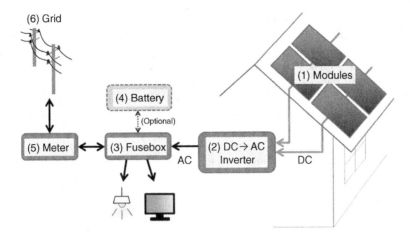

Figure 2.16 Typical simple solar PV installation (Source: http://commons.wikimedia.org/wiki/File:PV-system-schematics-residential-Eng.png)

Although historically batteries have played a very little role in commercial-scale electric power systems, with the increasing penetration of electric vehicles there is increasing interest in the possibility of small-scale electricity storage to smooth electricity production and consumption over time.

The proportion of electrical energy produced using each of these different approaches differs widely from one region to another and one country to another. Some countries, such as France, produce the bulk of the electricity from nuclear energy. Countries such as Norway and New Zealand produce the bulk of their electricity from hydroelectric sources. Australia produces the bulk of its electricity from coal.

2.5.1 The Key Characteristics of Electricity Generators

As we have seen above, different electricity generators differ greatly from one another. Yet, as we will see, many different types of generators will coexist simultaneously in an efficient electric power system.

Importantly, generators can differ from each other in the following dimensions:

- *Controllability* – The output of most thermal, geothermal or hydro generators can be varied by controlling the amount of steam produced and allowed to pass through the turbine (or the volume of water in the case of a hydro generator). In the case of many renewable generators, however, such as solar, wind or wave generators, the maximum output at any given point in time is primarily determined by weather conditions. These generators cannot usually increase their output in response to market conditions (although they may be able to reduce their output). Importantly, the output of these generators may drop off rapidly if weather conditions change – for example, if the sun passes behind a cloud, or if the wind drops suddenly. In many countries one of the most important current challenges facing the power system is the successful integration of large amounts of uncontrollable power sources.
- *Limits on rate of change of output* – Amongst those generators whose output is controllable, generators differ in the rate at which their output can be *ramped up or down* in response to

changes in the supply–demand balance in the wholesale market. Some sources, such as batteries, can ramp up quite rapidly. Hydro generation can also increase output rapidly over a period of seconds or minutes (the rate at which flow gates can be opened). However, many large fossil fuel or nuclear generators cannot increase or reduce their output very rapidly at all, taking many minutes to increase their output by a few megawatts.

- *Variable cost of production* – For most generators, the cost of producing electricity varies with the rate at which electricity is produced – the higher the rate at which electricity is produced, the more fuel is required and therefore the higher the cost. This cost depends on both the price of the input fuel (which may itself vary on a spot market) and the efficiency of the generator in converting the input fuel into electrical energy. The efficiency of a thermal generator is known as its heat rate. For a gas-fired generator, the heat-rate is measured as the ratio of the energy in the input fuel to the electrical energy produced. A generator that converts energy in a fuel source into electrical energy can be thought of as engaging in *arbitrage* between the price of the input fuel and the price of electricity. A generator's variable cost may also depend on the spot price of emission permits or renewable energy certificates, which may also vary over time. Renewable generators (solar, wind, wave, etc.) tend to have relatively low variable costs.
- *Limits on the rate of production* – All generators have some maximum rate at which they can produce energy, known as their generating capacity. The largest thermal generating stations may be capable of producing thousands of megawatts. However, a typical electric power system will also have many smaller generators, capable of producing a few hundred megawatts. The increasing penetration of wind and solar generation is resulting in hundreds of thousands of smaller generators capable of producing less than a few megawatts. The typical small rooftop PV system will be capable of producing only a few kilowatts of power.
- *Limits on the energy that can be produced in a period of time* – Although all generators have a limit on the rate at which they can produce electrical energy, a few generators are also limited in the amount of energy that they can produce in one session. For example, in the absence of new rainfall, a hydro generator situated below a lake can only operate continuously until its water supply is drained. The same applies to a generating unit that is drawing on some source of stored energy, such as a battery. Some generators have a limited stock of fuel on hand and limited ability to renew that stock. Such generators are known as *energy-limited* generators.
- *Fixed cost* – All generators require some physical plant. However, the relative importance of fixed and variable costs and the absolute size of the fixed cost differs across different generators. As we will see, amongst controllable generators, some generators have substantial fixed costs and relatively low variable costs. These are known as *baseload* generators. Other generators have relatively low fixed costs and high variable costs. These are known as *peaking* generators. A related idea is the notion of *economies of scale*. While some generation technologies (e.g. solar and wind) can be scaled down with little loss of cost-effectiveness, historically the trend amongst conventional generators has been to larger units to exploit economies of scale, with many generating plants capable of producing several thousand megawatts.
- *Environmental impacts* – Generators also differ, of course, in the nature and extent of their environmental impact. In many countries the electricity generation sector is a major contributor to emissions of greenhouse gases, such as CO_2. Many fossil-fuel generators are strongly affected by policies designed to mitigate CO_2 emissions. In addition, electricity

generators can be a source of other atmospheric pollutants. Many large generators also rely on large supplies of water for cooling. In urban areas there may even be concerns about noise levels associated with, say, a gas turbine.

- *Forecastability* – As we will see, efficient operation of a power system requires information on the likely supply–demand and network conditions in the future, both the near future (over the course of a day or so), and over longer terms, such as months or years. Predicting future power system conditions requires, amongst other things, predicting the output or the behaviour of generators in the future. Generators differ in the extent to which their future behaviour is predictable.

- *Reversibility* – Rarely, the process used to convert some other form of energy into electrical energy is reversible. For example, the drawing of electrical energy from a battery or a fuel cell is typically reversible. Some hydro generators have the ability to pump water uphill, to allow generation at a later time. Other forms of energy storage, such as heat, or flywheels, are theoretically possible but not yet in commercial operation. When the process is reversible the generator has the ability to also act as an electrical load – allowing the generator to arbitrage between the price of electricity at different time periods.

- *Relationship to related services* – Not all electricity generators have the production of electricity as the sole objective or even a primary objective. Some electricity generators also produce heat for a building or local area. For such generators, the need to provide heating services may partially override considerations in the electricity market. In other cases the production of electricity is ancillary to other objectives, such as the burning of waste products (e.g. as a by-product of sugar production in Australia). In the same way the charging or discharging of an electric vehicle is ancillary to other transportation needs and requirements.

- *Other characteristics* – There are a number of other characteristics of generators that affect generation operation and investment decisions in electric power systems, such as the following:

  - *Minimum loading* – The minimum safe and stable operating level. Many conventional thermal or nuclear generators cannot safely operate at very low levels of output. These generators prefer to be operating in a range near full capacity and would rather shut down than be forced to produce a low level of output.

  - *Start-up costs* – The costs that must be incurred before any output can be produced. Although some generators (such as solar or hydro) can switch on and off at will, many conventional thermal generators must heat water to produce steam before any electricity can be produced. This process can take several hours and consume a material amount of fuel. Such generators prefer to be producing for long periods of time, rather than undertake short periods of shutdown.

  - *Inertia* – The tendency of a generator to resist very short-run changes in the local supply–demand balance. Physically large generators with rotating turbine blades have a large amount of inertia and tend to naturally resist slowing down in response to a sudden increase in load, or speeding up in response to a sudden drop in load. Other generators, such as solar PV generation, do not exhibit such inertia at all.

  - *Ability to produce or consume reactive power* – As we will see, part of the active management of a power system is control of the level of reactive power at different parts of the network. Different generators have the capability to produce or consume reactive power, which can be a valuable service.

- *Black start capability* – Many generators require a source of electrical energy before they can start producing. This electrical energy may be necessary to operate pumps, cooling systems, and so on. This presents no particular problem as long as the generator is connected to a working electrical power system as the generator can draw a small amount of power from the grid to start operations. However, in the event of a major power system failure, the ability of a generator to start producing electricity without any external electrical input can be a very valuable service.

In modern economies, the demand for electric power is such that in all except small, isolated or remote systems, the number of different generating units that are required to meet total demand is typically quite large (often in the hundreds or thousands). Moreover, many of those generating units will be active at any one time. As a consequence, it is possible to envisage competition between rival generators. One of the key objectives of the reforms of the electricity industry, which we discuss in Chapter 3.2, is to foster competition between generators. As we will see, a large number of the chapters that follow explore the conditions under which competition between generators achieves socially desirable outcomes.

As we will see in later sections, in many electric power systems, generators compete with one another in a wholesale power market. However, not all generators (or consumers) will be required to participate in that wholesale market. Therefore, in such power systems generators also differ from one another in one further dimension: The extent to which they are integrated into and respond to wholesale market conditions. As we will see, most large generators are required to be directly involved with the wholesale market – submitting offer curves to the spot market and taking instructions from that market as to how much to produce. In addition, such generators are typically required to inform the market about scheduled outages and future expansion plans. These generators typically rely on sophisticated monitoring and communication facilities to communicate with the central market operator and to receive dispatch instructions and information on local network conditions, in return.

In contrast, at present, many smaller generators are not directly integrated into the wholesale market. These generators are not necessarily required to participate in the wholesale spot market and typically do not receive dispatch instructions from the wholesale market. In addition, many smaller generators, such as rooftop solar PV installations, do not respond to wholesale market prices at all. Although the output of these generators may be monitored, the extent of communication with the broader network is currently limited. As we will see, one of the themes of this textbook is the integration of smaller generation and consumption devices into the broader local and national electricity market.

2.6   Electricity Transmission and Distribution Networks

Although a small amount of electricity is produced where it is consumed, there are several reasons why electricity generators are often located a long way from loads, giving rise to the need for electricity transmission and distribution networks connecting generation sources to electricity consumers. These reasons are as follows:

- *Location of fuel sources* – Energy is only useful if it is available at the location where it is needed. It is often (but not always) cheaper to transport energy in the form of electricity than to transport the input fuel or energy source used to produce electricity. In addition, it is often cheaper to construct large power plants close to fuel sources than to construct large power plants close to electricity consumers. For both these reasons it is often cheaper overall (that is, the cost of production plus transportation) to locate an electricity generation plant close to energy sources and to construct transmission lines to transport that power close to consumers, than to locate the generation plant close to consumers and transport the fuel or energy source to the generation plant. This particularly applies for energy sources that effectively cannot be transported at all, such as geothermal energy sources, hydroelectric generation, or even much wind, solar or wave generation. Other energy sources, such as coal, can be transported but it often remains cheaper to locate an electricity generation plant at the mouth of a coal mine, and to transport the energy in the form of electricity, especially for low-grade coal. It is less clear whether it is cheaper to transport energy in the form of natural gas through pipelines, than over long-distance transmission lines.

- *Economies of scale* – Even where the fuel or energy source can be easily transported, many forms of generating technology are more cost effective at a large scale. In other words, it can be cheaper to build a single generating unit, even where that requires transporting electricity over longer distances, than to build multiple smaller generating units closer to the consumption sites.

- *Diversification of risks* – No generating sources are perfectly reliable. In practice, handling the sudden loss of large generating sources is costly to an electric power system. As a consequence (as we will see) it is common in power systems to take precautionary actions to reduce the cost of handling such contingencies when they occur. However, when the risks of generator outages are independent of one another, the cost of such precautionary actions do not scale proportionately with the number of generators in the system. As a consequence, it can make sense to connect two independent power systems to reduce the average cost of taking actions to handle contingencies.

Historically, the trend in electric power systems has been towards larger, more efficient generating plants, connected by networks spanning large distances. However, in recent years there has been some reversal of that trend. In particular, there has been an increasing take-up of small-scale solar and wind generation, often with generating units located close to consumers. Other technologies, such as small-scale fuel cells, may lead to further take-up of micro-generation facilities. This tendency, if it is pursued, will reduce the demand for long-distance transportation of electricity. Some commentators have even suggested that customers may increasingly choose to disconnect from the broader electric power system. However, disconnecting customers lose the benefit of electricity trade – importing power at times of local shortage and exporting power at times of local surplus. Although the distances over which electricity is transacted may diminish, it remains to be seen whether widespread disconnection will be an attractive option in the future.

As noted earlier, this large-scale network or grid is usually divided into two components: the high-voltage *transmission network* and the low-voltage *distribution network*. In practice there is no bright-line distinction between these two types of networks. The distinction will vary across different regions, across networks and across countries.

Figure 2.17 Typical high-voltage transmission lines (Source: http://commons.wikimedia.org/wiki/File: Electric_transmission_lines.jpg)

2.6.1 Transmission Networks

Transmission networks (as illustrated in Figure 2.17) typically carry large volumes of electric power over relatively long distances. A typical transmission network will have relatively few (usually hundreds) directly connected customers consisting of dozens or hundreds of large generating units that inject power into the network, and dozens or hundreds of points at which power is withdrawn from the network – either by directly connected large customers (such as aluminium smelters) or distribution networks, which themselves connect to a large number of smaller customers.

Since, as we saw in Section 2.1, the amount of power that is lost in the form of heat in electricity networks is inversely proportional to the square of the voltage, the higher the voltage, the smaller the losses. For this reason transmission networks tend to use very high voltages, with the highest voltages used for the longest and most heavily loaded transmission lines.

The voltages used for transmission around the world range from roughly 100 kV to as high as roughly 1000 kV, with 132, 220, 275, 330 and 500 kV being the most common.

Most of the largest generators in an electric power system are directly connected to the transmission network. Large industrial-scale generators usually produce electricity at a voltage in the range of 10 to 25 kV. Transformers are used to step the voltage up from this level to the voltage level on the local transmission network.

However, higher voltages bring their own problems. The higher the voltage, the greater the tendency for the voltage to break down the surrounding insulation. The surrounding insulation is often nothing more than air. When the voltage difference is high enough, a spark (or an *arc*) can form between two conductors, in effect creating a short circuit, which can damage the power system. To protect against this possibility, high-voltage equipment is typically taller and larger, with larger spacing between conductors, and between the conductors and the ground.

The electricity transmission network is a collection of several types of assets, including poles and wires, transformers, circuit breakers and other switching equipment, equipment for

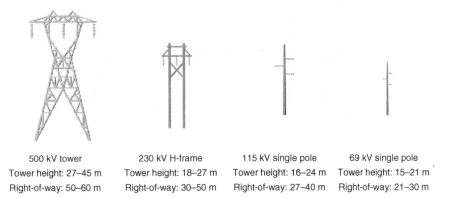

500 kV tower   230 kV H-frame   115 kV single pole   69 kV single pole
Tower height: 27–45 m   Tower height: 18–27 m   Tower height: 16–24 m   Tower height: 15–21 m
Right-of-way: 50–60 m   Right-of-way: 30–50 m   Right-of-way: 27–40 m   Right-of-way: 21–30 m

Figure 2.18 Dimensions of transmission towers of different voltages (Source: Minnesota Electric Transmission Planning at   http://www.minnelectrans.com/transmission-system.html)

controlling reactive power (including capacity and reactor banks), and communications and monitoring equipment.

Electricity is carried through the high-voltage transmission network in power lines suspended from *poles* and *towers*. The higher the voltage, the greater the distance required between the wires and between the wires and the ground. Figure 2.18 illustrates the different height and easement requirements for differing voltages. The easement is the area of land immediately underneath the power line over which there are strict limitations of use (no buildings, trees, and so on).

The transmission lines themselves usually consist of aluminium wires wrapped around a steel core. The steel core increases the strength of the transmission line and resists the tendency of the transmission lines to sag as the current increases. Aluminium is used for the conductor because of its low resistance to electricity (high 'conductivity').

Although most transmission lines are suspended from poles and towers, in some cases electricity is transported along insulated cables. Such cables can be buried underground in environmentally sensitive areas, and can even be submerged in water. Although cables tend to have higher reliability, they also have higher costs. For this reason they tend only to be used in areas where overhead transmission lines are not feasible (as when crossing a body of water) or would be unacceptable for environmental reasons.

Transmission lines are interconnected with one another at *switching stations* and *substations*. These substations house transformers and most of the other key network assets including switching devices, reactive power control devices, and monitoring, control and communications devices.

Of the switching devices, the most important are the *circuit breakers*. Circuit breakers must have the capability of disconnecting very high voltages and current flows. As we have already noted, at high-voltages the normal insulating properties of the air breaks down. Therefore, high-voltage circuit breakers need to be large in order to ensure that no arcing occurs when the switch is open. The process of opening and closing these circuit breakers can induce very large arcs as shown in Figure 2.19. Other devices, known as network protection devices, monitor the current flows on individual pieces of equipment and automatically switch the device out of service if the ratings on the equipment would be exceeded.

Figure 2.19   Arcing in a high-voltage switchyard (Source: http://www.lifeaftercoffee.com/2006/05/23/worlds-biggest-jacobs-ladder)

Substations also usually house a variety of equipment for regulating the level of reactive power in the manner discussed in Section 2.3. These include capacitor banks, synchronous condensers, static VAR compensators and other devices.

In addition to the equipment described above, power systems make substantial use of telecommunications equipment for remote monitoring and control of the power system. A large number of devices located at various points around the power system collect information and communicate that information (known as 'telemetry') back to the control centre. At the same time, the control centre can give instructions that are communicated back to the devices on the power system, opening or closing circuit breakers, switching capacitor banks in or out, increasing or decreasing the output of generators and so on.

For the reasons set out in Section 2.3, almost all AC transmission networks use three-phase power. That is, each transmission line is made up of three conductors. This also means that three circuit breakers are required, three sets of monitoring equipment, and often three transformers for each transmission circuit.

Figure 2.20 shows an extract of the schematic diagram of the power system in the vicinity of the Broken Hill region of New South Wales. This schematic diagram shows a substation connected to two generators through step-up transformers. The substation also includes a number of capacitor banks, static VAR compensators and reactors for regulating reactive power flows into this region. Each line on this diagram represents three conductors in the 'real world'.

Although the vast majority of modern power systems are alternating current networks, many electric power systems have a few DC network elements. DC transmission lines are commonly used for interconnecting two neighbouring power systems whose frequency is not otherwise synchronised. These lines are typically very high voltage, and are often abbreviated HVDC. In addition, high-voltage DC links are sometimes used for environmentally sensitive locations or crossing large bodies of water.

High-voltage DC links are either monopolar or bipolar. Monopolar links use a single conductor and rely on current flowing back through the ground to complete the circuit (these

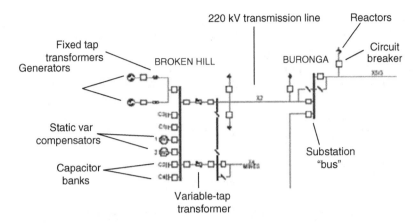

Figure 2.20 Schematic diagram of the electricity transmission network in the Broken Hill region (Source: From AEMO documents)

are also known as single wire earth return, as discussed later). Bipolar links use two conductors, with current flowing in opposite directions along each conductor – the residual current through the ground is negligible.

High-voltage DC links have three main advantages: (i) they typically require fewer conductors (only one or two conductors rather than three) and each conductor can be thinner than in an AC network; (ii) they can be used to connect two AC systems whose frequency is not otherwise synchronised; and (iii) the power flows are controllable. As we will see in due course, this last element can prove valuable as it implies that a DC link can be used to alleviate congestion on a parallel AC path. In addition, the loss of a parallel AC line does not overload a DC link. DC links also do not have the voltage stability problems of AC lines.

The primary disadvantage of DC transmission is that energy is lost in the process of converting from AC to DC and back again. A device for converting AC to DC is known as a *rectifier*. A device that converts DC to AC is known as an *inverter*. These devices are expensive and power is lost (and reactive power consumed) in the conversion process, making DC links uneconomical except in specific niche applications. In addition, the extra equipment required makes DC links less reliable than AC links.

In Australia there are currently three DC transmission links – the Murraylink project between South Australia and Victoria, the Terranora (Directlink) project between Queensland and New South Wales, and Basslink between Tasmania and Victoria. Murraylink and Directlink are bipolar HVDC links and Basslink is a monopolar cable. Murraylink consists of a 180 km underground cable operating at 150 kV. It is believed to be the world's longest underground transmission system. Terranora is a 59 km cable, operating at 80 kV. Basslink includes a 290 km stretch of undersea cable and is the second-longest undersea HVDC cable in the world.

2.6.2 Distribution Networks

Whereas transmission networks transport electrical energy long distances from relatively few points of power injection to relatively few points of power withdrawal, electricity distribution networks typically take power at a limited number of points of connection with the transmission network and deliver that power to a large number (hundreds of thousands) of geographically dense points in a given geographic region.

There is no clear dividing line between transmission and distribution networks. Distribution networks usually operate at voltage levels less than 50 kV, but some distribution networks are operated at voltages up to 132 kV and some transmission networks operate at voltages as low as 66 kV.

Distribution networks usually have a backbone based on three-phase AC power. As noted earlier, many commercial and industrial customers are directly supplied using the three-phase power at different voltages. In suburban areas transformers step the voltages down to 230 V. Most residential customers are supplied single-phase power at 230 V using a pair of cables.

In remote and rural areas, particularly in Australia and New Zealand, the final part of the distribution network connecting to remote farms might consist of a single wire, in a format known as single wire earth return (SWER). Australia and New Zealand were early pioneers and are large users of SWER in rural electricity distribution.

Distribution networks, like transmission networks, use a combination of conductors, poles, underground ducts, transformers, circuit breakers and switching equipment, power factor control equipment (capacitors, inductors) and communications equipment. Since the voltages are lower, these distribution assets are typically much smaller than the corresponding assets on a transmission network, and are often installed in equipment located along the side of a road, or on the distribution poles themselves, as illustrated in Figure 2.21.

In contrast to distribution networks, most customers of transmission networks are large producers or consumers of electric power. These customers typically have a substantial amount invested in producing or consuming electricity efficiently and are well-integrated into the wholesale electric power market. The production or consumption decisions of such customers

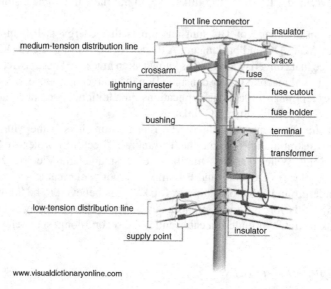

www.visualdictionaryonline.com

Figure 2.21 Typical distribution network assets (Source: http://visual.merriam-webster.com/energy/hydroelectricity/electricity-transmission/overhead-connection.php)

are closely metered at each moment of the day and they are typically willing and able to respond quickly to changes in supply and demand conditions in the market.

However, the situation has been very different for customers of distribution networks historically. Although there are some medium-scale producers and consumers of electricity connected at the distribution network level, most customers of distribution networks are very small scale producers and consumers of electricity. Historically, there was no capability to measure the production or consumption of such customers at any given point in time. Such customers were completely insulated from supply and demand conditions in the wholesale market. At the same time, distribution networks were operated in a way that did not take into account the potential for customers to respond to changing usage patterns on the local network.

However, as noted earlier, there is something of a revolution underway in the willingness and ability of small-scale producers and consumers of electricity to respond to wholesale market conditions. With the increasing penetration of microgeneration, local energy storage and smart appliances, there is greater ability and willingness of small customers to respond to wholesale market conditions. To use these assets efficiently will require changes in the way distribution networks are operated. In particular, mechanisms will need to be developed, along the lines of those that have been implemented for the transmission network, to reflect local supply and demand and network conditions back to customers – typically through local prices.

This is the goal of the Smart Grid concept – increasing the efficiency with which we use production and consumption assets, particularly assets connected to distribution networks, to increase the overall efficiency and effectiveness of the electric power system.

2.6.3 Competition and Regulation

It is almost universally recognised that electricity transmission and distribution networks represent what is called a *natural monopoly* – that is, that the features of these sectors are such that it is neither sustainable nor desirable to have competition between competing providers of transmission and distribution networks. At the present time it is not economically desirable to have competing power distribution lines running down every street. Moreover, for the reasons noted earlier, it is generally more efficient to connect together independent power systems, reducing the need for reserve capacity and thereby reducing the cost of ensuring reliable supply.

It is not sustainable or desirable to have competition between competing electric power systems. However, is it possible to have competition on just parts of the transmission or distribution networks? Could two or more firms compete to transport electricity from point A to point B? The problem is that the connection of a new AC transmission line to the existing network alters existing power flows in an important way. Paradoxically, the addition of a new line to an existing network can *reduce* the ability of some existing producers or consumers to exchange electric power. It remains an important research question (discussed further in Part VIII) whether or not it is possible to develop a system that rewards investors for financing a new transmission line.

However, the problems just mentioned only apply to AC transmission lines. It is possible to allow a third party to install a new HVDC link between any two points on the existing transmission network. In principle, such a link could make money by 'competing' with the existing network. However, there are serious questions about whether this can be made to work in practice (see Part VIII). Although in Australia some so-called merchant transmission links were established in the early 2000s, those links have subsequently applied for regulated status.

In short, the scope for conventional competition in distribution networks appears non-existent. The scope for competition in transmission networks is at best very limited. Transmission and distribution networks are in effect monopolies. As a consequence, transmission and distribution networks have substantial market power – that is, the ability (if left unregulated) to extract significant value from their customers (both producers and consumers).

As noted in Section 1.8, where a firm has substantial market power, customers are reluctant to make necessary sunk investments out of fear that the value of those sunk investments will be expropriated through an increase in prices. For this reason, around the world electricity transmission and distribution networks are subject to the range of ownership and/or price controls similar to those found in other public utility industries. Specifically, electricity transmission and distribution networks are typically either government owned, or cooperatives (owned by members), or subject to the regulatory controls that we know as public utility regulation.

2.7 Physical Limits on Networks

There are two types of physical limits or constraints on any electric power system. The first is the requirement that the supply of electricity must match the demand at every point in time. This is known as the *energy balance* requirement.

In practice, this requirement is closely related to the requirement to the keep the system frequency in a narrow band. Small imbalances in supply and demand cause the frequency on the power system to change – a slight excess of supply over demand causes the frequency to increase, while a slight excess of load over production causes the frequency to decrease. A large amount of the real-time activity of a power system operator is maintaining the system frequency within narrow bands. Figure 2.22 illustrates the typical allowable frequency range

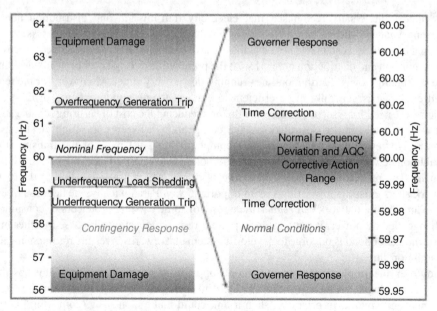

Figure 2.22 Normal frequency ranges for a 60 Hz power system (Source: Kirby, B., J. Dyer, C. Martinez, R. A. Shoureshi, R. Guttromson and J. Dagle, (2002), "Frequency Control Concerns in the North American Electric Power System", prepared for the U.S. Department of Energy, ORNL/TM-2003/41, available at: http://web.ornl.gov/~webworks/cppr/y2001/rpt/116362.pdf)

(and the impact of departures from this range) for a power system with a nominal frequency of 60 Hz.

The energy balance requirement must be satisfied at all times in all power systems, even in a (largely theoretical) power system that has no transmission or distribution network.

The second set of physical requirements or limits on power systems relate to the physical characteristics of the network itself. It is conventional to distinguish three categories of such limits:

- *Thermal limits* – which relate to the maximum amount of current that each network element can dissipate without overheating;
- *Voltage stability limits* – which relate to ensuring that there is an adequate supply of reactive power where it is needed in order to maintain the system voltage within prespecified limits;
- *Dynamic and transient stability limits* – which relate to the ability of the generators to remain in synchronism and for the system frequency to return to prespecified limits following a 'disturbance' to the system.

We will look in more detail at each of these in turn.

2.7.1 Thermal Limits

We have already seen how the power that is dissipated by a network element is proportional to the square of the current flow through that element. An increase in the current flow through a network element will therefore cause the element to heat up. The network element will continue to heat up to the point where either:

a. it establishes a new equilibrium with its environment – where the energy consumed in the network element is equal to the rate at which heat is passed to the environment; or
b. the element itself and/or the surrounding infrastructure is physically damaged. In some cases, violating the thermal limits can put human safety or the environment at risk.

Many network elements are specifically designed with devices that increase the rate at which they can dissipate heat into their environment. Cooling fans and radiating fins are common on many electrical appliances. Large transformers are often cooled by circulating oil or water.

In the case of high-voltage transmission lines, the primary risk from over-heating arises from the sagging of the lines themselves. As the wires heat up they tend to expand. As they expand the lines sag, reducing the clearance to the ground and other objects. As the wires approach grounded objects such as trees or buildings, the electric field between the wire and the grounded object increase. If the electric field becomes strong enough, the insulation between the wire and the grounded object (which is usually the air) breaks down and current flows in the form on an 'arc' or 'flashover'. These arcs can start fires or damage parts of the equipment itself. In most cases, the sudden increase in current flow will cause the protection devices on the transmission line to trip, causing the transmission line to be switched out of service. Sagging transmission lines also pose a risk to machinery, sailboats and humans.

In the case of transformers, over-heating can cause damage to the insulation surrounding the transformer or to the coils themselves. It is not unknown for transformers to overheat to the point where they explode. If the cooling oil overheats to the extent that it starts to boil, pressure release valves are opened, allowing the super-heated oil to escape. This oil may then catch fire, exploding in a ball of flame. This can cause the complete loss of a substation.

2.7.1.1 Static Versus Dynamic Thermal Limits

The amount of heat that a transmission line or a transformer can dissipate without overheating depends on factors, such as the ambient temperature, the degree of cloud cover (which affects the amount of solar energy absorbed by the lines), and the wind speed and direction. On a cold day with a brisk wind, a transmission line can carry substantially more current than on a very hot day with no wind.

To a very limited extent, this possibility is recognised by allowing for different static operating limits on transmission lines in summer and winter. For example, flows on the transmission lines from the Snowy region to NSW in Australia are limited to 3500 MW in winter but only 2800 MW in summer.

However, of course, although average temperatures are colder in winter, temperatures and wind speeds can vary greatly during winter or summer. These limits must therefore be set for the 'worst case scenario' (a day that is much hotter than average with no wind).

But since most days are not as bad as the worst case scenario, such a limit is well within what is technically feasible on most days. Therefore, one straightforward way to increase the allowable power flows is to install equipment that increases the information available to the system operator as to the instantaneous weather conditions. With additional information the system operator can set a limit that more closely matches the heat-dissipation conditions at the time, thereby allowing the operating limits to be set closer to the technical or physical limits of the network element. Since these limits vary frequently (often in real time) they are often known as dynamic limits or ratings.

For example, by observing the ambient temperature, the system operator could, in principle, set a thermal limit for a network element based on the ambient temperature. For example, the limit on Victoria to Snowy flows is nominally 1900 MW, but this reduces to 1700 MW when the temperature reaches 35 °C and 1600 MW at 40 °C.

In general, the more closely the system operator can measure the parameters that affect the physical limits, the lower the safety margin required and therefore the higher the average power flows the network can safely handle. This effect can be substantial. A New Zealand study of dynamic line ratings using devices for monitoring the tension on transmission lines found that the devices allowed a rating that was 43% higher than the static rating 60% of the time, 70% higher 40% of the time and 100% higher 20% of the time.[1]

Another approach is to directly measure the sag on the transmission line itself. Commercial products are available that allow transmission companies to measure the sag on transmission lines in real time.

2.7.1.2 Short-Term Versus Steady-State Thermal Limit

As noted earlier, an increase in the current flow through a network element will cause it to heat up. However, the rate at which a network element heats up for a given energy input depends on the physical characteristics of the element. The rate at which a given quantity of a substance heats up in response to a given energy input is known as its 'specific heat'. An object with a

---

[1] See Raniga and Rayudu (1999).

large specific heat and a large mass may take several minutes to heat up to its critical temperature, thereby allowing the system operator time to respond to the over-heating by taking corrective action.

For example, suppose that, on a certain day, a transformer can carry 1000 A of current continuously without exceeding a critical temperature. If the current suddenly increases to 1100 A the transformer will start heating but will not instantaneously exceed the critical temperature. Instead, there may be several seconds or minutes in which the system operator can take remedial action.

A transformer, which has a large iron mass and a large oil volume, may take 30 to 45 min to reach a critical temperature. On the other hand a transmission line, with a much lower mass of metal, may reach a critical temperature in 15 min.

As we will see later, this is important, because, as discussed in Part V, the ability of the system operator to correct a problem once it occurs has a large impact on the limits to which the system operator must operate the power system before a problem has occurred.

Operating limits that are allowed to be temporarily higher for a short period of time are known as 'short-term limits'. Manufacturers of large network elements, such as transformers, specify the short-term limits or short-term ratings of the equipment. A short-term limit consists of a higher power flow and a period of time for which that power flow can be tolerated after which the power flow must return to the normal, 'continuous' rating. A transformer with a continuous limit of 300 MVA might have short-term limits of 375 MVA for 15 min and 315 MVA for 30 min.

2.7.1.3 Increasing the Thermal Limits on a Transmission Line

We have already seen one way in which the thermal limits on a transmission line can be increased – through the use of 'dynamic' line ratings that vary with the ability of the line to dissipate heat. Thermal limits on transmission lines can also be increased by:

- *Reconductoring* – that is, replacing the existing conductors with new, larger conductors, or bundling new conductors with the existing ones. Whether or not this is feasible depends on the ability of the existing towers to carry the additional weight. In some cases, reinforcement of the towers or their concrete footings will be required.
- *Increasing ground clearance* – in some cases, especially where clearances are critical at only a limited number of spans, the tower height can be increased to restore sag clearances. In other cases it may be possible to re-tension the line to increase ground clearances. There are devices under development that are designed to automatically re-tension the line as the line heats up, offsetting the tendency for the line to sag as it heats up.
- *Increasing the number of circuits* – in some cases it may be possible to add another transmission line in parallel with an existing one, either by installing new lines on the existing towers or installing new towers and lines alongside the existing lines on the same easements. In other cases entirely new transmission paths and easements will be required.
- *Increasing line voltage* – there may be scope for a small increase in line voltage within a 'voltage class' without the need for substantial reconstruction of the power line or the substations at each end. Moving to a higher voltage class will usually require substantial new investment in towers, as higher voltages require greater clearances between the lines and

between the lines and the ground. The additional weight of the structures may also require reinforcement of the tower footings.

Furthermore, as we will see later, when there are multiple paths between a generator and a load, the power divides and flows over all the paths. In some cases, just one path will approach its thermal limit, thereby limiting the total power flows. In this case there may be some scope for changing the way the power flows over the system. This can be achieved through a device known as a phase-angle regulator (PAR). A phase-angle regulator changes the distribution of current flows in the overall network, without affecting the total power flow. Power flow can also be altered by changing the impedance of a transmission line by adding a series reactor or capacitor. Series capacitors reduce the impedance of a line while series reactors increase the impedance on a line. These devices also affect the level of reactive power, which we discussed earlier.

2.7.2 Voltage Stability Limits

We have already seen how, according to Ohm's law, the voltage drop across a network element is proportional to the current flowing through the network element. For a transmission line, this implies that the voltage at the 'receiving' end would normally be lower than the voltage at the 'sending' end.

While this is the case in a DC network, in an AC network, the system operator is typically required to keep the voltage and frequency at every point on the network within certain narrow ranges. In order to offset this tendency for the voltage to drop on a transmission line, devices for controlling the reactive power (inductors and capacitors as mentioned in Section 2.2) are installed at the receiving end of the transmission line to restore the receiving voltage to fall within an acceptable range.

However, restoring the voltage level in this way does not come without cost. Doing so introduces a phase difference between the voltage at the receiving end and the voltage at the sending end. The larger the current flow through the transmission line, the greater this phase difference.

When the phase difference approaches 90° it is no longer possible to maintain the voltage level using reactive power – instead, the voltage drops suddenly. This sudden voltage collapse would trigger the sudden disconnection of a large block of load. The resulting disturbance might be too large to be accommodated by the remainder of the power system, leading to a system blackout. The sudden voltage collapse can also do physical damage to generators attached to the power system.

In order to prevent this form of instability, the power system is operated in such a way that the power flows on the network are such that there is sufficient reactive power available to ensure that voltage lies within certain prespecified limits.

2.7.3 Dynamic and Transient Stability Limits

In an AC power system, virtually all the conventional generators on the interconnected part of the power system spin at exactly the same frequency. In the case of power systems that operate at 50 Hz, usually 3000 rpm (3000 revolutions per minute is the same as 50 cycles per second). Since large generators have very substantial mass (on the order of several tons), they have a sizeable amount of kinetic energy.

If there is a sudden increase in load on the power system, in the very short term this is not met by any explicit response from the system operator. Instead, an increase in load is met by a reduction in the kinetic energy of the generators. Since the kinetic energy is proportional to the square of the rotation speed, a small reduction in kinetic energy causes a slight reduction in system frequency. This reduction in frequency is detected by governing devices on the generators themselves, which increase the supply of fuel (usually steam) to restore the generator to its original frequency.

An increase in load at one point on the system does not affect all generators on the system equally. Those generators that are electrically 'closer' to the load will be placed under greater strain. It is possible that one or more of these generators will not be able to maintain synchronism with the overall system and will therefore be automatically disconnected. The resulting sudden loss in energy supply will increase the load on the remaining generators, which could lead to a cascading failure.

Even if the generators remain in synchronism, the system frequency does not necessarily immediately return to its desired level. The effect of a disturbance on the overall system frequency depends on both the inertia of the generators and the magnitude of the response of all the governing devices on all the active generators. If there is substantial inertia in the system, it may be that the regulating devices initially overcompensate for the increase in load so that the system frequency not only returns to its normal level but overshoots, leading to an increase in system frequency. This would cause the regulating systems to reduce the energy supply to the generators which might, in turn, lead to undershooting and so on.

In a stable power system, such oscillations around the target system frequency would quickly 'damp out' and the system would return to equilibrium. In an unstable power system, these oscillations could increase over time. This is illustrated in Figure 2.23 below.

Eventually, automatic protection devices would trigger, switching out blocks of generators or load. The resulting shock to the power system could cause more instability, leading to further tripping, resulting in a black-out as discussed earlier.

Whether or not the frequency is stable in response to disturbances in demand depends, in part, on the load flows. Power system operators carry out detailed analyses of stability under different power flow scenarios and under different network configurations to determine the stable 'envelope' within which the power system can be operated.

The dynamic and transient stability of the power system can be improved, amongst other things, through the use of gas-fired generators dispersed throughout the system. These generators have little inertia and can respond quickly to changes in load.

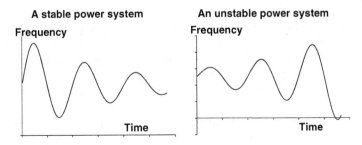

Figure 2.23 Frequency deviations in a stable and unstable power systems

Stability can also be increased by reducing the magnitude of the shocks to which the system is likely to be exposed. For example, the smaller the size of the generating units, the smaller the impact of any one generating-tripping event. In addition, adding switching stations in long transmission lines can also reduce the size of the shocks to which the system may be exposed. For example, suppose we have a pair of long transmission lines. If one of these lines fails, the impedance on the path doubles, which can have a substantial effect on the way in which power flows change around the rest of the interconnected network. On the other hand, if a switching station is installed in the middle of the pair of long transmission lines, the loss of one part of one of the lines increases the impedance of the path by only 50%, reducing the effect on power flows over the remainder of the network.

Result: Physical flow limits on networks come in three types: thermal limits, which relate to the amount of heat that can be dissipated through a network element; voltage limits, which relate to the level of reactive power support required on long transmission lines; and transient stability limits, which relate to the ability of the power system to respond to small shocks. The power system must be operated at all times to fall within these physical limits.

2.8 Electricity Consumption

Electrical energy has no value in and of itself. Instead, electrical energy has value because of the services provided by devices that consume electrical energy.

Electricity is used in so many industrial and commercial processes and in so many aspects of daily life that it is difficult to generalise the patterns of electricity usage. Nevertheless, we can make the following points: electricity consumption of individual entities (households, businesses, industrial enterprises) differ in the following characteristics:

- *Profile of consumption* – Although some industrial processes run for 24 h, consuming electric power at the same rate around the clock, most industrial and residential loads are cyclical, with greater consumption during working hours for industrial loads, and in the morning and evening for residential loads. Most loads also exhibit a cyclical pattern across the course of a week, or year, with lower consumption on weekends, public holidays, and lower consumption in mild seasons.
- *Flexibility* – Although some loads are relatively insensitive to incentives to alter consumption, other loads are able to be altered with relatively little inducement. For example, in the case where electrical energy is used for heating or cooling, where there is substantial thermal inertia, customers may be able to bring forward or defer some electricity consumption with little noticeable effect. In many countries hot water heating occurs at off-peak times, using the thermal inertia of the hot water cylinder to maintain hot water at other times. The same principles apply where electrical load is used to charge storage devices, such as batteries. In other cases the particular application may be able to be shifted in time or deferred with relatively little cost.

- *Responsiveness to wholesale market conditions* – When adequate metering facilities are available, electricity consumers are often partially exposed to wholesale market supply, demand and network conditions. There are two ways by which this is achieved:
  - By direct control of consumption by a third party, such as the retailer, or
  - By indirect control, by partially or fully exposing the customer to a time-varying retail price.

    For example, a customer might agree to have a swimming pool pump, air-conditioning, or refrigerator turned off for short periods by a third party when wholesale prices become very high. Alternatively, a customer might establish a device that responds directly to wholesale price signals, perhaps choosing the cheapest time to recharge an electric vehicle, for example.
- *Degree of integration with the wholesale market* – Even amongst those customers that are partially or entirely exposed to wholesale spot market conditions, only a few electricity loads participate directly in the wholesale market – submitting bid curves, and following directions as to how much to consume from the wholesale electricity market.

It is important to make a distinction between electricity consumption patterns at a point in time, and electricity consumption patterns over the long term.

In order to make use of electricity, customers have to install various customer-premises equipment, such as electrical wiring and electrical appliances, some of which may be customised to a particular use. In the short-term the electricity consumer has a stock of existing assets for converting electricity into valuable services. The patterns of electricity consumption, the controllability, and so on, depend on this stock of existing assets. However, over time these assets wear out and must be replaced. In making new investment decisions (or replacement decisions), the end user will consider alternatives, including other ways of organising production or consumption to alter or reduce electricity consumption, especially (where the end-user is exposed to a time-varying retail price), electricity consumption at peak times.

As a consequence, the responsiveness of retail electricity consumption to wholesale market conditions is different in the short-term and the long-term. In the long-term customers can change or replace large electricity consuming assets, or change production processes to consume less electricity, or switch consumption to off-peak times. Electricity consumers (large and small) tend to be more responsive to average wholesale market conditions in the long run than in the short run.

2.9 Does it Make Sense to Distinguish Electricity Producers and Consumers?

The previous sections (Sections 2.5–2.7) separately discussed the main features of electricity generators, networks and consumers. This follows the traditional structure outlined in Figure 2.13. In this traditional approach, electric power is generated by relatively few large generating units, and power flows over electricity networks predominantly in one direction to hundreds of thousands (or millions) of electricity consumers, most of whom are entirely passive and unresponsive to supply and demand conditions in the wholesale electricity market.

However, as we have repeatedly emphasised in this chapter, it is increasingly recognised that this traditional paradigm must change. There are emerging signs of something of a revolution in the way the electric power industry is organised.

Over the last decade or so there has been an increasing take-up of small-scale generation, such as solar PV, small wind turbines or fuel cells, often located on the site of the end-customer. In

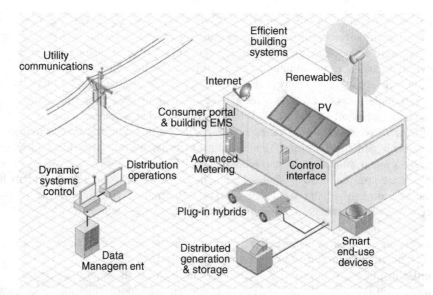

Figure 2.24 The range of controllable production and consumption devices in a possible electricity industry future (Source: EPRI, Smart Grid Demonstrations. http://www.smartgrid.epri.com/doc/EPRI% 20Smart%20Grid%20Overview.pdf)

addition, there is increasing use of so-called smart devices that are able to control their electricity usage in response to conditions in the electricity industry. Many customers are purchasing plug-in electric vehicles, which represent both a major source of load and, at the same time, a means of storing electrical energy, which could, in principle, be reinjected back into the grid when it is needed. The range of devices that may emerge in the future are illustrated in Figure 2.24.

The use of so-called smart devices and local generation facilities change the way the electric power system is organised and operated. Historically, as noted above, it made sense to draw a strong distinction between, on the one hand, the few, large, generating units, which were highly integrated into the wholesale electricity market, and the numerous, small, electricity consumers, which were largely insulated from conditions on the wholesale electricity market.

This distinction no longer makes sense. Instead, all customers are potentially generators or consumers, depending on preferences and local market conditions. Many commentators use the word 'prosumers' to reflect the fact that consumers of electricity are also often producers of electricity, and may switch from net consumption to net production according to changes in the wholesale price. All such customers must be at least partially integrated into the wholesale market and able to respond to local market conditions, particularly as reflected in changes in the local market price.

In this new environment the primary distinction is not between the type of customer (electricity producer versus electricity consumer), but between the size of the customer (large customers, who directly participate in the wholesale electricity market at some level, and small customers, who do not directly participate in the wholesale electricity market).

Proponents of this future vision of the electricity industry (which is also known as the *Smart Grid*) see several potential benefits from the increasing penetration of small-scale generation and smart devices and the integration of these customers into the wholesale electricity market. The integration of even small customers into the electricity market:

- Facilitates the integration of small-scale variable renewable generation (such as solar PV, wind and wave generation), thereby reducing the economic cost of achieving reductions in greenhouse gas emissions;
- Increases the resilience of the electricity network by increasing the range of production and consumption responses to supply, demand and network shocks; and
- Increases the utilisation of the network by reducing the need to build network capacity to cater for just a few peak hours per year, thereby reducing the need for costly network augmentations.

A major challenge for the electricity industry in the next few decades will be the effective extension of the wholesale market down to the level of small customers, and the effective integration of small customers into the wholesale market. This will require a change in the sophistication with which small customers approach their electricity consumption. Some customers will not wish to change. At this stage the extent of the benefits of this Smart Grid vision remain uncertain. However, there remains at least the potential for significant benefits. Achieving this Smart Grid vision will be a major theme of this text.

2.9.1 The Service Provided by the Electric Power Industry

What exactly is the service the electric power industry provides? Historically, answering this question would have been straightforward. Historically, to electricity consumers, the service provided by the electric power industry was a continuous, reliable supply of electricity, in the volume that the customers desire, at a given quality and at a fixed, reasonable price. (The quality of electricity supply depends on the consistency of the voltage, frequency and the purity or cleanliness of the sinusoidal waveform).

However, this response is no longer adequate. As we have seen, there is increasingly less distinction between electricity producers and consumers. The efficient use of and investment in assets such as smart devices or small-scale generation facilities cannot be achieved if customers face a fixed price. The concept of reliability makes no sense in a world in which electricity can be rationed through a wholesale spot price.

Therefore, what exactly is the service provided by the electric power industry? Instead of providing a reliable supply of electricity, the function of the new electric power industry is to provide a platform on which parties can trade and exchange electricity, with prices that reflect the supply and demand conditions at each different point on the network. In other words, the electric power industry exists not to provide a reliable supply of electricity but to provide a market, with market signals that allow customers (both generators and consumers of electricity) to make efficient decisions regarding the use of the existing set of assets and investment in new asset.

In our view the best way of summarising the service provided by the electric power industry is as follows:

> The electric power industry exists to provide a platform for efficient trade and transportation of electrical energy, resulting in efficient utilisation of and investment in electricity generation and consumption assets.

2.10 Summary

Electricity is a particularly versatile and useful form of energy that is ubiquitous in modern life. Electricity is the flow of electrons in conductors around circuits. Voltage is a measure of the force or pressure on those electrons. Current is a measure of the number of electrons passing a point on the circuit in a unit of time. The rate at which electrical energy is delivered, known as power, is equal to the voltage times the current, and is measured in Watts. The volume of electrical energy produced or consumed is measured in Watt-hours. In all circuits a small amount of power is lost in the conductors or other network elements. These losses are inversely proportional to the square of the voltage. For this reason, when large amounts of power must be carried long distances, very high voltages are used to keep the losses to a minimum.

In all large-scale power systems (with a few limited exceptions), the electricity flows in the form of an alternating current. In an alternating current the voltage and the current follow a sinusoidal pattern with a phase difference measured as an angle. When the voltage and current are not perfectly in phase, we can divide the total power flow into real and reactive power components. Only the real component of the power flow can be used to transfer electrical energy from one location to another. However, the reactive power still contributes to heating network elements and must be taken into account when assessing network limits. Reactive power is kept under control through capacitors, reactors and other related devices. Although two conductors are necessary to complete any circuit, it is often preferable and convenient to use three conductors with the voltage in each conductor 120° out of phase. Such an arrangement, known as three-phase power, delivers a constant power flow and is more convenient for many electric motors.

Electric power systems have three basic components: generators, networks and electricity consuming devices. Although there are a range of theoretical means for producing an electric current, by far the majority of electric power is generated by the conversion of kinetic energy into electrical energy. This typically occurs by means of a turbine, often driven by steam under pressure, using water heated by a heat source. In recent years wind generation has increased in importance. Solar cells are becoming increasingly important as a small-scale generation technology. Generators differ from one another in characteristics, such as their controllability, their cost of production, their capacity, ramp rate and environmental impact.

Although small-scale generation is growing in importance, historically there were significant cost benefits in centralising generation in relatively few large-scale generating plants. In addition, it is often cheaper to transport energy in the form of electricity than in other forms (such as coal). This focus on a few large generation stations results in a need to transport electricity long distances from where it is produced to where it is consumed. This service is provided by electricity networks. Electricity networks are typically divided into high-voltage transmission networks and low-voltage distribution networks. Both networks consist of a collection of assets including poles, wires, transformers, switching equipment, reactive power control equipment, and equipment for monitoring and control of the network.

Electricity is not consumed directly but is consumed in a huge range of devices that provide services to a range of commercial and industrial customers, small and medium-sized enterprises and households. Electricity consumers differ in whether or not they have meters capable of recording consumption at different points in time, and whether or not their consumption is responsive (directly or indirectly) to supply, demand and network conditions in the wholesale electricity market. With the increasing penetration of small-scale generation, electric vehicles and smart appliances, it is likely that consumers will increasingly have the willingness and ability to respond to wholesale market conditions.

The development of small-scale generation technologies, smart appliances, electric vehicles and other electricity storage technologies is changing the way we think about the role of the electricity industry. Historically, it made sense to distinguish between a few large generating units, delivering power in one predominant direction, to small passive electricity consumers. However, this distinction no longer makes sense. Instead it makes sense to think of all electricity customers (both producers and consumers) as being capable of producing, consuming and responding to wholesale market conditions. The primary distinction is no longer between producers and consumers, but between large customers (generators and loads), who directly participate in a wholesale electricity market, and small customers, who participate indirectly, perhaps via a retailer.

Similarly, it no longer makes sense to think about the electricity industry as providing a reliable supply of electricity to the location of the consumer. Instead, as the distinction between generators and consumers becomes blurred, it makes more sense to think of the electricity industry as providing a platform for the efficient trade and exchange of electricity. The primary service provided by the electric power industry is the promotion of efficient usage and investment decisions by electricity customers (both producers and consumers) by promoting efficient trade and transportation of electricity.

Questions

2.1 A generator is initially producing at the rate of 100 MW at the start of a dispatch interval. It is directed by the market operator to increase its output to 500 MW by the end of the dispatch interval (precisely 5 min later). It does so by increasing its output at a constant rate to just reach 500 MW at the end of the dispatch interval. How much electrical energy (MWh) does this generator produce in that 5 min period?

2.2 Why is electrical energy typically transported over long distances using high-voltage AC links? What are the pros and cons of using high voltages for transporting electricity? What are the pros and cons of DC as compared to AC links?

2.3 What are the primary elements of a transmission network? What is the role played by circuit breakers? What is the role played by capacitor banks, reactors and static VAR compensators?

2.4 Suppose that an open-cycle gas turbine (OCGT) has an efficiency of 30% (efficiency is given by the ratio of electrical energy produced to the energy in the fuel consumed), whereas a combined-cycle gas turbine (CCGT) has an efficiency of 50%. Assume that the variable cost of each generator is entirely comprised of the input fuel cost. On a day when the price of gas is $3/GJ, the variable cost of the CCGT generator is $50/MWh. What is the variable cost of the OCGT generator?

2.5 In market-based power systems there is always a market for real power, but there is seldom a market for reactive power. Instead, reactive power requirements are met through a combination of requirements on generators and loads and network assets (such as capacitors and inductors). What are the pros and cons of establishing a separate market for reactive power?

2.6   If the wholesale electricity market worked effectively, would there ever be involuntary load shedding? Why or why not?

2.7   What are the implications of having large amounts of electricity storage capability (e.g. through plug-in electric vehicles) connected to the network?

Further Reading

The introductory material in this chapter can be found in countless textbooks. If you want a textbook treatment try Wood and Wollenberg (2012), Saadat (1999), Grainger (2003) or Stevenson (1982).

Wikipedia is a good informal source of further information, with articles on Kirchhoff's Circuit Laws, Ohm's law, Transmission Losses, Direct Current, High Voltage Direct Current, AC Power, Three-Phase Electric Power, History of Electric Power Transmission, and the battle between DC and AC (known as the War of Currents), Solar Power, Electricity Generation, Electric Power Transmission, Electricity Distribution, Single Wire Earth Return, Murraylink, Terranora (Directlink), and Basslink, Smart Grids, demand response, demand side management, Value of Lost Load and Load shedding (rolling blackouts).

3

Electricity Industry Market Structure and Competition

The structure of the electricity industry varies greatly around the world. In many countries, electricity is provided by a single integrated monopoly provider. In a few countries, there is active competition in parts of the electricity industry, often coordinated through formal markets. This chapter seeks to provide an overview of the range of forms of markets in the electricity sector, with a more detailed look at those industries that rely primarily on market signals to drive efficient usage decisions and efficient investment decisions.

3.1 Tasks Performed in an Efficient Electricity Industry

One way to think about the task of designing an efficient electricity industry is to list all of the key tasks that must be performed, assign those tasks to different entities, and then to design frameworks, rules and incentives to ensure that those entities have incentives to carry out those tasks efficiently and effectively, and to coordinate and cooperate with the other entities in the execution of their respective tasks.

Let us begin by setting out the key tasks that an efficient electricity industry must perform. As explained in Section 1.1, it is useful to draw a distinction between short-term tasks and longer-term tasks. In addition we will distinguish 'risk-management' tasks, which make the bridge between the short-term outcomes and longer-term incentives.

3.1.1 Short-Term Tasks

An efficient electricity industry must perform the following short-term tasks:

- *Efficient short-run use of available generation resources.* For reasons that will become clear in Chapter 9, all efficient electricity supply industries have available a range of different generation technologies with different fixed costs and variable costs. Making efficient use of these resources implies ensuring that electricity is produced using the lowest-cost mix of generation resources, taking into consideration the limits of the transmission network, generator technological capabilities (such as startup costs and ramp rates) and resource

The Economics of Electricity Markets, First Edition. Darryl R. Biggar and Mohammad Reza Hesamzadeh.
© 2014 John Wiley & Sons, Ltd. Published 2014 by John Wiley & Sons, Ltd.

availability (such as hydro storage and wind and solar energy availability). This task also includes ensuring efficient maintenance practices, efficient levels of forced outages and efficient timing and duration of unforced outages. Where market-driven processes are used, ensuring efficient allocation of available generation resources may also involve controls or curbs on the exercise of market power.

Achieving efficient short-run usage outcomes may also involve intertemporal decisions, such as whether or not to startup a generator or load, when to charge or discharge electricity storage, such as an electric vehicle, or when to make use of limited energy resources (such as hydro storage). In these cases, achieving efficient outcomes is likely to require high-quality forecasts of likely future supply, demand and network conditions.

- *Efficient short-run use of available demand-side resources.* Efficient use of electricity involves electricity consumers adjusting their consumption in the light of changing supply/demand conditions. This adjustment may happen directly, through devices interacting with the wholesale market, or indirectly through customers responding to wholesale prices. In addition, when supply and demand conditions are at extreme levels, curtailing or rationing load in an efficient manner.

- *Efficient short-run use of the available network resources.* Efficient use of the available network resources includes (i) efficient redistribution or reallocation of generation and demand resources to accommodate transmission or distribution constraints, and (ii) efficient operation of the transmission and distribution network. Efficient operation includes efficient setting of network limits taking into account, say, weather conditions and credible contingencies; efficient maintenance and inventory practices, efficient levels of forced outages, and efficient decisions regarding the timing of unforced outages. Efficient operation may also include efficient switching decisions, bringing network elements into or out of service in such a way as to maximise overall economic benefits. These decisions must, to an extent, be coordinated with decisions as to, say, the use of generation resources. For example, it does not make sense for a generator to incur large startup costs early in a day if a key transmission line will be taken out of service later that same day, limiting the ability of that generator to get its power to the market.

- *Efficient response to very short run imbalances in supply and demand.* Every electricity industry, no matter how it is configured, must maintain a balance between supply and demand and must respect the physical limits of the underlying networks, at all points in time. However, supply, demand and network conditions change all the time. In the real world the power system is subject to a variety of shocks and contingencies, such as the loss of a large generator or network element. Overall efficiency in operation involves efficient response to contingencies including (as a special case of the points above) efficient reallocation of generation, efficient use of load resources and efficient use of network resources. In addition, the system must be operated in a manner that balances the costs of taking actions once a contingency has happened (so-called corrective actions) against the costs of inefficient operation before a contingency has happened (so-called preventive actions).

All of the tasks mentioned must be performed in an efficient electric supply industry, no matter how that industry is organised.

In addition, in a liberalised electricity market, there are many other related tasks that must be performed effectively, such as the metering of generation supplies and load consumption, the operation of billing and settlement systems, and the monitoring and enforcement of the market rules and processes.

3.1.2 Risk-Management Tasks

In a liberalised electricity market, as we will see, primarily reliance is placed on market-like processes to achieve many of the short-term tasks set out earlier. These market-like processes ensure efficient allocation of resources and efficient coordination of certain decisions, primarily through short-term market price signals. However, many long-term decisions (such as those set out below) require a degree of certainty or assurance as to the likely long-term average effect or incidence of these short-term outcomes. Short-term prices may reflect these long-term consequences on average, but potentially expose market participants to substantial risk. There is therefore a role for risk-management instruments to insulate market participants from short-term price volatility while retaining the long-term price signals.

In principle there may arise a need for risk-management instruments for each of the four primary tasks set out earlier. Specifically, a generator may require some sort of assurance as to the average price it will receive over a year or more, or a guarantee of the minimum amount of revenue it will receive under certain circumstances (such as when it decides to incur startup costs). Similarly, a trader who is procuring electricity at one location for supply to another location may require some sort of certainty or assurance as to the level of physical availability of the underlying network and/or a promise of financial compensation in the event that transmission or distribution congestion limits his or her ability to supply load. Similarly, relatively few downstream customers are willing to be exposed to the wholesale price for electricity. Most seek some form of assurance as to a long-term average price. As we will see, providing such a risk-management function is a key role of energy retailers. Finally, a generator whose primary function is to provide emergency balancing services may require some form of long-term assurance as to the revenue stream he/she will receive. Such a generator may seek a risk-management tool based on the prices for balancing services, where those are separately priced.

3.1.3 Long-Term Tasks

Achieving an overall efficient outcome in an electricity market requires more than just efficient short-term use of existing resources – it also requires efficient investment in resources over the long term, specifically:

- *Efficient investment in production resources.* Efficient investment in electricity generation includes efficient decisions as to the amount, the type, the location and the timing of new generation investment. This task includes both decisions regarding investment in large-scale generation, as well as small-scale microgeneration, such as roof-top solar, fuel-cells, or small-scale storage. This task also includes efficient decisions as to the retirement of existing generation plants as well as investment in very fast response plants, capable of responding to contingencies as they occur. These decisions must, of course, be closely coordinated with decisions as to the nature, timing and extent of any upgrades to the transmission network.
- *Efficient investment in consumption resources.* This includes efficient decisions as to the type, amount, location and timing of investment in electricity-consuming devices, including

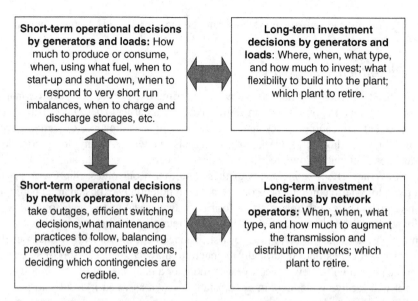

Figure 3.1   The key tasks to be performed in an efficient electricity supply industry

both large-scale investment in, say, an aluminium smelter, and small-scale investment in air-conditioning and electric vehicles. Closely related are decisions about the capabilities of these consumption resources including the ability to shift consumption across time periods, the ability to respond to dispatch instructions and/or their ability to respond on very short timescales. Again, decisions as to the nature, extent and timing of new load investment must be coordinated with decisions as to the nature, extent and timing of new generation and transmission investment decisions.

• *Efficient investment in network resources.* This includes the amount, type, location and timing of new transmission and distribution augmentation, and the retirement of existing transmission and distribution facilities. Again, this task must, of course, be coordinated with generation and load investment decisions.

Figure 3.1 highlights the different types of decisions involved in achieving overall efficient outcomes, and the interaction between them.

Having identified these key tasks, the next step is to decide who should perform the tasks and what governance arrangements or operating incentives will be put in place to ensure that the tasks will be performed in an effective manner.

3.2   Electricity Industry Reforms

As noted in Section 1.7, it is widely considered that electricity transmission and distribution networks are *natural monopolies*. It is not considered feasible or desirable to have multiple firms in the same location installing and competing in the provision of electricity network services. Where there are monopolies there may arise a risk that customers of those monopoly services resist making investments in reliance on the monopoly service out of fear that, once they have done so, the monopoly will raise its charges.

In the electricity industry both generators and final consumers of electricity are customers of the electricity networks. Both may fear making valuable sunk investments without some mechanism for protecting those investments.

As discussed in Section 1.7, there are two conventional approaches for protecting those investments: long-term contracts and vertical integration.

In the case of electricity generators, historically, vertical integration was the most common approach to solving the hold-up problem that would otherwise arise. Historically, electricity generators were almost always combined with the network businesses, to form the integrated electricity utility, which is common around the world.

However, vertical integration has several problems of its own. It is hard to achieve high-powered incentives for efficiency in a large organisation. Starting in the 1980s some countries started experimenting with long-term contracts for the purchase of electric power from independent generating companies. These long-term contracts are often known as power purchase agreements (PPAs). These and other approaches for facilitating competition are discussed further in Section 3.3.

The other major customers of electricity networks are the electricity consumers themselves. Occasionally, large consumers of electricity (such as large mines) will vertically integrate into the self-provision of electricity. More commonly large consumers of electricity will enter into long-term contracts. For example, an aluminium smelter might enter into a long-term 10–30 year contract for the purchase of electricity.

However, most smaller electricity consumers have no incentive or ability to vertically integrate into electricity production and distribution, and transaction cost makes it infeasible to enter into long-term contracts. Smaller consumers of electricity therefore require one of the public solutions to the monopoly problem: joint or government ownership or the long-term contract known as public utility regulation.

Historically, these solutions to the monopoly problem were common. In almost all countries around the world, electricity was historically provided by a single, integrated monopoly firm in each region. This firm was typically integrated from generation and transmission, through to distribution and retailing (i.e. the business of interaction with the end consumer). In many countries this integrated monopoly firm is owned by the government. As an institution, *government ownership* in the electricity industry has proven remarkably stable, starting from the earliest days of the electricity industry. Even today the majority of electricity utilities around the world remain government owned. In a few countries the electricity industry is organised as a customer-owned cooperative.

In a few countries – notably the United States – many local electricity providers were privately owned (known, in the US parlance, as an *IOU* or *investor-owned utility*). These privately owned monopolies were almost always subject to regulatory controls on their prices. This situation persisted for most of the twentieth century.

3.2.1 Market-Orientated Reforms of the Late Twentieth Century

In the 1980s and 1990s criticism of the performance of state-owned entities, combined with criticism of the effectiveness of monopoly price regulation and renewed interest in reliance on competition, led to a wave of regulatory reforms that had far-reaching consequences for the organisation and operation of the traditional 'public utility' sectors, such as electricity, gas and telecommunications.

These reforms had slightly different emphases in different sectors and different countries. However, there were a clear central set of ideas that were applied across many different industries. These central ideas included:

- The introduction of *commercial incentives* and operating practices into government-owned enterprises, primarily through corporatisation and, in some cases, privatisation. Corporatisation is the process of placing a government-owned entity on the same footing as any other commercial enterprise, with governance by a board, the same rights and responsibilities as other firms, and clear (usually profit-focused) objectives.
- The *introduction of competition* wherever feasible – specifically in those components of an industry that can sustain effective competition. This was achieved through the removal of entry barriers (such as statutory barriers to entry), the elimination of competitive advantages enjoyed by some government or private firms (so-called *competitive neutrality*) and the creation of arms-length market mechanisms.
- The *focussing of regulation or government ownership* on the remaining natural monopoly elements of the sector. This often implies a refocusing of regulation from the retail tariffs of the integrated firm to the wholesale tariffs of a component of the firm, such as the network businesses.
- Where economically feasible, *structural separation* of the natural monopoly components from the competitive segments. Structural separation of this kind eliminates the incentive otherwise faced by a vertically integrated and regulated incumbent to discriminate against its rivals in the competitive segments, by raising the price and/or lowering the quality at which it provides access to the natural monopoly component.

Applying these ideas to electricity, the reformers of the 1980s and 1990s viewed the generation sector and (to a lesser extent) the retail sector of the electricity industry as being potentially competitive. They therefore sought to promote competition, particularly in the generation sector. Some countries also pursued structural separation of the electricity industry – separating the monopoly network businesses from the competitive generation and retailing segments. At the same time, there was often a clarification and focusing of the monopoly price regulation on the network elements.

The objectives of these reforms differed slightly from one country to another, but one of the most important objectives was improving the efficiency and productivity of these industries. There was a common concern that government-owned vertically integrated electricity supply companies had become inefficient, over-manned, unproductive and unresponsive to customer desires. There was also often a concern that these companies had over-invested, in effect 'gold plating' the electricity supply industry.

In some countries, government-owned firms had under-invested over decades and lacked the ability to raise tariffs to sustainable levels. In some countries reducing the level of government subsidies was a primary objective. In a few cases the electricity supply business was politically powerful and reforms were, in part, to limit the challenges to central government authority.

In the United States, the primary concerns surrounded perceived weaknesses in the regulatory controls. Electricity utilities were often seen as inefficient, bureaucratic and unresponsive to customer desires, and changing customer priorities with respect to, say, renewable generation. Electricity market reforms were seen as a way to improve incentives for efficiency, to shift more investment risk away from consumers and on to investors, and to make the industry more adaptable to the introduction of new generation technologies.

3.3   Approaches to Reform of the Electricity Industry

Reforms to the electricity industry typically have focused on the use of competition to achieve efficient use of, and investment in, generation resources. However, different reforms differ in the degree of competition introduced on the 'buyer' side of the market. We can distinguish three broad categories of approaches:

- The *single buyer* approach, under which a single entity has responsibility for purchasing wholesale electricity.
- *Wholesale competition*, under which entities (such as distribution businesses) have a local monopoly over customers and negotiate on their behalf to procure electricity.
- *Retail competition*, under which any customer can, in principle, purchase electric power from any supplier.

Under the single buyer approach, a single entity – this may be the system operator or an integrated electricity company – has responsibility for procuring electricity at the lowest possible cost from different competing suppliers. Many less-developed countries have implemented a version of the single buyer model in order to stimulate competition in the generation sector.

The single buyer model may involve the establishment of a wholesale market for determining efficient short-run dispatch of generation. Where such a market is established, generators will usually be compensated based on the prices in that market and may even be expected to invest based on the prices in that market.

The single buyer model can be relatively straightforward to implement. Since the transmission network is usually integrated into the single buyer, there is usually no need for separate transmission pricing arrangements. The single buyer can choose to enter into a range of different risk-sharing arrangements with generators, taking on some or all of the generators' investment risk. Indeed, the single buyer model suffers from the same monopoly drawbacks we discussed earlier. The investment of generators remains subject to the hold-up problem – the risk that, once they have made a sunk investment, the single buyer will change the market arrangements to lower the prices paid to generators. Generators will typically be unwilling to invest without some form of long-term assurance of compensation, such as a long-term contract of some kind.

In the single buyer model, responsibility for operation and investment in the transmission and distribution networks is almost always the task of the single buyer. The assignment of tasks in this model is set out in Table 3.1.

Under the second approach, sometimes referred to as wholesale competition, local monopolies, typically called load-serving entities, have a monopoly over customers in a local area and take responsibility for purchasing electricity on behalf of their end-customers. There will typically be a wholesale market established to ensure efficient dispatch of generation resources and possibly efficient investment in generation resources.

Table 3.1 The assignment of key tasks in different approaches to reform

Task → Industry structure ↓	Efficient use of and investment in electricity consuming resources	Efficient use of and investment in electricity generating resources	Efficient use of and investment in network resources
Integrated industry	Customers (acting under price signals from integrated firm)	Integrated firm (no price signals)	Integrated firm
Single buyer	Customers (acting under price signals from single buyer)	Generation owners (responding to price signals) and Single buyer (through long-term contracts)	Single buyer
Wholesale competition	Customers (acting under price signals from load serving entity)	Generation owners and operators (responding to price signals or capacity prices)	Market operator (transmission); LSE (distribution)
Retail competition	Customers (acting under price signals from retailer)	Generation owners and operators (responding to price signals or capacity prices)	Market operator (transmission); market operator (distribution)

This approach has the advantage that there is enhanced competition between electricity buyers, reducing the hold-up problem, increasing the incentives for investment in generation capacity. The load-serving entity is often a local distribution business. Under this approach there need to be arrangements for pricing or allocating the capacity of the transmission network. Where the load-serving entity is also the distribution business it must develop mechanisms for efficient use of and investment in the distribution network.

Under the third approach, referred to as retail competition, any business may establish itself as an energy retailer and purchase electricity from any generator and on-sell it to an end user. This approach requires arrangements for pricing and use of both the capacity of the transmission and distribution networks. This is typically carried out by a market operator (although the market operator need not be the same for transmission as for distribution).

Very few industries place much focus on achieving efficient use of load resources. Decisions as to use of and investment in load resources are almost always delegated to electricity consumers, operating under price signals. In almost all cases those price signals are currently underdeveloped. For example, most consumers still face an electricity price that does not vary with wholesale market conditions. There is significant room for improvement in the efficient use and investment in small customer resources – as in the Smart Grid vision for the future discussed in the previous chapter.

Under all of the approaches to reform, the entity interacting with the end-customer may, in addition to price signals, also have direct control over the appliances and devices owned by the end-customer. For example, an integrated industry, or a single buyer, may enter into arrangements to control an end-customer's hot-water heating load, or air-conditioning unit. Under the retail competition model, competition between retailers should drive retailers to offer customers contracts involving combinations of price signals and direct load control that effectively control the resources owned by end-customers.

Under all of the approaches set out here, responsibility for efficient use of and investment in network resources is not driven by competitive market forces but is undertaken by an entity acting under regulatory controls or other general incentives.

As this table suggests, there is a clear conceptual distinction between (a) those tasks for which we rely on normal competitive market forces and price signals to deliver efficient operational and investment decisions, and (b) those tasks for which there is no role for competition or price signals.

In some markets nearly all of the operational and investment decisions of generators are driven primarily by decentralised profit-maximising objectives of generation entrepreneurs. These decisions are made in response to price signals – both the price signals that arise in competitive markets, and the prices for the regulated services, such as charges for the use of the transmission network. These decisions include most decisions as to how much to produce, when to start up and shut down, which fuels to use, and when and how to carry out maintenance.

On the other hand, even in the most liberalised electricity markets, primary reliance is placed on regulatory incentives and processes to deliver efficient decisions with respect to the operation and investment in distribution and transmission networks.

However, the boundary between regulatory decisions and competitive outcomes is not as crystal clear as it might at first appear. Some decisions are made through a combination of regulatory processes and competitive markets. For example, in the Australian energy market, the system operator determines, through administrative processes, how much of each of the so-called ancillary services to procure, but then uses a competitive-tendering process to select who should provide and how much they should provide for each service. In addition, it is common for the system operator to retain discretion to override the outcomes of the regular market processes and to issue orders ('directions') to a generator to produce more or less, or to operate in a particular way. In practice electricity markets involve a mix of reliance of market forces and central direction.

3.4   Other Key Roles in a Market-Orientated Electric Power System

In Chapter 2, we looked at the main components of an electric power system (generation, transmission, distribution and loads). These components are present in every power system no matter how it is organised or structured.

However, this textbook focuses on electric power systems that rely as much as possible on market processes to achieve the objectives set out earlier. In a liberalised or market-orientated electric power system, there are certain additional roles to be performed, which are worth highlighting. These include:

- *The market operator role* – Wholesale market price signals must be closely coordinated with the losses and physical delivery constraints arising on the electric power networks. With the current understanding of electricity market design, this coordination and integration can only be played through the operation of a centralised entity with knowledge of the bids and offers of market participants and physical network characteristics in its region. As we will see, the role of the market operator is, at each point in time, to take its knowledge of bids and offers and physical network characteristics and find a set of dispatch instructions for each generator and load in the region that maximises the economic value of the trade in electrical energy. This process produces a set of prices reflecting the value of electricity at each location. At present, it is not possible to envisage competition in the market operator role.

- *The retailer role* – Most small electricity market participants (whether consumers or producers) would prefer not to be exposed to volatile wholesale electricity prices. Instead, these small customers are prepared to pay for a service that insulates them from the wholesale electricity prices and packages those wholesale prices in a manner that they prefer. This is the role of the retailer. The retailer may also be involved in interfacing between the end-customer and the wholesale market, including aggregating end-customer demand, or end-customer generation, bidding in the wholesale market on behalf of the end-customer, and potentially controlling the generation or consumption devices owned by the end-customer in response to wholesale market signals. At present, it appears that there is scope for effective competition in this retail service.

- *The monopoly regulator role* – As noted above, there is no scope for competition in electricity network services. As a result, customers of electricity networks (generators and consumers of electricity) need some assurance that the prices they pay for network services will not increase in the future. This assurance could be provided through vertical integration (government ownership or cooperative ownership), or through an explicit long-term vertical contract. In many countries this assurance is provided through a form of government-administered long-term contract known as public utility regulation. The administration of this long-term contract requires a dispute-resolution body, which is commonly known as the public utility regulator.

- *The forward market trader role* – Even large electricity market participants (whether producers or consumers) would prefer not to be exposed to volatile wholesale electricity prices. Instead, these parties would prefer to offload their risk to other parties. To an extent, this can be achieved through vertical integration, or through tailor-made vertical contracts. However, many large electricity market participants also seek to hedge their risk using risk-sharing contracts. These are often traded between market participants or on organised securities exchanges. However, there is not necessarily always a match between the buyers and sellers of these hedge contracts, especially when the market participants are located at different geographical locations. In this case there can be a role for a forward market participant who intermediates between wholesale market participants, providing the contractual forms that each market participant require. This role has no standardised label, but we will refer to this role as the forward market trader. There appears to be scope for effective competition in this role.

3.5 An Overview of Liberalised Electricity Markets

All liberalised or market-orientated electric power systems have a centralised wholesale real-time or spot market (a 'smart market' of the kind introduced earlier) for energy in which a number of buyers and sellers participate and in which the wholesale real-time or spot price for energy is determined at different locations on the network. However, there still remain significant differences between liberalised electricity markets. This section introduces the key differences between liberalised electricity markets.

Liberalised wholesale electricity markets may differ in the following respects:

- The extent to which the physical network constraints are taken into account in the wholesale energy market process.

  In the presence of network constraints the dispatch of generation and load resources must be coordinated with the physical limits on the transmission and distribution networks. Many liberalised electricity markets carry out this coordination by integrating the physical network limits into the computation of the efficient use of generation and load resources in the manner set out in Part III. However, many other markets take into account physical network limits in an *ad hoc* manner. For example, the wholesale price may be computed ignoring network limits; where network constraints arise, the network operator may adjust generators and loads up and down to the extent necessary to relieve constraints.

  Even where network constraints are included in the computation of the efficient dispatch of generation and load resources, the market prices paid to generators and loads may not reflect those physical constraints, which can distort the incentives on generators and loads in offering their resources to the market.[1]

  Even in those markets that correctly include transmission constraints in the computation of the efficient generation and load resources, we are not aware of any markets that incorporate distribution network constraints. There remains substantial scope for improving the coordination of local generation and load resources with local network conditions.

- The time period between successive operations of the market process.

  In any electricity industry supply and demand must be kept in balance (and within the physical limits of the network resources) at all points in time. In a market-based electricity industry the primary mechanism for balancing supply and demand is the market mechanism and the associated price signals. However, for various reasons, the market mechanism is typically not operated very frequently. Some markets set wholesale prices every hour, some every half-hour, or every 10 min. A few set prices every 5 min. However, 5 min remains a long time in an industry in which supply and demand must be maintained on a scale down to milliseconds. In the periods between the operation of the market mechanism (i.e. during the 'dispatch interval') the market mechanism cannot be relied on to achieve a balance of supply and demand.

  All real electricity markets have some mechanism for handling events (known as contingencies) that occur during the dispatch interval. These mechanisms include mechanisms for procuring and dispatching of balancing services when they are required. The longer the period between the operation of the market process, the more important are these balancing mechanisms.

- The extent to which generation and load resources are able to or required to participate in the wholesale market.

  In all liberalised electricity markets large generation resources participate in the wholesale market. Participation in the wholesale market includes submitting offer curves and

---

[1] The Australian National Electricity Market (NEM) features a regional or zonal pricing model. There is only one wholesale price in each NEM region. This contrasts with those markets overseas that have chosen nodal pricing (such as PJM or New Zealand), on the one hand, and those markets that have chosen uniform pricing (such as the United Kingdom) on the other. Unlike in the United Kingdom, generators that are constrained off do not receive additional payments (known as constrained-on or constrained-off payments).

following dispatch instructions from the market operator, but may also involve submitting forecast offer curves and information on planned outages.

However, markets vary in the extent to which smaller generators and loads are able or require to participate in the wholesale market. Typically, smaller generators are not required to either submit offers to the wholesale market or follow dispatch instructions from the market operator. We are not aware of any market that has attempted to integrate small-scale generation and load resources into the wholesale market. This is a substantial potential area for future development.

Even where smaller generators and loads do not directly participate in the wholesale market, it remains necessary to forecast their production and consumption. Markets differ in who is responsible for providing information on net forecast load.

Many renewable generators, such as wind or solar generators, are not able to follow dispatch instructions from the wholesale market and are often given special treatment.

Another dimension on which markets differ is whether or not participation in the wholesale market is compulsory or voluntary. Many markets allow participants to sign supply contracts with each other separately from the spot market. However, this is usually accompanied with a requirement to notify the market operator the net position of the participants in those bilateral trades for the purpose of computing the market outcome.

- The range of other associated markets in addition to the primary real-time market in electrical energy.

As noted earlier, all liberalised electricity markets have mechanisms for handling contingencies that occur between successive operations of the market process. In many cases, the procurement of these balancing services itself occurs through a market process. In addition, some markets have mechanisms to provide an additional source of revenue to generators to reward them for maintaining a certain level of capacity, known as 'capacity markets'.

- The nature of any network risk-management tools provided by the market operator or network operator.

As we will explore further in Part VI, in markets that integrate network constraints, the resulting wholesale prices reflect those network constraints. Those price signals give rise to a revenue stream (known as the merchandising surplus or settlement residues) to the market operator, which reflects the nature and extent of those network constraints. Markets differ in the use to which this revenue stream is put. In some markets this revenue stream is used to offset other network costs. In many markets this revenue stream is used to provide a backing to risk-management instruments provided by the market operator or network operator.

Liberalised wholesale markets also differ in a range of other market design factors, including:

- Whether or not there is a formal, centralised day-ahead or hour-ahead market, in addition to the real-time or spot market. Where there is a short-term forward market, there are two short-term prices for electricity: the day-ahead price and the spot or real-time price. Since these markets settle transactions at two different prices, they are known as two-settlement systems (or sometimes multisettlement systems). Markets such as PJM and California have a formal day-ahead market. The Australian National Electricity Market (NEM) does not have a formal day-ahead market.
- The range of decisions that are under the direct control of the system operator and the extent of any financial commitments or promises made by the system operator. In some

markets, the system operator has control over whether or not a certain generating unit starts up or shuts down (known as the *unit commitment decision*). In some markets, generators that are required to start up by the system operator receive a minimum guaranteed revenue for doing so whatever the level of the wholesale spot price. Since the unit commitment decision requires knowledge of supply and demand conditions over a period of time into the near future, markets in which the system operator takes control of the unit commitment decision require a formal day-ahead market (that is, must be a two-settlement system).

- Whether market participants are paid a single market-clearing price or are paid an amount that depends on their actual bid (known as 'pay-as-bid'). Markets also differ in the nature of the bids and offers – are the bids and offers interpreted as a piece-wise linear function, or as a 'step function' (as in the Australian NEM)?
- The range of prices that can emerge in the wholesale spot market. Some markets place upper and lower limits on the range of allowance spot prices – that is, a price ceiling and/or a price floor.
- The freedom or flexibility of generators and loads in how they can offer their resources to the market. Some markets have strict requirements to ensure that generators and loads submit bid and offer curves that reflect their true costs. Other markets limit how frequently generators and loads may change their bids and offers prior to the real-time dispatch.

3.6 An Overview of the Australian National Electricity Market

Australia's National Electricity Market or NEM coordinates the use of the larger generators and largest loads connected to the interconnected transmission network on the eastern and southern part of Australia, including Tasmania (which is connected by an undersea DC link). At over 4000 km in length the NEM is one of the longest interconnected power systems in the world. The NEM is not connected to the transmission networks in the other parts of Australia (the Northern Territory or West Australia).

The NEM features a high degree of vertical separation. Generation and retailing are separated from transmission and distribution.

The NEM is organised around a 5 min compulsory gross pool wholesale spot market, which is operated by the system operator, the Australian Energy Market Operator (AEMO). All generators above a certain size, known as *scheduled generators*, are required to participate in this wholesale spot market. These generators are required to submit offers to AEMO the day before, although they are able to revise elements of these offers right up until the time of dispatch. These offers consist primarily of a schedule showing the amount that the generator is willing to produce at 10 different nominated prices.

Larger wind generators must also participate in the wholesale spot market although in this case the maximum output of the generator is not specified by the generator itself but by a separate system that forecasts wind speeds. Wind generators may be dispatched for an amount less than the maximum output determined by the wind forecasting system but will not be dispatched for more. Wind generators are referred to as *semischeduled*.

Large loads may also participate in the wholesale spot market as scheduled loads, although in practice direct participation is limited – the only regular demand-side participants in the wholesale spot market are the pumps associated with certain hydro generators.

Smaller generators and loads do not participate in the wholesale market directly and can produce or consume as much as they desire at the prevailing spot price. AEMO produces regular (5 min) forecasts of this uncontrolled net load.

The physical limits of the transmission network are represented in the dispatch process as a set of linear equations known as constraint equations. There are a large number of constraints (approximately 20 000) in the constraint library, but only a subset of these are ever invoked or active at any one time, according to the physical network configuration, the current state of outages and current operating conditions (such as say, the presence of lightning or bushfires). The constraint library is maintained by AEMO using information provided by the transmission network owners.

The dispatch process carries out a constrained linear optimisation that maximises the reported economic welfare subject to the physical limits of the transmission network. The output of this constrained optimisation is an output target for every scheduled generator and a consumption target for every scheduled load, which is consistent with the physical operating limits of the transmission network.

The market process produces five spot prices – one for each of the NEM zones or regions – every 5 min dispatch interval. These prices must fall between $-1000/MWh (the price floor) and a price ceiling currently around $14 000/MWh (this price ceiling is updated annually in the light of inflation). At present the NEM zones correspond to the five NEM states – Queensland, New South Wales, Victoria, South Australia and Tasmania.[2] At the end of every half hour the six preceding spot prices are averaged to find the settlement price. Generators are paid the corresponding settlement price in their region for their output and loads must pay the settlement price (adjusted for static loss factors).

In addition to the market for electrical energy, every 5 min there are also separate markets for balancing services, known as ancillary services. These markets are divided into two types: *regulation service* and *frequency control services*. If a generator is dispatched for the regulation service, it must increase or decrease output during the 5 min interval in response to automated requests from AEMO. If a generator is dispatched for the frequency control services, it must stand ready to offset a change in the observe system frequency (by increasing or decreasing its output) within a specified timeframe. The frequency control services are not centrally dispatched by AEMO but operate autonomously once enabled.

These markets for ancillary services are partly integrated into the same optimisation process as for electrical energy – in other words the provision of ancillary services is 'cooptimised' with the provision of electrical energy. A generator that is the cheapest provider of ancillary services may be backed off in the energy market in order to allow the generator to be available to provide ancillary services.

Transmission services are provided by six transmission companies – one in each state. These transmission companies are subject to revenue-cap regulation administered by a regulatory agency, the Australian Energy Regulator (AER). The AER has developed a set of service quality indicators that are designed to reward transmission companies for maximising the capability of their networks at times of peak load. Any major new investment carried out by these transmission companies is subject to a cost–benefit assessment known as the 'regulatory test'.

---

[2] At the commencement of the NEM there was an additional NEM region lying between Victoria and New South Wales but this region was abolished on 1 July 2008.

Generators pay the cost of any transmission assets necessary to connect them to the shared transmission network (so-called 'connection assets'). Otherwise, the costs of the transmission network are entirely borne by customers.

Transmission network service providers have the option of entering into contracts with generators at strategic locations for the provision of certain services, such as certain ancillary services, reactive power support, or the provision of energy at certain times.

There is an associated market in risk management instruments. The most commonly traded forms are swaps (also known as *contracts for differences*) and caps, but many other forms of risk management instruments are also available. This market is predominantly over-the-counter, but there is increasingly active trading on the Sydney Futures Exchange.

The presence of price differences between regions in the NEM gives rise to a stream of funds known as the 'inter-regional settlement residues'. These are auctioned off periodically by AEMO as an instrument for facilitating hedging of inter-regional trade.

There is a requirement on retailers to purchase a proportion of the electricity they sell from renewable sources. This requirement gives value to 'renewable energy certificates', which are provided by renewable generating units for each MWh of energy they produce. There has been substantial investment in wind generation (particularly in South Australia) as a result.

Generators in the NEM earn revenue from the sale of energy in the wholesale spot market, the sale of ancillary services, the sale of contracts in the hedge market and, possibly, through the sale of renewable energy certificates. There are no other sources of revenue for generators, such as payments for making capacity available or guarantees regarding the revenue stream the generator will receive in any given period of time. In particular, generators must cover their fixed costs through the revenue stream they receive.

3.6.1 Assessment of the NEM

Overall, the NEM is a relatively simple market design. There are no day-ahead markets, no centralised unit-commitment, no capacity markets and no market power mitigation measures. The NEM has proven relatively robust in that its fundamental design has not changed in 15 years.

However, there remains substantial room for improvement, particularly:

- *Improving the geographic differentiation of prices*. Although the NEM correctly prices certain nodes at the transmission network level, physical network limits within each nominal pricing region are not correctly priced giving rise to distorted bidding incentives for generators and loads (known in Australia as *disorderly bidding*). Several other liberalised electricity markets in the world have adopted nodal pricing. Some improvements in efficiency could be achieved through the adoption of nodal pricing in Australia.

  Even more importantly, like almost every other liberalised wholesale electricity market, there is scope for significantly improved efficiency through improved pricing of customers connected to the distribution network. At present, most smaller customers face a time-averaged tariff, which does not reflect the presence of congestion on the distribution network. This results in inefficient local production and consumption decisions, and results in the need to over-build the distribution network. Mechanisms for nodal pricing of distribution networks are discussed further in Part IX. This will require, amongst other things, improved metering of small customers.

- *Improving the temporal differentiation of prices.* At present, the spot prices in the NEM are computed every 5 min. However, these prices are averaged to form 30 min prices for settlement purposes. This distorts the 5 min price signals. Furthermore, increasing the frequency at which prices are computed would improve the signals for localised investment in fast-response generation and load resources for more efficient handling of contingencies. This is discussed further in Part V.
- *Improving the mechanisms for hedging network congestion price risks.* At present, the only mechanism for hedging inter-regional price risk are the inter-regional settlement residues. For various reasons these are not an effective hedging instrument. Consideration should be given to improving the products available for hedging congestion price risk, so as to facilitate trade across separately priced locations. This would become even more important once nodal pricing is extended down to the distribution network. This is discussed further in Part VI.

3.7   The Pros and Cons of Electricity Market Reform

Has electricity market reform been successful? To give a complete answer to this question is very difficult. It would depend on the objectives of the reform and the outcome in each case.

At the most basic level, with the exception of a few widely publicised incidents, the lights have mostly stayed on. There have been a couple of exceptions. In 2003, a blackout occurred in the northeast corner of the United States and Canada, affecting power to millions of homes. However, this appears to be due to a failure of coordination between power systems and not a failure of the reform process per se. During the California electricity crisis, power was rationed to some consumers in southern California. This was a direct result of some of the policies put in place during the reforms in California. Those lessons have now been well learned and probably will not be repeated elsewhere. In the majority of cases, however, electricity market reform has not obviously affected reliability in the short-term. Market reform has therefore passed that, most elementary, threshold.

Electricity industry reform has apparently improved the efficiency and productivity of the industry. In Australia there is evidence of greater efficiency, improved work practices and greater reliability amongst generating plants. Large parts of the sector are exposed to considerably greater commercial pressures than before. In addition, the scope for new entry by a variety of technologies is greater than ever before. On these dimensions, electricity industry reform has probably been a success.

The evidence for efficiency improvement is less obvious in the United States (where there was probably less to gain) and the cost of the reforms have been substantial, leading some to question whether the reforms have been worthwhile. It will probably take more time to establish this one way or another.

Electricity market reform has not been without its critics. Critics of the reform process have tended to focus on concerns about market power, investment and governance.

As regards *market power*, many commentators have noted that electricity markets are prone to the exercise of market power (discussed in more detail in Part VII). In particular, market power often arises in small geographic areas with high barriers to entry known as 'load pockets'. The exercise of market power may lead to prices that are politically unsustainable, or may lead to policy measures to mitigate market power, which themselves have undesirable side effects, such as acting as a barrier to investment. Even where barriers to entry are low, geographic markets may be small, especially where there are few or imperfect instruments for hedging basis risk.

Although there have been episodes of market power in the Australian NEM, at most times the market is broadly competitive. In recent years wholesale prices have been moderate. Market power may yet undermine the reform process, but to date its impact has been muted.

As regards *investment*, the primary concern is that investment in generation capacity will not be forthcoming or will not be forthcoming in a timely manner. Perhaps private investors will tend not to invest until the last possible minute, after prices have risen to politically unacceptable levels. Perhaps entry will tend to occur in waves, with periods of very high prices and high investment followed by long periods of low prices and little or no investment. Perhaps it will be difficult to coordinate generation and network investment. In many countries, incentives for investing in generation are increased through the use of reserve or capacity markets. However, there are concerns that these markets are not an effective substitute for private risk-taking investment decisions and will tend to become politicised. Even in countries without capacity markets, governments may still be able to directly affect levels of generation investment through, say, their control over state-owned generating enterprises. These actions may also politicise generation investment and tend to crowd out private investment.

Boom–bust cycles in generation investment could, in principle, be mitigated if retailers were prepared to purchase supply contracts for many years in the future. However, customers are typically reluctant to sign-up to multiyear contracts, preferring to purchase electricity on a month-to-month basis. The lack of demand for longer-term contracts may exacerbate the short-termism in generation investment cycles.

The experience of the Australian NEM is that there has been reasonably sustained and on-going investment in generation capacity. There has not yet been evidence of a boom–bust cycle in generation investment. However, these issues need to be kept under review.

Finally, there are concerns about the *long-term governance arrangements*. Electricity is a politically sensitive industry. Governments cannot easily commit to 'keeping their hands off'. For an energy-only electricity market such as the Australian NEM to operate effectively there must occasionally arise outcomes that are politically sensitive (such as episodes of high prices). If these outcomes are not allowed to occur, the market will not be credible and will not succeed in the long run.[3] As customers become more used to the outcomes in liberalised electricity markets their degree of concern may diminish over time. This remains to be seen.

3.8 Summary

In all electricity industries there are a range of tasks that must be performed efficiently if the industry is to achieve an overall efficient outcome. These tasks include efficient use of and investment in generation resources, load (consumption) resources and network resources. Electricity industries differ in who is responsible for each of these tasks and the incentives, information and governance on those entities.

Historically, the electricity industry in most countries was organised as a single, vertically integrated supplier in each region. This firm was often government owned. These arrangements were, in part, a response to the monopoly investment concerns identified in Chapters 1 and 2.

---

[3] The New Zealand Electricity Market, which relies heavily on hydro generation, has suffered from the on-going threat of dry years leading to significantly high electricity prices. This led to a Ministerial Review of electricity market performance.

However, over time, concerns about efficiency, productivity and responsiveness to customers led to pressure for reform – in particular, the introduction of competitive market processes to ensure efficient use and investment in generation and (to a lesser extent) load resources.

Several countries have placed primary reliance on a wholesale spot market for electrical energy to drive efficient use of and investment in generation resources. However, there remain some material differences in how these markets are organised, including differences in the extent to which network constraints are integrated into the energy market, the time period of operation of that wholesale market, the extent to which generation and load resources participate in that wholesale market, as well as various other factors.

The Australian National Electricity Market is an example of a liberalised market-based electricity industry. It is organised around a wholesale spot market that computes prices in each of six regions every 5 min. Retailers compete to offer contracts to smaller customers. This market design has proven simple and resilient. However, there remains scope for improvement, particularly in the geographic and temporal differentiation of wholesale prices.

Electricity market reform has produced some gains. There have been clear improvements in productivity in Australia and the UK. However, the reforms themselves have involved substantial transition costs. There remain some concerns about the ability of the existing wholesale electricity markets to deliver politically sustainable levels and timeliness of investment. The jury is still out on whether these reforms will deliver not just one-off benefits but long-term increasing gains for consumers.

Questions

3.1 How does electricity market reform change the set of tasks that need to be performed in an electricity supply industry? What new tasks must be performed in a liberalised electricity market? What new problems or challenges are created by the reform process?

3.2 What were the main objectives of electricity market reform? Has electricity market reform achieved those objectives? Are there any clear failing of electricity market reform (putting aside specific problem cases such as the experience in California)? Is there any clear benefit of electricity market reform?

3.3 What were the main groups of policies pursued in the liberalisation of public utility industries in the 1980s and 1990s?

3.4 In the Australian NEM generators offer their output in 10 price–quantity pairs where the quantity is the amount offered at the corresponding price. The resulting offer curve is a 'step function'. Would it be preferable for the offer curve to be a linear function between a set of price–quantity pairs? What are the pros and cons?

Further Reading

For more information on the design of wholesale electricity markets, you may find it interesting to read Stoft (2000). There are also texts by Murray (1998, pp. 554–561), Murray (2009), Harris (2011), Ilic, Galiana and Fink (1998), and (in the context of European electricity markets) Glachant and Finon (2003). For more information on the Australian NEM you may like to look at AEMOs *An Introduction to Australia's National Electricity Market*.

Part III

Optimal Dispatch: The Efficient Use of Generation, Consumption and Network Resources

Hastings power house, New Zealand, ca 1915 (Source: Neil Rennie)

Suppose we have an existing stock of generation, consumption and network assets. How should that stock of assets be operated so as to maximise economic welfare? This question must be answered whether the electricity industry is operated as a vertically integrated monolithic firm with no scope for competition or as a liberalised market with decentralised decision making by a large number of competing generators and consumers. The problem becomes even more complicated when the operation of generation and consumption assets must be coordinated with physical network limits. How should we utilise a stock of generation, consumption and network assets so as to deliver the greatest economic value? These are the issues we explore in this part.

4

Efficient Short-Term Operation of an Electricity Industry with no Network Constraints

Let us start by putting aside network issues. Let us suppose that we have a set of generators and loads that (for the moment) are all located at the same point on the network. Now imagine that you are an omniscient all-powerful operator, able to control all of the generation and consumption resources in the economy. How should you use this stock of generation and consumption resources in the most efficient way possible? We start by focusing on the question of the efficient use of a set of generation resources. Consideration of this question leads us to the notion of the merit order. In the next section, we shall ask whether it is possible to achieve this efficient outcome using decentralised decision-making and market signals alone.

4.1 The Cost of Generation

Let us focus first on the optimal use of a set of existing generation assets. As we saw in Part II, electricity is produced by physical equipment, which we will refer to as an electricity generator.

We will focus in this section on the short-run economic costs incurred by a generator in producing its output. In general, the costs incurred by a generator will depend on many parameters (such as the price of the input fuel and the cost of capital). We will focus on how the costs incurred by a generator might vary with the level of output of the generator (i.e. the rate at which the generator produces electrical energy). Let us call the rate of production of the generator Q (Q is measured in megawatts). Recall that this output represents a flow of electrical energy per unit of time. A generator that produces at the constant rate Q for a period of time T (measured in hours), produces QT MWh of energy.

As mentioned, we will assume that each generator can control the rate at which it produces electrical energy. While this assumption is true for most generators, such as thermal (coal or gas-fired) or hydro generators, it is not true for some other generators, such as wind- or solar-powered generators, which depend on the availability of resources. These generators are not

The Economics of Electricity Markets, First Edition. Darryl R. Biggar and Mohammad Reza Hesamzadeh.
© 2014 John Wiley & Sons, Ltd. Published 2014 by John Wiley & Sons, Ltd.

necessarily able to follow instructions from the system operator as to the rate at which they should produce electrical energy. These generators raise specific issues that will be addressed later.

Let $C(Q)$ ($/h) be the short-run cost function of the generator – that is, the rate at which the generator is incurring expenditure when producing energy at the rate Q MW. A generator that produces at a constant rate Q MW for a period of time T incurs costs of $C(Q)T$ in dollars (or some other unit of currency).

It is common to distinguish these costs into two components:

- The *fixed costs*, which are independent of the output of the generator. These costs include the costs of leasing the generating facilities (such as the land on which the generator is sited) and/or the costs of financing the purchase of the facilities. These costs also include the costs of any permanent operations and maintenance or management staff, which must be maintained whether or not the generator is in operation. These costs do not enter into the output decision calculus, which we will discuss below.
- The *variable costs* (also known as the 'production costs'), which vary with the output of the generator. These costs include the costs of any fuel consumed, any operating or maintenance costs, which vary with output, and the costs incurred in starting and stopping the generator.

In addition, a generator will typically have some minimum level of output below which it cannot physically operate effectively (without shutting down entirely), and some maximum level of output above which it cannot produce any more output.

Furthermore, there may be other important costs that might need to be taken into account, such as the *startup costs* of the generator. In many thermal generators energy must be consumed to heat the water in the boiler before electrical energy can be produced at all. For the moment we will put these costs to one side.

How might the variable costs of a typical generator vary with its output? At low levels of production (close to the minimum operating level), the average variable costs (that is the variable cost divided by the output of the generator) tend to be relatively high since there are often 'auxiliary' costs that must be incurred whenever the generator is in service and producing nonzero levels of output. The average variable cost then typically declines as the output of the generator increases, but may start to rise again as the output of the generator approaches the maximum operating level.

The cost function of a typical generator is sometimes approximated as a quadratic function of its output. For example, the cost function of a typical generator might be assumed to take the following form:

$$C(Q) = \begin{cases} F, & \text{where } Q = 0 \\ cQ + aQ^2 + b + F, & \underline{Q} \le Q \le \overline{Q} \end{cases}$$

where F is the fixed costs of the generator, a, b and c, are the parameters of the cost function, and $\underline{Q}$ and $\overline{Q}$ are the minimum and maximum operating levels of the generator, respectively.

As we saw in Section 1.3, in the short-run, economists usually focus primarily on the marginal cost function of producers such as generators. The marginal cost function is the

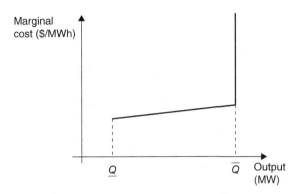

Figure 4.1　Marginal cost curve of a typical generator

derivative of the cost function with respect to the level of output, which we will denote $C'(Q)$. The marginal cost function of the typical generator referred to earlier would be

$$C'(Q) = \begin{cases} \text{undefined,} & \text{where } Q = 0 \\ c + 2aQ, & \underline{Q} \leq Q \leq \overline{Q} \end{cases}$$

The graph of the marginal cost of this hypothetical generator is illustrated in Figure 4.1.

Note that, as the output of each generator reaches its physical limit, it becomes very costly to increase its output further – at this point the marginal cost function turns sharply upward.

A couple of points should be borne in mind. First, the variable cost of a generator (and therefore the marginal cost) reflects the opportunity cost of using the generator's energy source in some other way or at some other time. This opportunity cost may change from moment to moment, shifting the marginal cost function up or down. For example, where there is a spot market for the fuel used in the generator, the price of the fuel on that spot market may change from one time interval to the next.

Even for those generators that have access to fuel under long-term contracts at fixed prices, as long as the generator has the opportunity to sell any fuel it does not use on the fuel spot market, or the opportunity to buy any additional fuel on the fuel spot market, the relevant opportunity cost for the generator is not the long-term or contract price for the fuel, but the short-run fuel spot price. When the fuel spot price is high the generator may find it more profitable to produce less electricity and to sell its input fuel on the spot market. Therefore, even for generators that have long-term fuel price contracts, if those generator have access to a short-term market on which they can buy and sell their input fuel, the marginal cost function will not be static but will depend on the spot price of the fuel.

Some generators, particularly some hydroelectric generators, have a limited stock of energy available to them. In the case of hydroelectric generators, the limited stock of energy is reflected in the limited stock of water in high-elevation lakes. Such generators are known as *energy-limited generators*. For energy-limited generators, the opportunity cost of producing at one instant in time is the foregone profit from not being able to produce at another point in time. Therefore, the marginal cost function of such generators depends on the potential value of the output produced at other times.

For these generators there is a very significant time-interdependence in their costs of production. To model the optimal use of these generators, we should consider the optimal production strategy over a period of time. If the price of electricity is expected to be higher tomorrow, every extra MW produced today implies there will be less energy available to produce electricity tomorrow. As a result, the variable cost of the electricity produced today is the foregone profit from production tomorrow. We will come back to this issue in Section 4.10.

In addition, as noted earlier, we have put to one side the issue of startup costs, which we will return to later in this chapter in Section 4.10.

It is worth noting that the cost function of a typical generator is not necessarily convex. A function is said to be convex if, when we take the weighted average of the function at two different points, the weighted average lies above the function evaluated at the weighted average of the two points. The cost function set out earlier is nonconvex due to (a) the presence of fixed auxiliary costs, which can be avoided when the output reduces to zero, and (b) the presence of a minimum operating level, which is greater than zero.

As we will see later, these nonconvexities complicate the task of finding the optimal combination of output of different generators. Even when the optimal combination of outputs exist, with nonconvexities some generators may not be able to cover their costs. For these reasons, it is common in electricity market analysis to assume these problems away. This is equivalent to assuming that $\underline{Q} = 0$ and $b = 0$ in the earlier cost function. This is precisely what we will do in the next section.

4.2 Simple Stylised Representation of a Generator

For our purposes it is useful, at least at the outset, to assume a particularly simple stylised shape for the marginal cost function of generators. In particular we will assume that there is no minimum operating level, no fixed auxiliary costs, and constant marginal costs of operation up to the maximum operating level.

In mathematical notation, we will assume that the minimum operating level is zero, and the parameters a and b in the cost function mentioned earlier are also zero. The maximum operating level will be said to be the generator's *capacity* and will be denoted K. In other words, we will assume $\underline{Q} = 0$ and $\overline{Q} = K$. The cost function of the generator is then simply: $C(Q) = cQ + F$ for $0 \leq Q \leq K$.

Under these assumptions the marginal cost curve for the generator is flat (horizontal) up to the generator's maximum operating level at which point the marginal cost curve becomes vertical. Since there is no ambiguity, we can refer to the generator's marginal cost in the flat part of the curve as the generator's *variable cost*. This marginal cost is illustrated in Figure 4.2.

Even in those cases where the marginal cost function of a generator is not flat over a wide range of output, it is often the case that the marginal cost function can be approximated by a 'step function' – that is, a function that is flat over a range of output. Mathematically this is equivalent to breaking a single generator up into smaller units each of which has a constant marginal cost, allowing us to use the simple stylised representation of a generator as set out in Figure 4.2. In principle, this approximation can be made arbitrarily accurate.

Figure 4.3 shows how the cost function of a generator might be approximated using three hypothetical generators each with a constant marginal cost.

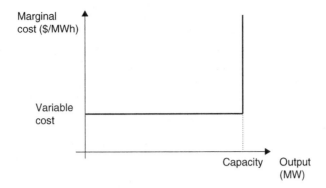

Figure 4.2   Stylised representation of the marginal cost of a generator

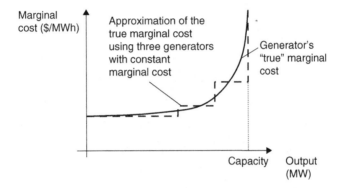

Figure 4.3   Approximation of an arbitrary marginal cost curve using generators with constant marginal cost

In practice, many liberalised electricity markets require generators to submit an offer curve (that is, a statement of how much they are willing to produce at a given price) that is a step function of just this kind.

4.3   Optimal Dispatch of Generation with Inelastic Demand

Let us consider first a power system with no physical limits except one – the constraint that total power produced must equal total power consumed. We will ignore the transmission and distribution networks entirely – in effect, we will assume that all production and consumption of electricity takes place at a single location.

In addition, in this section, we will ignore any controllable consumption assets. We will consider these further in Section 4.4. Finally, to keep things simple, let us put to one side, at least temporarily, the nonconvexities mentioned earlier. Specifically, we will assume that the cost function of every generator $C_i(Q_i)$ is convex. This implies, amongst other things, that there is no minimum operating level. We will also ignore startup costs.

Let us consider the task of an omniscient system operator who has a portfolio of N generators over which it has perfect information and total control. The system operator is able to direct each generator as to how much it should produce. If generator i is directed to produce energy at the rate Q_i (MW) it incurs cost at the rate $C_i(Q_i)$ ($/h). The total demand for electricity is Q (MW).

The set of all the orders given by the system operator to the generators is known as the *dispatch* (also often spelled 'despatch'). The dispatch is a vector $Q = (Q_1, Q_2, \ldots, Q_N)$, which specifies for each generator the total rate at which it is to produce energy.

The task of the system operator is to find the dispatch that minimises the total cost of generation subject to the constraint that the total amount of generation must equal the total load. Let us assume for the moment that the total demand is less than the total capacity of the generators to produce so we do not have to worry about rationing.

4.3.1 Optimal Least Cost Dispatch of Generation Resources

The task of the system operator is to find a dispatch that minimises the total overall cost of meeting demand. In other words, the task of the system operator can be written as follows:

$$\min \sum_{i=1}^{N} C_i(Q_i)$$

$$\text{subject to } \sum_{i=1}^{N} Q_i = Q \leftrightarrow \lambda$$

$$\text{and } \forall i, \ Q_i \leq K_i \leftrightarrow \mu_i \text{ and } Q_i \geq 0 \leftrightarrow \nu_i$$

The KKT conditions for this problem are as follows:

$$\forall i, \ C_i'(Q_i) = \lambda - \mu_i + \nu_i$$

In addition,

$$\forall i, \ \mu_i \geq 0 \text{ and } \mu_i(Q_i - K_i) = 0$$

$$\forall i, \nu_i \geq 0 \text{ and } \nu_i Q_i = 0$$

From these expressions we can see that for each generator there are three possibilities: Either the generator is dispatched to an intermediate level (between its maximum and minimum output, so that $\mu_i = 0$ and $\nu_i = 0$) in which case the generator is dispatched to a point where its marginal cost is equal to some common value:

For each i for which $0 < Q_i < \overline{Q}_i$, $C_i'(Q_i) = \lambda$ for some constant λ.

Alternatively, the generator is dispatched to its maximum output $Q_i = K_i$, in which case its marginal cost is smaller than the common value $C_i'(Q_i) \leq \lambda$, or the generator is dispatched to its minimum output $Q_i = 0$ in which case the marginal cost is greater than the common value: $C_i'(Q_i) \geq \lambda$.

The common marginal cost λ is known as the *system marginal cost* or SMC. Note that the SMC is equal to the marginal cost of producing an additional unit to meet an additional unit of demand.

We can define the total cost of the optimal dispatch as $C(Q) = \sum_{i=1}^{N} C_i(Q_i)$. If the cost function of each generator is convex, the total cost is convex in the total load.

Result: Given a set of controllable generators with convex cost functions and upper and lower bounds on production, the least cost dispatch (ignoring network constraints) has the following characteristics:

a. For every generator that is despatched to a rate of production that lies between its minimum and maximum operating level, the marginal cost of each generator is the same; this common value is known as the 'system marginal cost' or SMC;
b. All generators that are despatched to a target of zero have a marginal cost that is above SMC, and all generators that are despatched to their maximum operating level have a marginal cost that is below the SMC.
c. The sum of the output of all generators is equal to the total demand.

In addition, it is straightforward to demonstrate that (i) the total cost of generation is convex in the total system load Q, (ii) the SMC is monotonically increasing in the system load Q, and (iii) the output of each generator is monotonically increasing in both the SMC and load.

4.3.2 Least Cost Dispatch for Generators with Constant Variable Cost

The results just mentioned hold whatever the shape of the marginal cost function of each generator. Let us focus now on the special case where the marginal cost curve of each generator takes the simple stylised form of Figure 4.2.

Specifically, let us assume that when there are K_i units of capacity of generation of type i, and when that type of generation is producing at the rate Q_i, costs are incurred at the rate

$$C_i(Q_i) = c_i Q_i \text{ for } 0 \leq Q_i \leq K_i$$

The social optimisation problem is as follows:

$$\min \sum_i c_i Q_i$$

$$\text{subject to } \sum_i Q_i = Q \leftrightarrow \lambda$$

$$\text{and } \forall i, Q_i \leq K_i \leftrightarrow \mu_i \text{ and } Q_i \geq 0 \leftrightarrow \nu_i$$

The KKT conditions for this problem include the following condition:

$$\forall i, c_i = \lambda - \mu_i + \nu_i$$

From the KKT conditions we find that the least cost dispatch has the following properties: Each generator can be ranked in order, from the generator with the lowest variable cost to the generator with the highest variable cost. This is known as the *merit order*. Intuitively, the system operator can then work its way up the merit order dispatching each generator up to its maximum capacity in order until all demand is satisfied.

Result: In the least cost dispatch of a set of controllable generators with a constant variable cost and ignoring network constraints, each generator is dispatched according to the merit order. Generators are dispatched in order from the lowest-variable-cost to the highest, working up the merit order until all demand is satisfied.

More formally, for each generator:

a. if the variable cost of the generator is below the SMC, the generator is dispatched for its full available capacity;
b. if the variable cost of the generator is above the SMC, the generator is not dispatched at all;
c. if the variable cost of the generator is equal to the SMC, the output of the generator is indeterminate, but the total output of all generators is equal to the total load.

Intuitively, the optimal dispatch can be found as follows. We can sum the marginal cost curves for each generator horizontally to find the *industry supply curve*. When the marginal cost curves take the simple stylised form of Figure 4.2 the industry supply curve is a simple step function. We can then find the intersection between this industry supply curve and the (vertical) demand curve. This intersection yields the SMC.

The optimal output for each generator is where the marginal cost of each generator is equal to the SMC. The output of each generator is either at its maximum capacity, if the generator's variable cost is below the SMC, or zero, if the generator's variable cost is above the SMC, or at some intermediate level when the generator's variable cost is equal to the SMC (Figure 4.4).

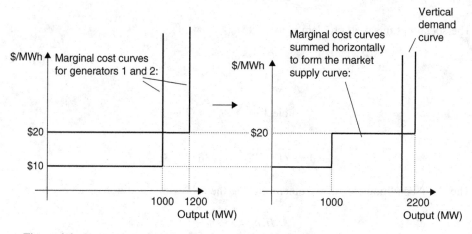

Figure 4.4 Intersection of market supply and market demand gives market-clearing price

Table 4.1 Cost data for a simple three-generator electricity supply industry

Generator	Variable Cost ($/MWh)	Capacity (MW)
A	$10	1000
B	$25	500
C	$100	200

As we will see later, in a liberalised electricity market, the SMC will correspond to the wholesale spot price for electricity. However, nothing we have said so far relies on the existence of a wholesale spot market. The same principles would apply for a vertically integrated electricity industry as for a liberalised market.

We can say that the SMC is always equal to the marginal cost of every generator that is dispatched for a positive amount. Strictly speaking, however, the marginal cost of some generators is indeterminate at their maximum capacity (indicated here by the vertical marginal cost curve at this point). However, mathematically we can define a left-hand marginal cost as the marginal cost saving of a small reduction in output and a right-hand marginal cost as the marginal cost addition from a small increase in output. It is perhaps more correct to say that the SMC lies between the left-hand marginal cost and the right-hand marginal cost for all generators. However, if we allow the marginal cost to take a range of values at this point (all values between the left-hand marginal cost and the right-hand marginal cost) we can make the assertion that the SMC is always equal to the marginal cost of every generator that is dispatched for a positive amount.

The SMC is sometimes said to be equal to the variable cost of the last generator to be dispatched – which is also known as the *marginal generator* or sometimes as the *price setter* (this last term is something of a misnomer since, of course, it is the output of all generators combined, together with the level of demand, which sets the level of the price).

4.3.3 Example

For example, suppose we have an electricity industry with three generators with constant variable cost and capacities as shown in Table 4.1.

Suppose that load varies between 500 and 1700 MW. Given this information about supply and demand conditions we can work out the dispatch of each generator for each level of demand, as shown in Table 4.2. For example, when the load is 1100 MW, generator A is dispatched up to its maximum capacity (1000 MW) and generator B is dispatched for the remainder (100 MW). Since generator B is the marginal generator, the SMC (the wholesale spot price) is $25/MWh.

Table 4.2 The optimal dispatch and SMC for different levels of demand in a simple three-generator industry

Load (MW)	Dispatch (MW)			SMC (Price) ($/MWh)
	Gen A	Gen B	Gen C	
$500 \leq Q < 1000$	Q	0	0	$10
1000	1000	0	0	$10 \leq P \leq $25
$1000 \leq Q < 1500$	1000	Q−1000	0	$25
1500	1000	500	0	$25 \leq P \leq $100
$1500 \leq Q < 1700$	1000	500	Q−1500	$100
1700	1000	500	200	$P \geq $100

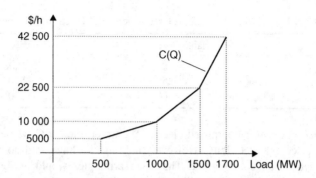

Figure 4.5 Cost function for the simple three-generator case

We can also express the total cost of the optimal dispatch in this example as follows:

$$C(Q) = \begin{cases} 10Q, & 500 \le Q \le 1000 \\ 25Q - 15,000, & 1000 \le Q \le 1500 \\ 100Q - 127,500, & 1500 \le Q \le 1700 \end{cases}$$

As we noted before, this function is convex in the total demand, Q, as illustrated in Figure 4.5.

4.4 Optimal Dispatch of Both Generation and Load Assets

Section 4.3 focused on the optimal use of a set of generation assets to serve a given quantity of load. However, this analysis is incomplete. Both production and consumption assets can, in principle, respond to wholesale market conditions. Overall efficiency requires that we consider efficiency in the use of both production and consumption assets.

As in Part I, we will assume that each electricity customer has a utility function from the consumption of electricity $U_i(Q_i)$. Each customer is assumed to be a price-taker on the electricity market. As we know from Part I, if each consumer faces a simple linear price for electricity, we can derive a downward sloping demand curve for each customer.

In Section 4.3, the task of the market operator was to choose a set of production rates (one for each generator), known as the dispatch, which minimises the overall cost of meeting demand.

Now the market operator must choose a combination of a rate of production for each generator and a rate of consumption for each consumer, which maximises the total economic welfare. In other words, the dispatch now has two components. We can write the optimal dispatch as (Q^S, Q^B) where, as before, $Q^S = Q_1^S, Q_2^S, \ldots$ is a vector that specifies for each generator the total rate at which that generator is to produce energy. In addition, $Q^B = Q_1^B, Q_2^B, \ldots$ specifies the rate at which each electricity consumer is to consume electrical energy.

For simplicity, let us temporarily set aside the generator production bounds (this is without loss of generality since these bounds are embodied in the generator cost function in any case).

The problem of finding the efficient use of a set of production and consumption assets was considered in Section 1.4. Here we apply that analysis to the case of the electricity market.

Given a set of generation and consumption resources, the task of the system operator is to find a dispatch that minimises the total economic surplus. In other words, the task of the system operator can be written as follows:

$$\max \sum_i U_i(Q_i^B) - \sum_i C_i(Q_i^S)$$

Subject to

$$\sum_i Q_i^B = \sum_i Q_i^S \leftrightarrow \lambda$$

This problem has the KKT conditions

$$\forall i, \ U_i'(Q_i^B) = \lambda \text{ and } \forall i, \ C_i'(Q_i^S) = \lambda \text{ and } \sum_i Q_i^B = \sum_i Q_i^S$$

In other words, this problem has the following solution: the point at which the demand and supply curves intersect yields a price. All generators produce at a rate where their marginal cost is equal to this common price or SMC. In exactly the same way, loads are dispatched to the point where their marginal valuation is equal to the SMC.

Result: The welfare-maximising dispatch of a set of controllable electricity production and consumption assets has the following characteristic: There is a common system-wide marginal cost. Each generator produces at a rate where the marginal cost is equal to the common system-wide marginal cost. Each consumer consumes at a rate where the marginal value of consumption (the point on the demand curve) is equal to the common system-wide marginal cost.

For example, it might be that there are three different types of consumers in the market. Consumers of type A are prepared to consume at the rate of up to 200 MW of electricity and value that electricity at $200/MWh. Consumers of type B are prepared to consume at a rate of up to 300 MW of electricity and value that electricity at $100/MWh. Consumers of type C are prepared to consume up to 50 MW of electricity and value that electricity at $20/MWh. The resulting market demand curve (the sum of the individual demand curves) is illustrated in Figure 4.6.

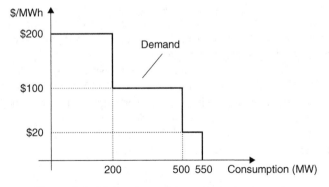

Figure 4.6   Market demand curve for a simple market

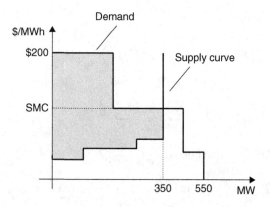

Figure 4.7 Illustration of optimal dispatch in a simple electricity market

Given a market supply curve such as the one in Figure 4.7, we can determine the optimal dispatch. From Figure 4.7 we see that the SMC is equal to $100/MWh. The generators are dispatched according to this SMC as before. In addition, customers of type A consume at the rate of 200 MW. Customers of type B consume at some intermediate rate, while customers of type C choose not to consume at all. The total economic surplus is the shaded area in Figure 4.7.

4.5 Symmetry in the Treatment of Generation and Load

4.5.1 Symmetry Between Buyer-Owned Generators and Stand-Alone Generators

In the previous analysis we have assumed that consumers (loads) have some controllable consumption for electricity given by the utility function, without enquiring further into how that controllable consumption comes about.

This controllable consumption could arise from an interruptible production process or from the deferment of consumption in cases where there is substantial thermal inertia (i.e. when the electricity is used for heating or cooling). However, in addition, an important source of controllability in the level of load arises when the buyer has an on-site controllable generator. In this case, even if the underlying load of the buyer is fairly stable, the net load as it appears to the market (that is, the underlying load less the on-site production) can appear controllable.

As we noted in Part II, one of the key potential future developments in electric power systems is the increasing penetration of small-scale generation including roof-top solar PV, small wind generation and local storage. A key question therefore, is whether the earlier analysis should treat this generation differently.

Does the analysis distinguish in any significant way between stand-alone generation and generation integrated with a buyer? The simple answer is no. We can treat the utility function of an integrated load-generator as equal to the underlying utility from the total electricity consumption less the cost of electricity produced on site. Specifically, if the underlying utility function is denoted $V(Q)$ and the local cost of production is $C(Q)$, then the utility reported to the central market operator is as follows:

$$U(Q) = V\left(Q + Q^S\right) - C\left(Q^S\right)$$

If the buyer operates the local generator in such a way as to maximise his/her utility then he/she will choose the rate of production for which the marginal value of consumption is equal to the marginal cost of production:

$$V'(Q + Q^S) = C'(Q^S)$$

Moreover, this common marginal value and marginal cost is equal to the common marginal value and marginal cost for the other generators and loads in the market. This is the condition for the overall efficient outcome above.

In other words, there is no difference in the treatment of integrated generators and stand-alone generators. We can combine a generator with a load and the combined entity can be treated the same way in the market as a separate load and a stand-alone generator.

4.5.2 Symmetry Between Total Surplus Maximisation and Generation Cost Minimisation

We are focussing on the question of the efficient use of a given set of production and consumption assets. We have viewed this as the problem of finding the dispatch that maximises total surplus, but we can also view this problem as a minimisation problem very similar to the problem of minimising the total cost of generation as we saw in Section 4.3.

Let us suppose that, at the price of zero, the demand for electricity by customer i is finite, and takes the value $Q_i^{B \ max}$. This implies that the utility function $U_i(Q_i^B)$ is bounded above. Let the maximum of this function be U_i^{max}. Let us define a new variable $Q_i^{Sdr} = Q_i^{B \ max} - Q_i^B$ and define a new function $V_i(Q_i^{Sdr}) = U_i^{max} - U_i(Q_i^{B \ max} - Q_i^{Sdr})$. This function $V_i(\cdot)$ is positive and upward sloping (and has a positive second derivative), just like a generator cost function.

In fact we can think *any* demand-side responsiveness to the wholesale price as being exactly equivalent to a generator. We can imagine that the demand for each customer is represented as a single, fixed, demand of Q_i^{Bmax}, which is valued at the amount U_i^{max}. Any responsiveness of the customer to the wholesale market is reflected in the form of a generator with output Q_i^{Sdr} and cost function $V_i(Q_i^{Sdr})$. The task of the market operator can then be viewed as simply minimising the cost of generation.

Result: In the economic analysis of power systems there is no need to draw a distinction between controllable generation, controllable loads, or sites that include both generation and consumption. Economically, we can treat the market as consisting of all controllable generation or all controllable loads. Similarly, we can allow colocated generation and load assets to be treated collectively (as either a generator or load) or separately as a generator separate from a load. (Later we will see that this result depends on the assumption that generators and loads face the same market price).

4.6 The Benefit Function

We can express the mathematical problem of finding the optimal dispatch of generation and consumption resources in a slightly different way. This alternative way is mathematically identical to the problem set out in Section 4.5. However, this alternative formulation proves more convenient when we introduce network constraints in Chapter 5.

This alternative formulation works as follows: Consider a partition of the set of producers and consumers.[1] The jth partition will be labelled N_j. Let us define the net injection of the producers and consumers in the jth partition as the total rate of production of all the producers in this partition less the total rate of consumption of all the consumers:

$$Z_j = \sum_{i \in N_j} Q_i^S - \sum_{i \in N_j} Q_i^B$$

Now let us define the benefit function for each partition as follows: The benefit function for the jth partition for the net injection Z_j is the maximum level of economic surplus that can be achieved for the producers and consumers in the jth partition while holding constant the net injection Z_j. In other words, the benefit function is defined as follows:

$$B_j(Z_j) = \max_{Q_i^B, Q_i^S} \sum_{i \in N_j} U_i(Q_i^B) - \sum_{i \in N_j} C_i(Q_i^S)$$

$$\text{subject to } Z_j = \sum_{i \in N_j} Q_i^S - \sum_{i \in N_j} Q_i^B \leftrightarrow \alpha_j$$

As before, at the optimum, the marginal valuation of each customer in the jth partition and the marginal cost of each supplier in the partition are the same and equal to α_j. Furthermore, an increase in the net injection for this partition by a small amount increases the benefit function by the amount α_j:

$$B_j'(Z_j) = \alpha_j$$

Importantly, the optimal dispatch task can be rewritten more simply in terms of the benefit function. The optimal dispatch task is now:

$$W = \max_{Z_j} \sum_j B_j(Z_j)$$

$$\text{subject to } \sum_j Z_j = 0$$

4.7 Nonconvexities in Production: Minimum Operating Levels

In the previous sections we assumed, for convenience, that the cost function of each generator was convex. However, in practice, many generators have a minimum level of output below which they cannot physically operate. This introduces a new set of constraints into the optimal dispatch problem. How do these new constraints affect the optimal dispatch?

It turns out that the presence of minimum operating levels affects the optimal dispatch in four ways:

- It may not simply be feasible to meet certain levels of demand;
- As load increases, some generator's output may need to be *reduced* (rather than increased) in the optimal dispatch;

---

[1] A partition of set A is a set of mutually dis-joint subsets with a union equal to the original set A.

Table 4.3 Key cost data for a simple illustration of the impact of nonconvexities

Generator	Variable Cost ($/MWh)	Minimum Operating Level (MW)	Maximum Operating Level (Capacity, MW)
A	$10	250	500
B	$20	260	500

- The marginal or price-setting generator may not be the 'last' generator in the merit order to be dispatched;
- The total cost of the optimal dispatch may not be convex in the load.

These results can be illustrated using the following simple example. Let us suppose an electricity industry has just two generating units, each with a capacity of 500 MW. Unit A has a constant variable cost of $10/MWh, and a minimum operating level of 250 MW. Unit B has a constant variable cost of $20/MWh and a minimum operating level of 260 MW, as summarised in Table 4.3.

Clearly, when demand is below 250 MW, there is no feasible dispatch – that is, there is no dispatch that satisfies the operating constraints of the generators.

If demand is larger than 250 MW but less than 500 MW, the load is met by unit A. However, what about if load increases to 501 MW? This load exceeds the capacity of unit A to supply alone, so we need to increase the output of unit B. However, the minimum output of unit B is 260 MW. If we dispatch unit B to 260 MW, we would have to reduce the output of unit A to 241 MW, which is below its minimum operating level. We see that for loads between 500 and 510 MW there is, again, no feasible dispatch.

For loads larger than 510 MW and up to 760 MW, we can dispatch unit B for its minimum (260 MW), and then dispatch unit A for the remainder. Note that unit A is, in this case, the marginal generator, even though unit B is also dispatched. For loads larger than 760 MW, up to 1000 MW, unit A is dispatched for 500 MW, and unit B is dispatched for the remainder.

Figure 4.8 illustrates the shape of the total cost of optimal dispatc as a function of system load. As we can see, this function is no longer convex.

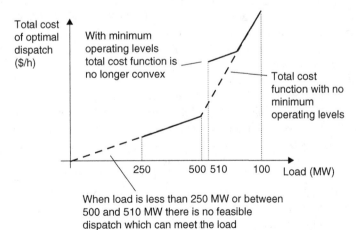

Figure 4.8 Cost of dispatch in the presence of nonconvexities

4.8 Efficient Dispatch of Energy-Limited Resources

Some generators have only a limited stock of their fuel or key energy input on hand, and cannot easily obtain more at short notice. These generators are known as *energy limited*. As we will see, it will typically not make sense to put an energy-limited generator in the merit order at the variable cost of obtaining its key input. Instead, as we will see, we have to take into account the *opportunity cost* of not being able to produce at other times.

The classic example of an energy-limited generator is a hydro-generator with a limited stock of water held back in a dam. Once the dam is full, any additional in-flows must be spilled down the river. Since this water cannot be stored and used at any other time, the opportunity cost of using any further inflows of water is zero – the hydro-generator may as well use this water for generating electricity provided the spot price for electricity exceeds the variable cost of converting that water into electricity (which is typically close to zero for a hydro-generator).

Let us focus on the case where there are no more inflows forecast over the period in consideration so that generating electricity at the current time involves consuming water that could otherwise be used to generate electricity at some other time. What is the most efficient way to use this resource of stored water?

If the water in the dam is used for generating electricity at a particular time, it cannot be used for generating electricity at another time. The opportunity cost of using the water for generation at a particular point in time is the value of that water in generation at some other time.

Let us suppose, for simplicity, that the variable cost of the energy-limited generator is close to zero. This is a reasonable approximation for hydro-generators.

It turns out that the most efficient use of the energy-limited resource is to use the resource to generate at those times when the SMC is the highest over the period in question. In other words, if a hydro-generator captures enough water to produce, say, 1000 MWh of output each week, the hydro-generator should generate those 1000 MWh precisely at those times when the SMC is the highest.

Note that this implicitly requires a degree of *future price forecasting and intertemporal optimisation*. In deciding whether or not to generate 1 MWh from a hydro resource, an efficient market operator must look at the generation that will be displaced by that 1 MWh of production today and the generation that will be displaced by that 1 MWh of production at some point in the future. It makes no sense to use an energy-limited resource to displace 1 MWh of low-cost generation today if that same resource could be used to displace 1 MWh of very high cost generation at a time in the near future.

Let us modify the optimal dispatch problem set out earlier, to incorporate the possibility of energy-limited resources. Since this is an intertemporal optimisation problem we now have to keep track of the time dimension of each variable. As before, let us suppose we have N generating units, operating over T periods. Suppose that the duration of period t is represented by σ_t (with units of time). The total demand in period t is Q_t (MW). The output of generating unit i in period t is Q_{it} (MW). Let us assume that the physical characteristics of each generating unit remains the same over time – that is, the marginal cost of generating unit i is $C_i(Q_{it})$ ($/MWh) over the range $0 \leq Q_{it} \leq K_i$. In addition, the Nth generator is assumed to face an energy constraint, that its total output over the T periods does not exceed an energy limit E_N.

The task of the system operator can now be written as follows:

$$\text{Min} \sum_{t} \sigma_t \sum_{i} C_i(Q_{it})$$

$$\text{Subject to } \forall t, \sum_{i=1}^{N} Q_{it} = Q_t \leftrightarrow \sigma_t \lambda_t$$

$$\text{And } \forall i, t, Q_{it} \leq K_{it} \leftrightarrow \sigma_t \mu_{it} \text{ and } Q_{it} \geq 0 \leftrightarrow \sigma_t \nu_{it}$$

$$\text{And } \sum_{t} \sigma_t Q_{Nt} \leq E_N \leftrightarrow \gamma$$

The KKT conditions are very similar to the problem solved earlier (see Section 4.3) except for the Nth generator. The KKT condition for the Nth generator is as follows:

$$C'_N(Q_{Nt}) = \gamma + \lambda_t - \mu_{it} + \nu_{it}$$

In other words, if the energy constraint on the energy-limited generator is not binding, the optimal dispatch is exactly as before (recall that every generator is dispatched according to merit order; all generators that are dispatched for an intermediate level of output have the same marginal cost). However, if the energy constraint is binding, the optimal dispatch is the same as before except that the energy-constrained generator is dispatched as though it has a marginal cost that is above its true or underlying marginal cost (by the amount γ). In effect, the energy-constrained generator is dispatched according to its opportunity cost, not its true marginal cost.

Result: When a generator is limited in the amount of energy it can produce in a given period of time, the optimal dispatch for that generator should reflect not the generator's true marginal cost of production, but a higher value reflecting the opportunity cost of not being able to produce at other times.

4.8.1 Example

An example will make this clearer. Let us suppose we have an energy market with three conventional thermal generators, each with a capacity of 500 MW. These generators have a variable cost of, say, $10/MWh, $50/MWh and $100/MWh, respectively. In addition there is a single energy-limited generator. Let us suppose each day is divided into six periods of equal length. The demand over these six periods is, say, 400, 800, 1200, 900, 700 and 200 MW.

What is the optimal way to dispatch these generators? The solution to this problem is illustrated in Figure 4.9. As before, we construct a merit order, from the lowest cost to the highest cost. The lowest-cost generator is dispatched first, and then the next-to-lowest and so on. However, when should the energy-limited generator be dispatched?

If the energy-limited generator has 500 MWh of energy available, the efficient dispatch is for the energy-limited generator to only produce at the peak period (period 3). At this time it

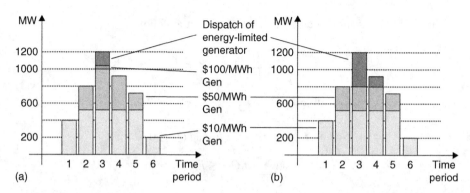

Figure 4.9 (a) Energy-limited generator has 500 MWh of energy available. (b) Energy-limited generator has 2000 MWh of energy available

displaces the highest cost generator. It can produce at the rate of 125 MW over this 4 h period, producing 500 MWh. This is illustrated in Figure 4.9a.

If the energy-limited generator has 2000 MWh of energy available, an efficient dispatch is for the energy-limited generator to produce at both the peak period (period 3) and the next-highest peak (period 4). It produces at the rate of 400 MW in period 3 and 100 MW in period 4, for a total of 2000 MWh of energy produced. In this case the energy-limited generator completely displaces the highest-cost generator in the peak period and partly displaces the $50/MWh generator in the next highest period. This is illustrated in Figure 4.9b.

4.9 Efficient Dispatch in the Presence of Ramp-Rate Constraints

Generators vary in how quickly they can increase or decrease their output. Some generators, such as large coal-fired generators, can only increase their output by a few MW every minute. On the other hand, some other generators, such as hydro-generators, are able to respond much more quickly to changes in the supply–demand balance. In practice, these *ramp rate constraints* must also be taken into account in the optimal dispatch.

In the presence of ramp rate constraints, it may be necessary to dispatch generators out of merit-order for a period of time. This can lead to large swings in the wholesale spot price.

Let us consider again the optimal dispatch task we have seen above, but this time let us explore the impact of ramp-rate constraints. Let us suppose that we have a power system with a number of generators. Let us suppose that the output of each generator is initially in some steady state Q_{i0}, and the total demand is Q_0. The power system then evolves over time $t = 1, \ldots, T$, following the demand Q_t. During this evolution ramp rate constraints may be binding. Let us suppose that at time $t = T$ the ramp rate constraints are no longer binding. The optimal dispatch task is now formulated as follows:

$$\min \sum_{i,t} c_i Q_{it}$$

$$\text{subject to } \forall t, \sum_i Q_{it} = Q_t \leftrightarrow \lambda_t$$

and

$$\forall i, t, Q_{it} \leq K_{it} \leftrightarrow \mu_{it} \text{ and } Q_{it} \geq 0 \leftrightarrow \nu_{it}$$

and

$$\forall i, t, \ Q_{it} - Q_{it-1} \leq R_i \leftrightarrow \tau_{it}$$

Here, R_i is the (upward) ramp rate constraint on generator i.
The KKT conditions for this problem include the following:

$$\forall i, t, c_i = \lambda_t - \mu_{it} + \nu_{it} - \tau_{it} + \tau_{it+1}$$

Paradoxically, when ramp-up constraints will be binding in the future, the wholesale spot price (i.e. the SMC) can drop to very low levels. Intuitively, the reason is that when ramp-up constraints will be binding in the future, adding more production today reduces the duration of the ramping period, reducing costs in subsequent periods as long as those ramp rate constraints are binding.

4.9.1 Example

To illustrate this let us suppose we have an electricity industry with 1000 MW of generation with a marginal cost of $10/MWh, with a low ramp rate of, say, 100 MW every 5 min. In addition, let us suppose we have 500 MW of generation with a marginal cost of $20/MWh with a ramp rate of, say, 75 MW every 5 min, and 500 MW of generation with a marginal cost of $100/MWh and a ramp rate of say 500 MW every 5 min. (This fast-ramping generation could be energy-limited hydro-generation in which case, as we know from the discussion in Section 4.8, this marginal cost is the 'opportunity cost' of generation rather than the true marginal cost). For simplicity, let us ignore other startup costs or minimum operating levels. The generation assets are summarised in Table 4.4.

Let us suppose that demand is initially at 500 MW and in a steady state. This load can be supplied entirely by the low-cost generator. The price is $10/MWh. However, let us suppose that it is known that at a particular point in time (say $t = 3$) demand will increase over 5 min to (just under) 1000 MW (later, in Section 12.2, we will deal with the case where demand varies in an uncertain manner).

At time $t = 3$, as demand increases, the lowest-cost generation cannot increase from 500 to 1000 MW in a single 5 min period. It can only increase from 500 to 600 MW. We can also increase the output of the next-lowest-cost generation from 0 to 100 MW in the 5 min period, increasing total output to 700 MW. We must use the highest-cost generator to provide the remaining 300 MW of output.

Table 4.4 Key cost data for a simple illustration of the impact of ramp rate limits

Generator	Variable Cost ($/MWh)	Maximum Ramp Rate (MW/5 min)	Capacity (MW)
A	$10	100	1000
B	$20	75	500
C	$100	500	500

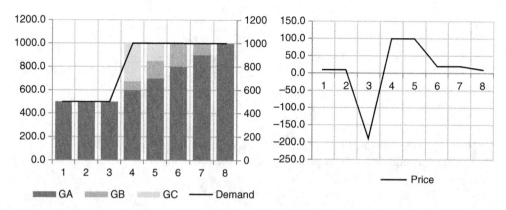

Figure 4.10 Efficient pricing and dispatch outcomes in the presence of ramp rate constraints

Importantly, the spot price at time $t = 3$ (before the increase in demand occurs) is $-190/MWh. The reason is that any increase in production by generator A at this point increases the cost of dispatch by $10/MWh in that period, but reduces the dispatch cost over the next several intervals – by $10/MWh in periods 6 and 7 and by $90/MWh in periods 4 and 5.

In period 4 any additional output must be met by the $100/MWh generation, so the price in this period is $100/MWh. The price remains at this level until the ramp rate constraints on generator B are no longer binding, at which point the price drops to $20/MWh. These results are illustrated in Figure 4.10.

The key result here is the counter-intuitive variation in the price. The price drops to a low level (here the price is negative) in the period *before* the ramp rate constraints start to bind.

In principle, a similar outcome can arise in the event of a rapid decline in demand. Again, certain generators, especially large thermal generators, may not be able to efficiently reduce their output rapidly. In this case the price may temporarily increase to significantly above its normal level.

Even more importantly, it turns out that if a period of binding ramp rate constraints is anticipated in the future, the optimal dispatch may require that this episode be anticipated. Specifically, it may be efficient to ramp up a low-ramp-rate plant in advance in order to reduce the cost of transition to the new steady-state equilibrium.

Actions that are taken by the power system to adjust to a new steady-state equilibrium after an event occurs are known as *corrective actions*. Actions taken in advance of an event occurring in order to reduce the cost of corrective actions *ex post* are known as *preventive actions*. These are discussed in more detail in Part V. Here we can observe that the presence of ramp rate constraints limits the ability of the power system to adjust to changing demand *ex post*, thereby raising the cost of corrective actions. It may be efficient for the power system to take actions in advance (preventive actions) to reduce the costs of those corrective actions *ex post*. In this case, the preventive action consists of ramping up the low-ramp-rate plant in advance of the increase in demand.

To see this consider the earlier example again, but this time let us suppose that demand starts at 1200 MW. In period 4 this demand ramps up to (just under) 1500 MW. If no preventive actions were taken, following the increase in the demand, generator B would ramp up slowly to 500 MW. However, this requires substantial use of the $100/MWh generator. This is not the

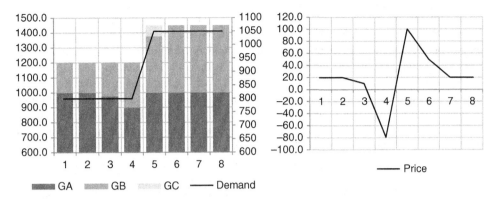

Figure 4.11 Efficient dispatch in the presence of ramp rate constraints may require taking preventive actions

least cost dispatch. Instead, it is efficient to take preventive actions – to ramp up generator B in advance of the increase in demand.

As shown in Figure 4.11, the price first drops in anticipation of the ramp-rate-binding episode and then increases during the episode. Intuitively, the reason is that, in this example, an increase in demand before the episode reduces the cost of adjustment. This occurs because the burden of the adjustment falls on the generator with the lowest ramp rate. In this example, if the overall demand were lower (as in the previous case) or higher, the burden of adjusting to the new steady state would fall on generators with a higher ramp rate, and the adjustment cost would be lower.

Result: When the supply, demand or network conditions change rapidly, the change of production or consumption targets may exceed the ramp-rate limits of some assets. In this case some assets may need to be dispatched out of merit order. This raises the cost of adjustment to the new steady state, known as corrective actions.

Where the change in supply/demand conditions is anticipated it will often be efficient to take actions in advance of the change in conditions, known as preventive actions, to reduce the cost of corrective actions. This may involve dispatch out-of-merit-order before the change in conditions occurs.

During the period of preventive actions the price may move in a counter-intuitive manner. Specifically the price will be lower than otherwise expected when an increase in the load in that period reduces the cost of taking preventive actions in subsequent periods.

4.10 Startup Costs and the Unit-Commitment Decision

Many generators must incur material costs before they are able to produce any output at all. For example, the generator might need to boil water to create steam. The cost of the energy required to get the boiler up to normal operating temperature must be incurred whether or not the generator operates for 1 h or 1 year. These costs are known as *startup costs*.

Up until now we have assumed that we have a stock of generating assets that are ready and willing to produce at a moment's notice. In the presence of startup costs, however, there is another decision to make – whether or not to bring a generating unit into a position where it is ready to produce output. This decision is known as the *unit commitment decision*.

Whether or not it is socially efficient to incur startup costs depends, amongst other things, on the length of time that the output of that generator will be required by the market. If a generator is only expected to be operating for an hour or less, it will typically not be worthwhile incurring substantial startup costs. On the other hand, if the generator is expected to be running continuously for months, the startup costs are largely irrelevant. Therefore, we cannot look at the effect of startup costs on the optimal dispatch in a single dispatch interval. Instead, we need to look at the effects of startup costs over a period of time, as in the earlier discussion on energy-limited generators.

In general, whether or not it will be worthwhile to incur the startup costs of a particular generating unit involves a comparison of (a) the magnitude of the startup costs of that generating unit and (b) the additional costs incurred by the market in the absence of that generating unit. Suppose a generator is expected to generate, say, 1000 MWh of electricity over the course of a day. Suppose that the startup cost of this generator is, say, \$50 000. If this generator is not started, then some other more expensive generator must be called on to supply this 1000 MWh of electricity. This will tend to increase the total cost of generating electricity. In addition there may be a need to incur startup costs from some other generators. Whether or not to incur startup costs depends on a comparison of these startup costs and the additional costs incurred when this generator is not started.

Where startup costs are material they can be incorporated into the optimal dispatch task – but, as with ramp-rates and energy limits, we need to consider an intertemporal dispatch task.

Up until this point we have modelled the dispatch task as a set of linear equations. These can be solved easily using standard algorithms. The introduction of startup costs requires that we introduce binary variables to the constrained optimisation. A binary variable can only take one of two values: zero or one.

As before, let us suppose that we have a power system with a number of generators. The output of generator i at time t is given by Q_{it}. At each time the state of the generator is given by the binary variable S_{it}. $S_{it} = 1$ means that the generator has started and is in an operational state. $S_{it} = 0$ implies the generator is in the state of being shut down. We can model startup costs as the costs of transitioning between states. For example, the transition from shut-down to started can be modelled as a binary variable: $T_{it}^{\text{off} \to \text{on}} = S_{it} - S_{it-1}$. The transition from started to shut-down can be modelled as a binary variable: $T_{it}^{\text{on} \to \text{off}} = S_{it-1} - S_{it}$. Let us suppose that the startup cost is $C_i^{\text{off} \to \text{on}}$ and the shut-down cost is $C_i^{\text{on} \to \text{off}}$.

The optimal dispatch task is now formulated as follows:

$$\min \sum_{i,t} c_i Q_{it} + \sum_{i,t} T_{it}^{\text{off} \to \text{on}} C_i^{\text{off} \to \text{on}} + \sum_{i,t} T_{it}^{\text{on} \to \text{off}} C_i^{\text{on} \to \text{off}}$$

$$\text{subject to } \forall t, \sum_i Q_{it} = Q_t \leftrightarrow \lambda_t$$

and

$$\forall i, t, \quad Q_{it} \leq K_{it} \leftrightarrow \mu_{it} \text{ and } Q_{it} \geq 0 \leftrightarrow \nu_{it}$$

Table 4.5 Key cost data for the illustration of the impact of startup costs on optimal dispatch

Generator	Variable Cost ($/MWh)	Capacity (MW)	Startup Cost ($)
A	10	1000	
B	20	500	20 000
C	50	500	

and

$$\forall i, t, \ T_{it}^{\text{off}\to\text{on}} - T_{it}^{\text{on}\to\text{off}} = S_{it} - S_{it-1} \text{ and } T_{it}^{\text{off}\to\text{on}} + T_{it}^{\text{on}\to\text{off}} \leq 1$$

and

$S_{it}, T_{it}^{\text{off}\to\text{on}}$ and $T_{it}^{\text{on}\to\text{off}}$ are binary variables.

To illustrate this let us consider a simple network with three generators as set out in Table 4.5.

Initially, generators A and C are operating; B is shutdown. Let us suppose that there are 8 periods. Demand is 800 MW in the first two periods, 1000 MW in the next two periods, 1300 MW in the next two periods and 750 MW in the last two periods. With this demand pattern, the efficient dispatch outcome involves generator B remaining off. Generator A produces the first 1000 MW and generator C produces the rest.

Now suppose that demand is 300 MW higher in each period (1100 MW in the first two periods, then 1300 MW, then 1600 MW and then 1050 MW). With this demand pattern it is efficient to incur the startup costs for generator B at the outset. The generators are then dispatched in their merit-order.

Result: Where there are material costs associated with the transition from shut-down to operational or vice versa, these can be taken into account in an intertemporal optimal dispatch task. The decision to incur startup costs depends on the contribution of the generator to reducing the cost of dispatch once it is operational.

4.11 Summary

A core task in achieving an overall efficient electricity industry is achieving efficient use of a given set of generation and consumption assets. This requires information about the cost function of generators and the utility function of consumers. In the modelling of electricity markets it is common to assume that generators have a marginal cost function that is constant up to some fixed capacity.

The optimal least-cost dispatch of a set of generators (i.e. the outcome that minimises the cost of generating sufficient electricity to meet demand) is where each generator produces up to the point where its marginal cost is equal to the common industry-wide marginal cost, which is also sometimes known as the system marginal cost or SMC. In the case of constant marginal cost, the efficient outcome involves the construction of a merit order of generators. The efficient outcome is often also referred to as optimal dispatch.

Optimal dispatch of load (consumption) resources occurs in a similar manner. The basic task is to choose the rate of production for each generator and the rate of consumption of each customer in such a way as to maximise total economic welfare or surplus. The optimal dispatch has the characteristic that each customer consumes at a rate where his/her marginal valuation is equal to a common system marginal cost, and each generator produces at a rate where his/her marginal cost is equal to the same common system marginal cost. We can treat customer responsiveness to price as equivalent to a hypothetical generator, converting the task of surplus maximisation into a task of cost minimisation.

Nonconvexities in production, such as minimum load levels, or startup costs significantly complicate the task of finding the optimal dispatch. There may be no feasible solution to the optimal dispatch problem.

Where a generator is limited in how much energy it can produce, the task of finding the least-cost dispatch or the welfare-maximising dispatch is an intertemporal problem involving choosing the rate of production or consumption over time. It will typically make economic sense to hold back the output of the energy-limited generator at times of low system marginal cost in order to increase the output of the generator at times of high system marginal cost.

The optimal dispatch task also requires an intertemporal optimisation where generators are limited in the rate of change of their output or where generators must incur startup costs to transition from shutdown to operational. Where generators are subject to ramp-rate constraints it may be necessary to dispatch generators out of merit order in order to balance supply and demand. It may make sense to adjust the power system *ex ante* (before the ramp rate constraints are binding) in order to reduce the cost of adjustment *ex post*. This illustrates the principle that it may make sense to take preventive actions to reduce the cost of taking corrective actions *ex post*. When ramp rate constraints are binding, the system marginal cost (which is also the price in a liberalised market) may move in a counter-intuitive manner.

Questions

4.1 What does it mean for the cost function of a generator to be nonconvex? What characteristics of a generator might give rise to a nonconvex cost function?

4.2 Under what conditions is there a monotonic-increasing relationship between the SMC (or market price) and the quantity of electricity supplied (i.e. under what conditions does higher demand lead to prices that are equal or higher)?

4.3 True or false: In a general least cost dispatch with upward-sloping generator marginal cost functions, every generator always produces to the point where its marginal cost is equal to the SMC?

4.4 Suppose that all customers have an inelastic demand for electricity up to a marginal value V (at which point demand for electricity drops to zero). Show that the problem of total surplus maximisation is equivalent to a problem of generator cost minimisation by including a hypothetical generator with a marginal cost equal to V.

4.5 An energy-limited generator has a very low marginal cost of $1/MWh. Should this generator be classified as a 'baseload' generator in the merit order?

4.6 True or false: In the presence of ramp rate constraints, price spikes can occur at times of off-peak demand? Explain why or why not.

4.7 In the presence of startup costs, is it still correct to say that generators should be dispatched according to the merit order? Why or why not?

Further Reading

For more on the unit-commitment issue, see Padhy (2004).

5

Achieving Efficient Use of Generation and Load Resources using a Market Mechanism in an Industry with no Network Constraints

In Chapter 4, we explored the characteristics of efficient dispatch of a set of generation and consumption assets. In this chapter, we explore whether or not that outcome might be achieved using decentralised competitive markets in which market participants act independently to pursue their own ends. To keep things simple, in this chapter we will ignore network constraints.

5.1 Decentralisation, Competition and Market Mechanisms

Chapter 4 focused on the question of how an omniscient system operator would efficiently utilise a set of generation and consumption assets. In principle, no matter how the electricity industry is organised, given enough information and enough control powers, an omniscient system operator could achieve an efficient dispatch of generation and consumption resources, and thereby achieve an economically efficient short-run outcome.

However, in the longer term, achieving economically efficient outcomes is not just a matter of information flows and control powers. Instead, achieving economically efficient outcomes is a matter of getting the incentives right – including incentives for efficiency, investment, innovation, cooperation and coordination.

Experience shows that achieving effective incentives in large vertically integrated businesses is difficult. These problems are significantly more severe when the business is government owned. Over time, large vertically integrated businesses struggle to maintain efficiency, productivity and customer responsiveness.

The Economics of Electricity Markets, First Edition. Darryl R. Biggar and Mohammad Reza Hesamzadeh.
© 2014 John Wiley & Sons, Ltd. Published 2014 by John Wiley & Sons, Ltd.

Incentives tend to be much sharper in competitive markets. Competitive markets provide strong incentives for productive efficiency, innovation and customer responsiveness. Furthermore, competitive markets adjust autonomously to changing supply and demand conditions, without any need for regulatory oversight or involvement. As set out in Chapter 3, a key element of the reforms that occurred in the late twentieth century was a shift towards greater reliance on competitive market forces wherever possible.

In the context of the electricity industry, this shift towards greater reliance on competitive market forces was reflected in a greater reliance on competition between generators to determine prices and dispatch outcomes. However, is there a market mechanism that allows us to achieve efficient outcomes – both in the short-run and in the longer-run?

Let us focus first on the short run. Can we achieve the efficient short-run dispatch outcomes discussed in Chapter 4 using some form of competitive market mechanism?

As discussed earlier, all competitive electricity markets require careful integration of individual production and consumption conditions with the physical limits on the network. As a consequence, competitive electricity markets are organised around a centrally coordinated *smart market*, introduced in Section 1.5.

Specifically, this smart market will be assumed to operate as follows:

a. First, each market participant sends key supply and/or demand information to a central operator, known as the *system operator* or *market operator*. Generators send to the system operator information on key operating characteristics of the generating plant in question. This includes information on, say, variable cost, maximum and minimum operating limits, and may include information on, say, startup costs, ramp rate limits and so on. Consumption resources, on the other hand, send to the system operator information on their willingness to pay for different rates of consumption.

b. The system operator uses this information on supply and demand to compute the dispatch outcome that maximises total economic welfare, subject to any physical operating or network constraints. The output of this process is a set of prices and a rate of production or consumption for each market participant.

c. The system operator communicates back to each market participant key information, such as the level of output they are to produce or consume, whether to start up and the price the market participant will pay or be paid. Each market participant then makes a decision as to how much to produce or consume.

Since, in such a mechanism, the actions of one participant can affect the outcomes of another, this is a situation of strategic interaction (see Section 1.8), which is best modelled as an economic 'game'. We will look for the *Nash equilibrium* of this game. A set of actions is a Nash equilibrium if, for each player of the game, his/her action maximises his/her payoff given the actions of the other players. In this market mechanism, where each market participant reports information to the system operator, a Nash equilibrium has the property that no player has an incentive to change the information it reports to the system operator given the information reported by the other players.

The market mechanism will be said to *support optimal dispatch* if and only if (a) when all market participants truthfully report their key operating characteristics, the resulting dispatch is optimal (that is, welfare-maximising) and (b) for each market participant, truthfully reporting their key operating characteristics is a Nash equilibrium.

5.2   Achieving Optimal Dispatch Through Competitive Bidding

Can we achieve an efficient dispatch of electrical energy through a market mechanism?

Let us focus on the case where there are no intertemporal constraints such as startup costs or energy-limited plant. In addition, let us suppose that there are no constraints on the level of the market price. Let us suppose that each generator is able to offer to the system operator an upward sloping supply curve, which is interpreted as representing, for each price, the rate of production the generator is willing and able to produce. In addition, let us suppose that every consumption resource is able to offer to the system operator a downward sloping demand curve, which represents, for each price, the rate at which the consumer is willing to consume electricity.

Let us assume that the system operator treats these reported supply and demand curves as reflecting the true marginal cost curve for each generator and the true demand curve for each consumer. The system operator is assumed to naively compute the dispatch that maximises the total economic surplus given the reported information. As we have seen, this constrained optimisation problem yields a common marginal cost, which we have referred to as the system marginal cost (SMC). We will now take this SMC to be the market price. Each generator and consumption resource will be paid (or pays) an amount of revenue equal to the market price multiplied by the corresponding rate of production (or consumption).

Importantly, as we noted in Section 1.5, provided there is sufficient competition between generators so that no generator or consumer has any influence over the market price, this market mechanism induces each generator to truthfully reveal its marginal costs of production and each consumer to reveal its true marginal value of electricity, at least in the region immediately around the expected market price.

To see this, consider the position of a generator considering at what price to offer in a small amount of its capacity. Let us suppose that this generator is sufficiently small that it has no practical impact on the market price.

Let us suppose that at a given price the generator offers a total volume to the system operator that is less than the corresponding volume on the generator's marginal cost curve. In the event that price arises, the generator will be dispatched for a price–quantity combination that lies above the marginal cost curve. As long as there is a positive probability that that price will occur, the generator can increase its expected profit by increasing the volume that it offers to the market at that price. This is illustrated in Figure 5.1.

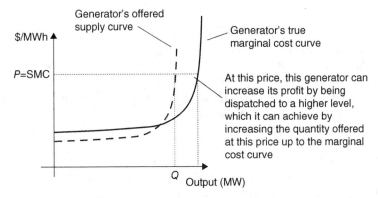

Figure 5.1   At any given price a price-taking generator has an incentive to offer a volume of output given by the marginal cost curve

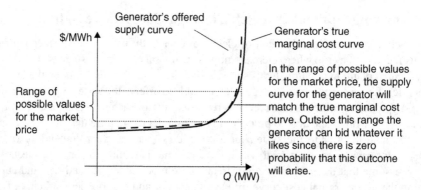

Figure 5.2 A price-taking generator has an incentive to submit an offer curve equal to its marginal cost curve

On the other hand, if at a given price, the generator offers a total volume to the system operator that is greater than the corresponding volume on the marginal cost curve, there is a chance the generator will be dispatched for a price–quantity combination that lies below the marginal cost curve. Again, the generator can increase its expected profit by reducing the volume that it offers to the market at that price.

We can conclude that at least in the region of prices that may arise with positive probability the generator has an incentive to submit an offer curve equal to its marginal cost curve. (Outside this region the generator can essentially bid any price or quantity since the corresponding price will never occur.). This is illustrated in Figure 5.2.

The same results also apply to consumers who are bidding in this market. Provided the consumer is a price taker, the consumer has an incentive to submit a bid curve that reflects his/ her true demand curve for electricity (at least over the range of prices that arise with positive probability).

We can conclude that, provided there is adequate competition between generators, on the one hand, and between consumers on the other, it is possible to decentralise short-run production decisions to a market mechanism. Competition between generators and between consumers will ensure that generators and consumers truthfully reveal their true marginal cost and marginal value to the system operator who will therefore ensure an optimal dispatch at each moment in time.

Result: If (a) each generator is a price-taker and is able to submit an upward-sloping supply curve to the system operator; (b) each consumer is a price-taker and is able to submit a downward-sloping demand curve to the system operator; (c) if the system operator computes the total-surplus maximising dispatch (and market price) on the assumption that each generator's reported supply curve matches its marginal cost curve, and each consumer's reported demand curve matches its true demand curve; and (d) there are no binding intertemporal constraints (ramp rates and energy limits) or other constraints on the market prices, then each generator will have an incentive to offer a supply curve that is equal to its marginal cost curve, and each consumer will have an incentive to offer a demand curve that is equal to its true demand curve over that range of prices that arise with a nonzero probability and the resulting dispatch will be efficient.

This result shows that, at least in principle, competitive market mechanisms can yield the efficient short-run outcome. However, real-world markets never perfectly achieve this theoretical ideal. In particular, the following problems may arise in practice:

- *Market power* – one or more market participants may have some influence over the wholesale market price. As we will see in Part VII, electricity markets are prone to the exercise of market power. The exercise of market power leads to inefficient dispatch outcomes and has a chilling effect on the hedge market.
- *Limits on prices* – many real-world markets have price ceilings or price floors. To the extent that the market price reaches these extreme levels, the market mechanism will no longer result in efficient outcomes.
- *Limits on the frequency of the market process* – In any electricity industry supply and demand must be kept in balance at all times. Yet the market process mentioned earlier takes time to operate. During the time intervals between successive runs of the market process, the market outcomes will not necessarily be efficient. This is discussed further in Part V.
- *Nonconvexities* – such as startup costs and minimum production levels.
- *Limits in the way that demand and supply information can be communicated to the central market operator* – for example, in the Australian electricity market, supply and demand information is communicated to the market operator in the form of a simple step function with 10 price–quantity steps. Representing supply and demand information in this way may require an approximation of the true underlying cost and demand information.

5.3   Variation in Wholesale Market Design

Section 5.2 emphasised that, in principle, a simple mechanism involving the communication of supply and demand curves to a central market operator could achieve an efficient dispatch of generation and consumption resources. In practice, however, there is a reasonable amount of variation across different wholesale electricity markets. Different electricity markets differ in, amongst other things:

a. The period of time between successive operations of the market process.

   The market process mentioned earlier (involving the communication of information to the market operator, the computation of the optimal dispatch, and the communication of price and dispatch information back to the market participants) takes time. There is therefore inevitably a period of time between successive operations of the market process. In practice, this period may be as long as 1 h, or as short as 5 min or less.

b. The range of decisions made by the central market operator.

   The market process mentioned focussed on a single decision – the rate of production or consumption of electricity. However, there may be other decisions involved in the efficient operation of a generator, such as whether or not to start up or shut down. There is a question whether these decisions should also be determined by the market operator or whether they should be left to the market participants. This issue is discussed in more detail in the next section.

c. How revenue to be paid by (or paid to) market participants is calculated.

   The same market process assumed that market participants would be paid (or would pay) each time period an amount equal to the market price multiplied by the electricity produced (or consumed) in that time period. However, there are other theoretical possibilities, such as basing the amount of revenue paid on the participant's bid or offer curve. This is sometimes known as 'pay-as-bid' and is discussed further later.

d. Whether or not market participants are permitted to make their own bilateral arrangements outside the centralised market process, or whether all transactions must occur through the centralised market process. This is the issue of *gross versus net pool*, and is discussed further later.

e. The range of associated markets. As noted earlier, the market process only operates every few minutes at best. This leaves open the question of how to balance supply and demand in the interval between successive operations of the market process. In many markets, the system operator is responsible for ensuring balancing on very short time scales. In many cases the system operator procures balancing services from generators and loads through a related market process. This is often known as the market for ancillary services.

f. The range of other information required to be provided. In addition to a wholesale spot market, in most cases a market operator will also provide short-term price forecasts for the next 24 h or so. In order to provide such forecasts, the market process may require that generators and loads submit forecast demand and supply curves for a period of time into the near future.

A few of these issues are discussed in more detail next.

5.3.1 Compulsory Gross Pool or Net Pool?

Should market participants be required to participate in the centralised market process described earlier, or should we allow market participants to make their own bilateral or multilateral arrangements? In fact, should we allow groups of participants to set up their own sub-market process, along these lines?

The market mechanism described earlier represents a simplified form of a smart market. In particular, we have ignored network constraints. If we ignore network constraints there is no particular need to operate a *centralised* market process. Participants may, if they choose, enter into bilateral transactions with other participants or groups of participants. (In fact a smart market is not needed at all; a conventional decentralised market process would work just as well).

However, as we will see in Chapter 7, the coordination of generation and consumption with network resources requires that the physical network limits be incorporated in the smart market process. This requires a single centralised market process. At a minimum therefore, participants who make bilateral arrangements must notify their net position (i.e. net injection or withdrawal at each node in the network) to the central market operator for the purpose of computing the optimal dispatch.

With that proviso, there is no strong economic reason for requiring market participants to transact through the centralised wholesale market. Market participants can choose to purchase or sell all of their requirements through the wholesale market, or can carry out whatever bilateral or other arrangements they would like.

Even where market participants are required to purchase or sell all of their requirements through the wholesale spot market, market participants usually hedge their exposure to the spot price risk through hedge contracts so that, in effect, they are only selling or buying a small proportion on the spot market.

Similarly, there are no strong economic reasons for discouraging market participants from transacting through the centralised wholesale market. The United Kingdom market seems designed to encourage bilateral arrangements. Participants who are in 'imbalance' (i.e. net buyers or sellers in the spot market) pay a penalty that is designed to discourage such imbalances. In effect, this forces market participants to enter into arrangements beforehand that closely balance their expected supply and demand. Such arrangements may encourage a high level of forward contracting, thereby discouraging market power. However, where the market is sufficiently competitive there seems to be no particular reason for favouring centralised over bilateral transactions or vice versa.

5.3.2 Single Price or Pay-as-Bid?

The usual practice in economic markets is for there to be a single market clearing price. If the market is competitive, this price, as we have seen, reflects the common marginal cost of the producers and the common marginal utility of the consumers. However, this marginal cost may of course be well above the variable cost of most of the generators in the market. For example, a generator with a variable cost of $10/MWh might be prepared to offer nearly all its output to the market at a price a little above $10/MWh. However, the wholesale market price may be much higher, such as $1000/MWh. Some might consider this level of compensation to the generator to be, in a sense 'unfair'. After all, the generator indicated in its offer to the market that it would be willing to produce at full capacity if the price was just $10/MWh. Why should it then be paid $1000/MWh for each unit of output – 100 times the amount at which it indicated that it was willing to produce that output?

Perhaps for reasons such as these, it is sometimes suggested that instead of a single market-clearing price, market participants should be paid an amount that reflects their bid or offer curve. For example, let us suppose a generator offers 100 MW at a price of $10/MWh, and another 100 MW at a price of $50/MWh, and the wholesale spot price happens to be $100/MWh. Under a single price market, this generator would be dispatched to produce at the rate of 200 MW, and would receive income at the rate of $20 000/h. However, under a pay-as-bid market it will only be paid $10 for the first 100 MW of output and $50 for the second 100 MW of output – in other words, it will receive income at the rate of $6000/h.

The same principle would apply to the buyer side of the market. If an electricity customer offered to pay, say, $1000/MWh for the first 50 MW of consumption and $500/MWH for the next 50 MW of consumption, at a time when the spot price is $100/MWh, this consumer would be instructed to consume at the rate of 100 MW. In a single-price market, this consumer would pay at the rate of $10 000/h. In a pay-as-bid market, this consumer would pay at $50 000/h for the first 50 MW and $25 000/h for the second 50 MW, for a total of $75 000/h.

Importantly, however, if there were a change from a single-price to a pay-as-bid market the bids and offers of the market participants would not remain the same.

Let us suppose, following the previous example, that the market participants believed that the wholesale spot price would be exactly $100/MWh. Then, would the generator in the example submit an offer curve as set out earlier? The answer is no: the generator can raise the revenue it receives without changing the amount for which it is dispatched by increasing its offer – it could offer up to just under $100/MWh for its entire 200 MW capacity and still be dispatched. By doing so this generator could bring its income arbitrarily close to $20 000/h – the amount it will receive in a single-price market.

In the same way, if the same consumer believed that the wholesale spot price would be exactly \$100/MWh, it could lower its bids to just above \$100/MWh for the 100 MW it desires to consume. By doing so it could bring its charges down to \$10 000/h – the amount it would pay in a single-price market.

As these examples show, in a pay-as-bid market, it is not a Nash equilibrium to submit a bid or offer curve that reflects the true underlying marginal cost of a generator or the marginal utility of a consumer. Instead, if the uncertainty about the market price is small enough, the market participants will submit bids and offer curves that reproduce the outcome in a single-price market.

Result: In a pay-as-bid market, participants no longer have an incentive to submit bids or offers that reflect their true marginal cost or demand. Instead market participants have an incentive to submit bids or offers under which they produce or consume at the same rate as in a single-price market, but the price they are asked to pay (or are paid) is close to that which would arise in a single-price market. In other words, market participants have an incentive to nullify the effects of a change to pay-as-bid.

This example focuses on the case where there is little uncertainty about the market price. Where there is some uncertainty about the market price, the analysis will change slightly. Nevertheless, the current consensus amongst economists is that a change to pay-as-bid is neither necessary nor desirable.

5.4 Day-Ahead Versus Real-Time Markets

In many countries, there are, in effect, two centralised market processes: a day-ahead market and a spot market. Such markets operate in two stages.

In the first stage, which usually occurs in the afternoon before the day of actual dispatch, market participants submit bids and offers to the day-ahead market, the market operator calculates the optimal dispatch over a period of, say, 24 h into the future. Market participants are then provided a forecast price and forecast dispatch and are paid according to the forecast price and forecast dispatch.

In the second stage, which occurs during the day of actual dispatch, the market operator calculates the optimal dispatch again using any updated information on supply and demand conditions. This results in a real-time price, and real-time dispatch. Market participants are paid the real-time price for the difference between their real-time and day-ahead dispatch.

Many liberalised electricity markets around the world feature a day-ahead market, or sometimes an hour-ahead market, in addition to a spot market. Since the production or consumption of each market participant is settled at two different prices – a spot price and a day-ahead price – these markets are known as *two-settlement systems* (or multisettlement systems).

Efficient operation of an electricity system requires a real-time or spot market, balancing supply, demand and network conditions at a given point in time. Do we need an additional market process operating a day in advance of real time? Organising and operating a day-ahead market involves some additional complexity and cost. Yet, a day-ahead market remains part of the Standard Market Design required by the United States Federal Energy Regulatory Commission. What might be the advantages of a day-ahead market?

First, it is worth emphasising that although a market participant may be settled at the day-ahead price in the day-ahead market, it is only the real-time price that determines the decision as to how much to produce or consume. All prices (day ahead, month ahead, etc.) in advance of real time do not affect production or consumption decisions.

To see this, let us suppose that the day-ahead price and quantity for a market participant is P^{DA} and Q^{DA} and the real-time price and quantity is P^{RT} and Q^{RT}. In the day-ahead market the market participant is paid $P^{DA}Q^{DA}$. In the real-time market the participant is paid, in addition, the real-time price for the difference in dispatch between the real-time and day-ahead markets: $P^{RT}(Q^{RT} - Q^{DA})$. The total revenue received by the market participant can be written as

$$P^{RT}Q^{RT} + (P^{DA} - P^{RT})Q^{DA}$$

From this we can see that it is only the real-time price that affects the real-time production or consumption decision. The effect of the day-ahead market is to act as a short-term hedge or forward market.

In many markets the central market operator makes short-term startup or *unit commitment* decisions. For many midmerit or peaking generators, these startup decisions occur over a 24 h timeframe, based on forecast prices over the next 24 h.

In order to make centralised unit commitment decisions, therefore, the market operator needs effective forecasts of supply/demand conditions over the next 24 h. In other words, where the market operator makes unit commitment decisions, some form of day-ahead market is essential. However, should the market operator be making unit commitment decisions? Or should these be left to the market participants themselves?

There seem to be two main advantages of a day-ahead market:

a. First, a day-ahead market could, in principle, improve the forecasts of prices and market conditions over the subsequent 24 h. Improvements in price forecasts, in turn, have the potential to enhance reliability and to improve the efficiency of the unit commitment decision. This is discussed further later.
b. Second, a day-ahead market may be able to reduce the exercise of market power.

These issues are discussed in more detail in Sections 5.4.1 and 5.4.2.

5.4.1 Improving the Quality of Short-Term Price Forecasts

Let us focus first on the scope for a day-ahead market to improve the quality of short-term forecasts of prices and market conditions.

As we saw in Chapter 4, the efficient short-run operation of a set of generation and consumption resources often requires accurate knowledge of supply and demand conditions over a period of time. Such knowledge allows efficient decisions to be made regarding whether or not to incur startup costs (the unit commitment decision), whether or not to ramp a generator up to a higher level of output (in anticipation of future binding ramp rate limits), or the best timing as to when to consume limited resources.

For example, suppose a hydro-generator receives enough water to generate for a few hours each day. If the price forecasts are inaccurate and under-forecast prices later in the day the hydro-generator may use up its water during the lower-priced periods. This reduces the profitability of the generator and the overall efficiency of resource use. As another example, a particular industry might lack flexible plant able to respond quickly to changes in demand. Based on preliminary forecasts, the owners of a slow-response plant may consider it not worthwhile to ramp their plant up slowly to cater for a forecast evening peak. If the evening peak demand is higher than expected the generator may not be available, reducing the profitability of the generator and the overall efficiency of resource use. Imperfect price forecasts can reduce the efficiency of dispatch.

Imperfect price forecasts may also reduce the efficiency of the unit commitment decision. A forecast of a particularly high price later in the day may inefficiently induce a number of generators to incur startup costs to be ready to exploit that high price. Improved price forecasts may also improve the extent of demand-side responsiveness. This might be the case if, for example, customers require a few hours notice to put other arrangements in place before changing their consumption of electricity.

All efficient wholesale electricity markets need the ability to accurately forecast short-term prices. However, does this justify the establishment of a day-ahead market? The Australian National Electricity Market (NEM) has a mechanism for forecasting short-term prices, known as the predispatch process. In the Australian NEM, generators are required to submit their offer curves the day before each trading day. This information is combined with demand forecasts by AEMO and submitted to the dispatch engine to produce forecasts of likely market outcomes for each half-hour of the subsequent trading day. This process produces forecasts of prices, flows on interconnectors and binding transmission constraints.

The predispatch process is, in some respects, similar to a day-ahead market. However, there is one important difference. Generators in the NEM are not paid on the basis of the predispatch prices. Generators in the NEM are able to change the amount they offer in each price band up until 5 min before real time, in a process known as 'rebidding'. The price forecasts produced in the predispatch process would have no meaning if the offer curves were not a reasonable reflection of each generator's expected behaviour. For this reason, there are requirements in the Australian market for scheduled generators and loads to submit their day-ahead bids and offers in 'good faith'. A bid is made in good faith if a generator has a 'genuine intention to honour the bid or offer if material conditions and circumstances on which that bid or offer was based remain unchanged'.

Despite these rules, under the current market arrangements in the Australian NEM there is relatively little to prevent a generator from submitting misleading bids in the day-ahead predispatch process. The reason is that only the bids submitted in real-time affect market outcomes. Although the day-ahead bids are supposed to be submitted in good faith (with a 'genuine intention to honour the bid') in practice market conditions can change in a myriad of ways from day-ahead to real time (demand may be higher or lower than forecast, different generating units or network elements may be in or out of service than was forecast, operating conditions of a particular generator may be different from expected) so that it is in practice very hard to prove that an original bid was not made in good faith.

If a generator submits misleading bids in the predispatch process, the short-term price and market forecasts will be distorted. This can affect the efficiency of dispatch. However, does a day-ahead market achieve more accurate price forecasts? The day-ahead market is very similar to a short-term (day ahead) hedge market. Hedge markets exist for many periods in advance of real time, from one day to several years. Competitive forces in the hedge markets ensure that the day-ahead price reflects the corresponding expected future real-time or spot price. However, the forecast real-time price will depend on factors, such as which generators to choose to start up, which depend on a host of market conditions. By centralising this decision with the market operator, the market operator may be able to provide better price forecasts than individual market participants. This could improve market outcomes. However, the practical significance of this effect is unclear.

5.4.2 Reducing the Exercise of Market Power

As explained further in Part VII, electricity markets tend to be prone to the exercise of market power. In practice, one of the primary constraints on incentives to exercise market power is the degree to which a generator is contracted in the forward or contract markets. The higher the degree to which a generator has 'presold' its output in the forward markets, the lower its incentive to exercise market power in real time.

A day-ahead market is just like any other forward market. The greater the proportion of output that a generator presells in the day-ahead market, the lower the incentive to exercise market power in the real-time dispatch process.

Why might a day-ahead market be preferable to a short-term forward market for mitigating market power? In the absence of a centralised unit-commitment a generator may be uncertain as to its future production, thereby limiting the extent to which it is prepared to commit to forward sales 24 h in advance. The centralised unit-commitment might increase the certainty of the future sales of the generator allowing it to commit to higher levels of hedging, thereby resulting in lower levels of market. This argument is plausible, but again the practical significance of this observation is unclear.

In our opinion the jury is still out on whether or not day-ahead markets are essential to achieving overall efficient outcomes in a liberalised wholesale electricity market.

5.5 Price Controls and Rationing

We have seen that in a typical wholesale electricity market, the price is determined through the optimal dispatch process and is equal to the common marginal cost (the system marginal cost or SMC), which determines the rate at which all generators produce and all customers consume.

Provided the demand of customers tends to zero for high enough prices, there always exists a price that balances supply and demand.

Provided the demand of customers is properly integrated into the wholesale market, the optimal dispatch task determines a price, and the resulting dispatch is efficient no matter what level of demand arises. In particular, it is important to be clear that in an electricity market with an effective demand side there is never any need for nonprice rationing of electricity or involuntary load shedding. As in any conventional market, in the event of a shortage of supply relative to demand, the price would rise to the point where supply and demand are again in balance (this is known as market clearing).

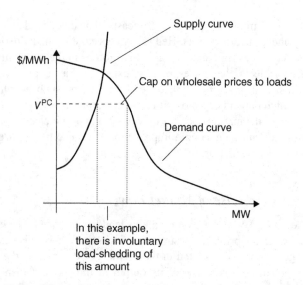

Figure 5.3 A price-cap may lead to involuntary load-shedding

However, where the wholesale market is not allowed to operate properly there is a chance that the wholesale market will not be able to clear, and some involuntary rationing of electricity will be required. For example, it may be that there is some maximum amount that consumers are allowed to report as their valuation, or some limit on the level at which the wholesale price is allowed to rise, such as a *market price cap*.

In the presence of a market price cap, consumers no longer have an incentive to report their true marginal cost or marginal value for electricity when that marginal cost of marginal value is above the price cap. Specifically, a generator has no incentive to offer more output than it is willing to produce at the market price cap (even if it technically has the capability to produce more). Similarly, a customer has no incentive to reduce its demand below what it is willing to consume at the market price cap (even if it would reduce demand if the price were to rise to a higher level).

In this context the market mechanism may not balance supply and demand and involuntary load shedding may be required.

Let us call the highest allowed price charged for wholesale loads V^{PC}. If, at this price, consumers desire to consume at a faster rate than producers are willing to produce, some load must be shed involuntarily. This is illustrated in Figure 5.3.

Result: In an effective wholesale electricity market the price adjusts to balance supply and demand. Concerns about the reliability of supply (where reliability is interpreted as continuity of service) do not arise (although there may be concerns about price volatility). However, where there are market distortions, such as a market price cap, the market may not clear and involuntary load shedding may be required.

5.5.1 Inadequate Metering and Involuntary Load Shedding

However, price caps are not always imposed without good reason. In many electric power systems many electricity consumers simply cannot participate in the wholesale market due to inability to monitor their rate of electricity consumption.

In order for electricity customers to participate in the wholesale market, their rate of electricity consumption must be metered at least at the same frequency as the wholesale market itself. This requires so-called *interval meters* or *smart meters*. Historically, most small customers do not have meters capable of recording consumption at different times. As a consequence, most small customers cannot participate effectively in the wholesale market at all.

In the absence of adequate metering it is simply not possible for customers to participate in the wholesale market. Even if they are willing to curtail their consumption in response to changing prices, without adequate metering, there is no way of determining that they have done so.

There is a limited exception to this principle. Perhaps the electricity company cannot meter individual customers on a time-of-use basis, but perhaps the electricity company retains some control over the usage of customers – perhaps it can set the usage of some customers to a low level or zero. In other words, perhaps the rate of consumption of a customer cannot be measured, but it is possible to determine whether or not that customer is consuming at all.

In this context it is still possible to implement an imperfect form of rationing. Customers could, in principle, communicate their willingness to be curtailed to the market operator (or more likely a market intermediary such as a retailer). This would then create a merit order for curtailment, exactly the same as a merit order for generation. As supply/demand conditions in the wholesale market tighten, the electricity company could work its way up this 'supply curve for curtailment', reducing the usage first for those with a low disutility for curtailment, and so on.

However, in many countries, even the capability for selective disconnection of customers does not exist. In this case it is not possible to make customers responsive to wholesale market conditions at all. In effect, the apparent demand curve is perfectly inelastic – that is, perfectly unresponsive to market conditions.

How should a market operator behave when faced with a perfectly inelastic demand curve? The market operator could simply seek to match supply with this inelastic demand at all times. However, this is inefficient. There could arise times where the marginal cost of the last unit of generation required to match supply and demand is so high that it exceeds the likely implicit marginal valuation of electricity for all except a very few customers.

In other words, even if customers cannot directly communicate their preferences to the wholesale market, this does not mean we should treat their desire for electricity to be infinitely valuable at the margin. Instead a point will be reached where it no longer makes sense to continue to supply electricity – where it would be socially preferable to curtail consumption rather than call on additional extremely expensive generation.

However, where that point is reached requires difficult trade-offs. In particular it depends strongly on the cost of involuntary curtailment or load shedding. Involuntary curtailment or load shedding is a highly economically inefficient process. Customers cannot usually be disconnected individually. Instead large groups of customers are disconnected in blocks. In any large group of customers there will be some with a high marginal valuation for electricity and some with a low marginal value for electricity. Ideally, as we have seen, customers with the lowest marginal value for electricity would be curtailed first. However, the market operator does not have such information. At best the market operator has information on the broad

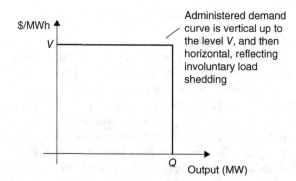

Figure 5.4 Effective demand curve in the presence of unresponsive load and an administered price cap

nature of each customer. It may be possible to separate different classes of customers – for example, disconnecting industrial customers before residential households, or schools before hospitals, but even this might not be possible.

Ideally, the market operator would seek to replicate as closely as possible the outcomes that would arise if it had information on the actual preferences of the customers. This would involve disconnecting low-marginal-value customers ahead of high-marginal-value customers. If the market operator had at least some information on the marginal value of customers, the market operator could in principle substitute an *administered demand curve* for the true underlying demand. The administered demand curve will not replicate the true demand curve, but imposing some load curtailment is more efficient than providing a power system capable of meeting all demand.

In practice, one approach (and the approach used in the Australian NEM) is to set a price level above which the market operator will invoke involuntary load shedding. This is a highly imperfect outcome, but it is preferable to not setting such a price.

In other words, in markets where there is a complete absence of effective metering, so that electricity customers are completely insulated from the wholesale market conditions, it makes sense to, in effect, assume that the demand curve is only vertical up to some price cap V. The administered demand curve is in effect horizontal at this point. This is illustrated in Figure 5.4.

How should the level of the price cap be determined? To answer this question, let us suppose that all electricity consumers face a constant price ($/MWh) for consumption of electricity, independent of the wholesale price. Let us assume that there is an underlying unobservable demand curve. The market can only observe the total demand of customers at that fixed price.

At off-peak times, this constant price is likely to be above the correct wholesale market price (which would arise if the true demand curve could be reflected to the market), inefficiently deterring some consumption at the margin. The actual or out-turn market price is inefficiently low. On the other hand, at peak times, this constant price is likely to be significantly below the correct wholesale market price. There is too much consumption (the marginal value of consumption is below the marginal cost of production). This results in an inefficiently high wholesale spot price and substantial deadweight loss. This is illustrated in Figure 5.5.

As demand increases, eventually a point will be reached where the economic loss from supplying an extra unit exceeds the economic harm from involuntary load shedding. At this point it makes sense to curtail demand rather than continue to supply.

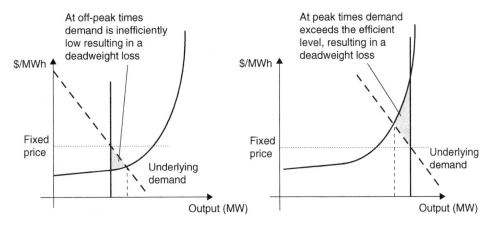

Figure 5.5 Determining the efficient level of the price cap

The economic harm from load shedding will depend on how that load shedding is carried out (e.g. whether it is possible to curtail low-marginal-value customers before high-marginal-value customers). The economic harm from load shedding is sometimes referred to as the Value of Lost Load or VoLL. This analysis suggests that the price cap should be set equal to the value of lost load plus the fixed price of consumption. Since the latter is much smaller than the former, the wholesale price cap is in practice set at the value of lost load.

Result: in a situation where customers cannot participate in the wholesale market, so that the wholesale demand curve is, in effect, perfectly inelastic, it makes sense to limit the short-run economic harm from inefficient dispatch by imposing a price cap above which involuntary load shedding will be imposed. The price cap should be set at the level where the economic loss from supplying another unit is equal to the economic harm arising from involuntary load shedding.

5.6   Time-Varying Demand, the Load-Duration Curve and the Price-Duration Curve

When talking about demand for electricity we need to be careful. In a normal market there is not a single quantity that is 'the demand' at a point in time. Rather, the demand is itself a function of the market price, as reflected in the demand curve.

However, in many wholesale electricity markets, as we have just noted, customers are not directly exposed to the wholesale market price. Instead they pay a simple fixed, time-averaged price. In effect, as we have seen, the demand curve is vertical over a wide range of prices. In this case it is possible to talk about 'the demand' or 'the load' at a point in time.

In practice, the total demand for electricity in any real electricity supply industry varies continuously. This variation in load is partly cyclical (over the course of a day, week, or year), partly dependent on exogenous factors (particularly the weather) and partly random. Over time load tends to grow. The growth rate depends on factors such as economic growth, the price of electricity and the take-up of air-conditioning. In recent years the total demand for electricity in Australia has been declining, due in part to the increasing use of on-site generation by customers (particularly solar PV) increasing take-up of energy-efficiency measures and the changing manufacturing landscape in Australia.

Mathematically, we could model the load on an electricity network as being a function of many different parameters, such as the wholesale spot price, time, and external factors, such as the ambient temperature or a random variable. In subsequent sections we will need to distinguish between these different ways of modelling demand.

For the moment however, let us ignore the dependence of demand on observable factors, such as the wholesale spot price or temperature. Instead, let us just treat demand as a random variable.

One key way to represent random demand in a simple diagram is through the device known as a *load–duration curve*. A load–duration curve sets out, for each possible level of load, the probability that the load will exceed that level.

Mathematically, we can model the load as a random variable, Q. This random variable is assumed to have a probability density function of $f(q)$ and a cumulative density function of $F(q) = \Pr(Q \leq q) = \int_0^q f(q)\mathrm{d}q$. Let $Q^{LD}(z)$ be the inverse of the cumulative density function – that is $Q^{LD}(z) = F^{-1}(z)$ where z lies between 0 and 1. $Q^{LD}(z)$ is said to be the *load–duration curve*.

Note that $f(Q^{LD}(z))\mathrm{d}q = \mathrm{d}z$ so the average load is the area under the load-duration curve:

$$E(Q) = \int qf(q)\mathrm{d}q = \int Q^{LD}(z)f(Q(z))\mathrm{d}q = \int_0^1 Q^{LD}(z)\mathrm{d}z$$

The shape of the load–duration curve depends on the probability distribution of the load. If the distribution of load is 'uniform' (that is, if the load is equally likely to take any value in a range), the load–duration curve will be a simple straight line. Typically, the load–duration curve for a normal market will be 'S-shaped'.

Figure 5.6 shows the load-duration curve for the Victorian region of the NEM for the calendar year of 2003. As shown, during that year, the total demand in this region exceeded 6210 MW 20% of the time.

As the load varies over time, the wholesale spot price will also vary. We can treat the price as a random variable. As with load, we can represent this variation in the wholesale price through the device known as the *price–duration curve*. The price-duration curve shows, for each price level, the probability that the wholesale spot price will exceed that level.

Figure 5.7 shows the price–duration curve for a simple industry in which there is 1000 MW of capacity with a variable cost of $10/MWh and 1000 MW of capacity with a variable cost of $40/MWh. Demand is assumed to be inelastic up to the price $1000/MWh, at which point demand drops to zero. As the demand varies from 800 to 2200 MW, the price increases from $10/MWh to $40/MWh and (when the load is above 2000 MW) to the point at which demand is willing to curtail ($1000/MWh).

Notice that in this simple electricity supply industry with no risk of outages, there is a simple monotonic relationship between load and price. This simple relationship does not hold in more complex markets in which there is both demand and supply uncertainty.

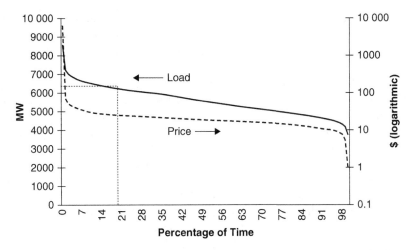

Figure 5.6 Load– and price–duration curves for the Victorian node in the NEM, calendar year 2003

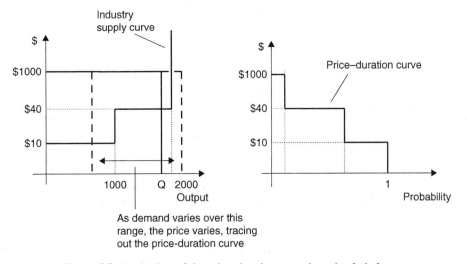

Figure 5.7 Derivation of the price–duration curve in a simple industry

The price–duration curve for the Victorian node of the NEM in the calendar year 2003 is illustrated in Figure 5.6 As with the load–duration curve, the area under the price–duration curve is equal to the average price.

As we will see in a later part of this book, the shape of the price–duration curve provides important signals as to the need for investment in different types of generators.

5.7 Summary

In Chapter 4, we examined how to achieve efficient use of a set of generation and consumption resources when there are no network constraints. In principle this efficient outcome could be achieved in a vertically integrated electricity company, but experience shows that establishing effective incentives in large vertically integrated entities is difficult, especially when those businesses are government owned. A primary objective of the reforms of the late twentieth century was to introduce competition into those segments of the electricity industry where it is feasible.

In this chapter we explored whether we can achieve efficient use of a set of generation and consumption resources using a competitive market mechanism. The advantage of a market mechanism is that it allows production and consumption decisions to be decentralised to competing market participants, significantly strengthening incentives for productive efficiency, innovation and customer responsiveness.

Anticipating the issues that will arise when we take in network constraints, we considered a centralised market mechanism with a system operator. Market participants communicate supply and demand information to the market operator who computes the optimal dispatch and communicates price and dispatch information back to the market participants. Provided there is effective competition between market participants, so that each participant behaves as a price-taker, and provided there are no price caps or other distortions, market participants have an incentive to submit their true supply and demand information. The resulting outcome from the market process is efficient.

There is a degree of variation in wholesale market designs in the electricity industry practice. These variations include the extent to which market participants are able to enter into bilateral or multilateral arrangements outside the centralised market process, and whether or not the centralised market process incorporates the startup or unit commitment decision of generators. Anticipating the issues that will arise in future chapters, as long as individual market participants communicate their net position (net injection or withdrawal at every node of the network) to the central market operator, there is no reason to prevent bilateral arrangements.

Many countries operate a two-stage market, with settlement occurring at both the day-ahead price and the real-time price. A day-ahead market is technically very similar to a short-term (day ahead) forward market. However, by incorporating the unit commitment decision, it is theoretically possible that a centralised day-ahead market might yield better price forecasts than a decentralised hedge market and may encourage higher levels of hedging, perhaps leading to lower levels of market power. However, the practical significance of these effects is not clear in practice.

In a functioning and effective wholesale electricity market there is no concern with reliability (where reliability is understood as involuntary curtailment of supply) since the price always rises to a level that will clear the market (there could, however, be problems with volatility of prices). In the presence of price caps the market may fail to clear, resulting in the need for involuntary shedding of load. In practice mechanisms for involuntary shedding of load are typically very economically inefficient resulting in substantial economic harm.

The involvement of customers in the wholesale market requires adequate metering facilities. Where such facilities do not exist customer demand is inelastic and unresponsive to the wholesale spot price. In this context introducing a price cap can improve welfare by limiting the level of economic losses at very high levels of demand. The price cap should be set at a level where the economic loss from supplying an extra unit is equal to the economic harm that arises from involuntary shedding of load.

Questions

5.1 What is a market mechanism? Why do all wholesale spot markets integrate the energy market with the physical limits of the transmission network?

5.2 True or false: A price-taking generator will submit an offer curve that matches its short-run marginal cost curve.

5.3 What are the arguments for and against the establishment of a day-ahead market?

5.4 An electricity market features a price floor of $-1000/MWh. Many generators offer a proportion of their output to the market at the price floor. Is this evidence of a lack of competition in the market?

Further Reading

For more on the single price versus pay-as-bid debate, see Baldick (2009). On whether or not day-ahead markets are required, you might look at Irastorza and Fraser (2002) or Counsell and Evans (2004). Chao and Wilson (1999) have written on issues in the design of wholesale electricity markets.

6

Representing Network Constraints

The previous chapters have considered how to achieve efficient short-run use of a set of generation and consumption resources in the absence of network constraints. As we will see, the issues become more complex and more interesting when we seek efficient use of generation and consumption resources in the presence of network constraints.

The first step in the process of introducing network constraints into our analysis is to show how the physical limits of the network can be represented as a set of equations known as constraint equations. These constraint equations, which define the technical capability of the network, depend on the net power injection at each node on the network. The first step in the process, therefore, is an understanding of how the physical limits of the electricity network can be expressed as a set of constraints on the net power injection at each node of the network.

6.1 Representing Networks Mathematically

In Part II of this book we discussed all of the elements of the power system – the transformers, the transmission lines, the switches and so on. It turns out that for the purposes of modelling the power flows on an AC network we can abstract away from much of that detail. In fact, it turns out that we can model an AC network as simply a collection of *nodes* (also called 'buses') and *links* between the nodes.

Figure 6.1 illustrates a number of different network configurations. The power flow on each network can be adequately modelled once we know the network configuration and the electrical characteristics (specifically, the reactance and impedance) of each of the transmission links.

It is common to distinguish two different kinds of networks, according to whether or not the network features *loop flow*. A network has loop flow if there is more than one path between two nodes on the network. A network with no loop flow is sometimes called a *radial network*. By definition, a radial network has the property that there is only one path between any two nodes. Figure 6.1 shows examples of radial networks.

Radial networks are also sometimes called linear or *hub-and-spoke* networks.

The opposite of a radial network is a network with loop flow or a meshed network. In a meshed network there are at least two nodes for which there are at least two different paths between those nodes. Figure 6.2 illustrates some meshed networks.

The Economics of Electricity Markets, First Edition. Darryl R. Biggar and Mohammad Reza Hesamzadeh.
© 2014 John Wiley & Sons, Ltd. Published 2014 by John Wiley & Sons, Ltd.

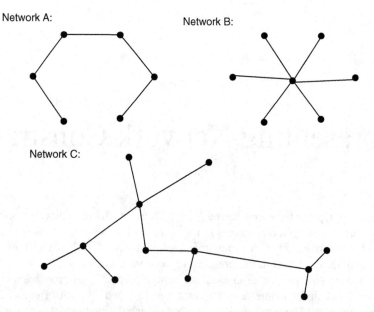

Figure 6.1   Examples of radial networks

Definition: A transmission network is *radial* if there is only one path between any two nodes on the network. A transmission network is *meshed* if there exist two nodes for which there are two or more paths between those nodes.

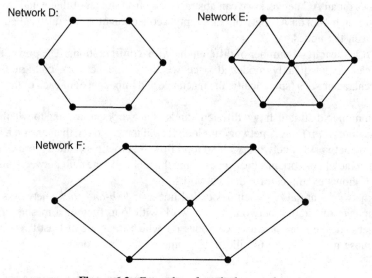

Figure 6.2   Examples of meshed networks

Suppose we have a network with N nodes and M links between the nodes. We can represent the network configuration mathematically using the *network incidence matrix* A. This is an $M \times N$ matrix. If there is a link from node i to node j in the network, the row of this matrix corresponding to the link $i \rightarrow j$ takes the value 1 in the ith column and the value -1 in the jth column and is zero elsewhere.

For example, the following simple network has the network incidence matrix set out as shown.

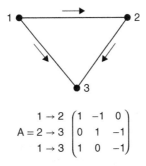

$$A = \begin{matrix} 1 \rightarrow 2 \\ 2 \rightarrow 3 \\ 1 \rightarrow 3 \end{matrix} \begin{pmatrix} 1 & -1 & 0 \\ 0 & 1 & -1 \\ 1 & 0 & -1 \end{pmatrix}$$

Definition: The *network incidence matrix* has the following entries:

$$A_{li} = \begin{cases} 1, & \text{if } i \text{ is the originating node for link } l \\ -1, & \text{if } i \text{ is the terminating node for link } l \\ 0, & \text{otherwise} \end{cases}$$

6.2 Net Injections, Power Flows and the DC Load Flow Model

Let us suppose that we have a set of network links. A link l in this set connects node i with node j in the direction from node i to node j. Node i is said to be the origination node for link l, (denoted $org(l)$) and node j is said to be the termination node (denoted $term(l)$).

It is useful to make a distinction between *bidirectional* links and *unidirectional* links. In much of the analysis that follows we will ignore losses. In the absence of losses the power flow from node i to node j is equal and opposite to the power flow from node j to node i. In this case we need only keep a list of bidirectional links (since the flow in the opposite direction is just the negative of the flow in the given direction). However, in the presence of losses, the flow from node i to node j is not equal-and-opposite to the power flow from node j to node i. In this case we must keep track of each direction separately.

Let us define B to be the set of bidirectional network links, where flow in the opposite direction is just taken to be the negative of the flow in the given direction (e.g. in a three-node, three link network we could have $B = \{1 \rightarrow 2, 1 \rightarrow 3, 2 \rightarrow 3\}$). Similarly, let us define U to be the set of one-directional links, where the flow is not equal and opposite (e.g. in a three-node, three link network we could have $U = \{1 \rightarrow 2, 2 \rightarrow 1, 1 \rightarrow 3, 3 \rightarrow 1, 2 \rightarrow 3, 3 \rightarrow 2\}$).

Normally, in what follows, we will not need to make a distinction between the set of bidirectional and unidirectional links. However, in a few places this distinction is important.

With few exceptions the physical limits on an electricity network can be expressed as limits on the power flows over or through particular network elements, such as a transformer or a transmission line. In other words, the physical limits on networks can be expressed in the form:

$$F_l \leq K_l$$

where F_l is the power flow over (unidirectional) network link $l \in U$, and K_l is the physical limit of the power flow on link l.

In a typical AC power network the system operator has relatively little control over the way the power divides and flows over the elements of the network. In principle, a system operator could control power flows in one or more of the following way:

- By switching passive devices such as capacitors or reactors in or out of the network;
- By the use of active devices (known as FACTS devices) to control power flow; or
- By switching network elements in or out.

There is increasing interest in how system operators could use these tools to optimise overall economic outcomes (or possibly even to mitigate market power). We will discuss some of these issues further below. However, in practice, the use of such tools has to date been limited. In practice the primary means by which a system operator controls power flows over the electricity transmission network is through control over the production and consumption decisions of generators and loads.

If we put aside reactive power (i.e., if we assume that there is sufficient reactive power resources at each location on the network to keep the power factor close to one), the power flows over the transmission network depend on the *net injection* of power at each node of the network. The net injection of power at a node on the network at a given point in time is the total amount of generation (MW) less the total amount of load (MW) at that node.

Every producer and consumer in an electric power system is located at one and only one node. The set of producers and consumers at each node in the power system is therefore a partition of the set of all producers and consumers. Let us suppose that the set of producers and consumers at node j is denoted N_j. We can therefore define the net injection of electric power at each node as the total production of the producers at that node less the total consumption of the consumers at that node:

$$Z_j = \sum_{i \in N_j} Q_i^S - \sum_{i \in N_j} Q_i^B$$

Definition: The *net injection* of power at a given node of the network at a given point in time is equal to the total generation at that node less the total consumption at that node.

Let us define the net injection at node i as Z_i. By Kirchhoff's law, the net injection at any given node must be equal to the net flow out of that node:

$$\forall i, Z_i = \sum_{l \in U : \text{org}(l)=i} F_l$$

Here, org(l) is the originating node for link l.

If we ignore losses, then the flow from i to j is the same as the flow from j to i (except for a minus sign) $F_{i \to j} = -F_{j \to i}$. We can therefore write the net injection at each node as follows:

$$\forall i, Z_i = \sum_{l \in B : \text{org}(l)=i} F_l - \sum_{l \in B : \text{term}(l)=i} F_l$$

Here, term(l) is the terminating node for link l.

Finally, we can observe that if we ignore losses, the sum over all of the net injections of each node of the network must be zero:

$$\sum_i Z_i = \sum_{l \in B} F_l - \sum_{l \in B} F_l = 0$$

Result: The net injection of power at a node is equal to the net flow of power away from that node on network links. If we ignore losses, the sum of the net injections across the whole network must be zero.

The system operator uses its control over the controllable generation and load resources in such a way that the net injection of power at each node on the network corresponds to a set of power flows over the network that is within the feasible limits.

The set of all net injections at each node of the network can be written as a vector $Z = (Z_1, Z_2, \ldots Z_n)$. The constraint equations can therefore be written in the form

$$F_l(Z) \le K_l$$

As we saw earlier these operating limits could vary in real time according to the changing physical limits on the network. In principle all of the physical limits of the network (i.e. the thermal limits, voltage stability limits and transient stability limits) can be represented in this way.

Definition: The *constraint equations* are a set of functions of the net injections at each node of the network that define the boundaries of the set of secure operating states.

6.2.1   The DC Load Flow Model

In order to make progress, we need to understand these functions that map the net injections at each node of the network to the power flows over elements of the network.

Modelling power flows on AC power networks accurately requires a moderately complicated set of equations. Although accurate modelling of power flows is essential in some applications, for the purposes of this book, we can use a significant simplification of the full set of equations, known as the DC Load Flow model.

The DC Load Flow model is a reasonable approximation where there are sufficient reactive power resources available around the network to ensure that the power factor is close to one. One of the key characteristics of the DC Load Flow model is that the constraint equations are linear in the net injections.

The derivation of the DC Load Flow model from the full AC power flow equations is set out below. The key result is that the power flow over any network element is a linear function of the net injections at each node:

$$F_l = \sum_i H_{li} Z_i$$

Let us assume we have a network with N nodes and M links between the nodes. The exact equation for the real power flow from node i to node j (measured at node i, in the direction of node j) is

$$F_{ij} = G_{ij}\left[V_i^2 - V_i V_j \cos\left(\delta_i - \delta_j\right)\right] + \Omega_{ij} V_i V_j \sin\left(\delta_i - \delta_j\right)$$

where $G_{ij} = R_{ij}/(R_{ij}^2 + X_{ij}^2)$, $\Omega_{ij} = X_{ij}/(R_{ij}^2 + X_{ij}^2)$, R_{ij} is the resistance on the line from i to j and X_{ij} is the reactance on the line from i to j. V_i and V_j are the voltages measured at node i and node j and δ_i and δ_j are the phase angles of the voltage relative to some node known as the reference node.

We can simplify this equation by making certain assumptions. Specifically, let us normalise the voltages to be equal to one (i.e. $V_i \approx 1$ and $V_j \approx 1$). Similarly, let us assume that the phase differences across the transmission line $\delta_i - \delta_j$ are small so that $\cos(\delta_i - \delta_j) \approx 1$ and $\sin(\delta_i - \delta_j) \approx \delta_i - \delta_j$. Then the equation reduces to

$$F_{ij} = \Omega_{ij}\left(\delta_i - \delta_j\right)$$

We can express this equation in matrix form using the $(M \times N)$ network incidence matrix A introduced in Section 6.1. Let Ω be the $(M \times M)$ diagonal matrix corresponding to Ω_{ij}. Let δ be the (length N) vector of voltage phase angles relative to the reference node and let F be the (length M) vector of flows on each line. Then the equation above can be written as

$$F = \Omega A \delta$$

Since, by Kirchoff's first law, power flows into a node equal power flows out, it must be that $Z = A^T F$. Hence, combining these equations, we have that $Z = A^T \Omega A \delta$. Solving for δ yields $\delta = (A^T \Omega A)^{-1} Z$. Hence, we can write

$$F = HZ$$

where $H = \Omega A (A^T \Omega A)^{-1}$.

Result: In the DC Load Flow model, the flow on any network element can be represented as a linear function of the injections at the nodes of the network.

6.3 The Matrix of Power Transfer Distribution Factors

Let us suppose that we have a network with N nodes and M transmission links between the nodes. Let Z_i denote the net injection of power at node i.

As we will see later in Section 7.9, in general, the sum of the net injections across all nodes of the network will be equal to the losses on the network elements. However, for the moment let us ignore the losses so the energy balance equation takes the form

$$\sum_i Z_i = 0$$

Note that, as a result of the energy balance equation, the system operator cannot independently choose the net injection at all the N nodes of the network. Instead, the system operator can only choose the net injection at $N - 1$ nodes, and then the net injection at the Nth node is determined automatically by the energy balance equation. The Nth node is known as the *swing bus* or the *reference node*.

As emphasised in Section 6.2, under the DC Load Flow model that we are using, the relationship between the power flows on the lth transmission line and the net injections on the $N - 1$ nodes other than the reference node can be written as

$$F_l = \sum_{i=1}^{N-1} H_{li} Z_i$$

Here, H is an $M \times N - 1$ matrix. The matrix H is sometimes referred to as the matrix of shift factors, *power transfer distribution factors* (PTDFs or PDFs) or just distribution factors.

H_{li} is the amount by which the flows over the link l vary with a change in the injection at node i (and a corresponding offsetting withdrawal at the reference node N).

These distribution factors are fundamental for the analysis of electricity transmission networks. For example, consider the following simple three-node, three-link network that will appear throughout this book:

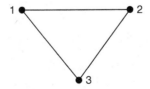

If each link on this network has the same impedance, this network has the following matrix of distribution factors (taking node 3 as the reference node):

$$H = \begin{matrix} 1 \to 2 \\ 2 \to 3 \\ 1 \to 3 \end{matrix} \begin{pmatrix} 1/3 & -1/3 \\ 1/3 & 2/3 \\ 2/3 & 1/3 \end{pmatrix}$$

In other words, $F_{2 \to 3} = (1/3)Z_1 + (2/3)Z_2$ and so on.

6.3.1 Converting between Reference Nodes

Let us define H_{li}^n to be the matrix of distribution factors when the reference node is n. Then, by definition, the column of the matrix corresponding to node n is all zeroes: $\forall l, H_{ln}^n = 0$.

Given a matrix of distribution factors for a given reference node n, it is easy to convert to the matrix of distribution factors for any other reference node m, simply by subtracting the value H_{lm}^n. This can be proved as follows:

$$F_l = \sum_i H_{li}^n Z_i = \sum_i H_{li}^n Z_i - \sum_i H_{lm}^n Z_i = \sum_i H_{li}^m Z_i$$

Therefore, for any nodes n and m

$$H_{li}^m = H_{li}^n - H_{lm}^n$$

Result: Given the matrix of distribution factors with respect to any one reference node n, we can easily find the matrix of distribution factors with respect to any other reference node m simply by subtracting the column for node m from every other column.

6.4 Distribution Factors for Radial Networks

Suppose that we are given an arbitrary network and two nodes i and j on that network. Consider the effect of a small increase in the injection at node i and a small and equal increase in the withdrawal at node j. This change causes a change in the power flows on every path between node i and node j. If there is only one path between node i and node j, the flow on that path increases by precisely the amount of the increased injection. Flows that are not on a path between node i and node j are completely unaffected by this change in the injections.

We can use this information to learn more about the properties of the matrix of distribution factors. In the previous sections we saw that the flows on a network are related to the injections on that network in the following way: $F_l = \sum_{i=1}^{N-1} H_{li} Z_i$. From this equation we know that if the injection at node i increases by 1 unit and the injection at node j decreases by 1 unit, the flow over any link l in the network changes by precisely $F_l = H_{li} - H_{lj}$.

If the link l is not on some path between node i and node j, the flow on link l is not affected by an increase in the injection at node i and a corresponding increase in the withdrawal at node j. Hence, we have that $F_l = H_{li} - H_{lj} = 0$ if link l is not on a path between node i and node j.

In the case where there is only one path between node i and node j, an increase in the injection at node i and a corresponding withdrawal at node j causes a proportional increase in the flow on any link that lies on the path between node i and node j. Hence, we have that

$$
H_{li} - H_{lj} = \begin{cases} 1, & \text{if } l \text{ lies on the path from } i \text{ to } j \\ -1, & \text{if } l \text{ lies on the path from } j \text{ to } i \\ 0, & \text{otherwise} \end{cases}
$$

For example, consider the following simple linear network:

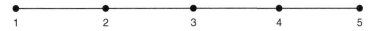

Ignoring losses, the matrix of distribution factors must therefore be as follows (taking node 5 as the reference node):

$$
H = \begin{matrix} 1 \rightarrow 2 \\ 2 \rightarrow 3 \\ 3 \rightarrow 4 \\ 4 \rightarrow 5 \end{matrix} \begin{pmatrix} 1 & 0 & 0 & 0 \\ 1 & 1 & 0 & 0 \\ 1 & 1 & 1 & 0 \\ 1 & 1 & 1 & 1 \end{pmatrix}
$$

As another example, consider the network illustrated here.

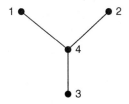

This network has the matrix of power transfer distribution factors as shown (node 4 is the reference node):

$$
H = \begin{matrix} 1 \rightarrow 4 \\ 2 \rightarrow 4 \\ 3 \rightarrow 4 \end{matrix} \begin{pmatrix} 1 & 0 & 0 \\ 0 & 1 & 0 \\ 0 & 0 & 1 \end{pmatrix}
$$

6.5 Constraint Equations and the Set of Feasible Injections

As before, let us suppose that we have a matrix with N nodes and M links. Let us suppose that there are physical limits on the flows on each link. Let us suppose that the physical limit on link l is denoted K_l. The system operator must control the net injection at each node in such a way that the flow on each network element l is less than the physical limit. As in Section 6.2, this can be expressed as

$$\forall l \in U, F_l \leq K_l$$

However, the flow on each transmission link is directly related to the net injection through the equation $F = HZ$; hence, we can deduce that the system operator must operate the system in such a way that the following equations hold:

$$\forall l, \sum_{i=1}^{N-1} H_{li} Z_i \leq K_l$$

These are the *constraint equations* for this network. The set of all injections that satisfy these constraint equations and the energy balance equation is known as the *set of feasible injections*.

Definition: The *set of feasible injections* is the set of all possible combinations of injections at each node on the network that simultaneously satisfy the constraint equations and the energy balance equation and are therefore within the physical limits of the network.

The set of feasible injections is convex. This arises because the constraints that define the set of feasible injections are linear.

Let us suppose we have any two vectors of feasible injections Z^A and Z^B. Let λ be a real number between zero and one. Consider the new injection defined as follows: $Z = \lambda Z^A + (1 - \lambda)Z^B$. This injection satisfies the energy balance constraint (using vector notation):

$$1^T Z = \lambda 1^T Z^A + (1 - \lambda)1^T Z^B = 0$$

Here, 1 is a vector of ones (so that $1^T Z = \sum_i Z_i$). The injection also satisfies the generic transmission constraints (again, using vector notation):

$$HZ = \lambda HZ^A + (1 - \lambda)HZ^B \leq \lambda K + (1 - \lambda)K = K$$

This proves that the set of feasible injections is convex. In addition the injection $Z = 0$ (i.e. balanced supply and demand at every node) is also feasible.

Furthermore, let us suppose we take a network and we augment the capacity of one or more of the transmission lines on the network, without adding new lines or changing any connections and leaving the electrical characteristics of the lines unchanged. This always increases the size of the set of feasible injections.

We can express this formally as follows. Suppose we have a network described by the PTDF matrix H and flow limits K_l. Suppose we augment this network so as to yield new flow limits K_l' where $\forall l, K_l' \geq K_l$ while leaving the PTDF matrix unchanged, then the set of feasible injections of the original network is a subset of the set of feasible injections of the augmented network: If an injection Z is feasible on the original network, it will remain feasible on the augmented network.

Result: The set of feasible injections is convex and includes the special case of zero injection at each node. Moreover, augmenting the capacity of transmission lines while leaving the PTDF matrix unchanged increases the size of the feasible set.

To see how this works in a simple example, consider the following three-node, two-link network:

Taking node 3 as the reference node, and assuming both links have identical electrical characteristics, this network has the matrix of power transfer distribution factors as shown:

$$H = \begin{matrix} 1 \to 3 \\ 2 \to 3 \end{matrix} \begin{pmatrix} 1 & 0 \\ 0 & 1 \end{pmatrix}$$

In general, it may be the case that the maximum flow (the flow limit) on a transmission line in one direction may be different from the maximum flow in the reverse direction. However, let us assume for the moment that the maximum flow on each transmission line is the same in both directions. Specifically, let us assume that the maximum flow on the lines on this network are given as $K_{1\to3}$ and $K_{2\to3}$.

Therefore, the set of feasible injections is the set of injections $z = (z_1, z_2, z_3)^T$, which satisfy $-K_{1\to3} \leq z_1 \leq K_{1\to3}$, $-K_{2\to3} \leq z_2 \leq K_{2\to3}$ and $z_1 + z_2 + z_3 = 0$. Since there are only two independent injections, we can illustrate this feasible set on a two-dimensional diagram as in Figure 6.3.

Note that increasing the capacity on existing links (i.e. increasing the limits K_l) without changing the configuration of this simple network always increases the size of the feasible set.

Now consider what happens when a new link is constructed between nodes 1 and 2. If we assume that all the links have identical electrical characteristics, we have the three-node, three-link network described earlier.

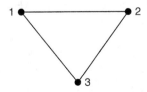

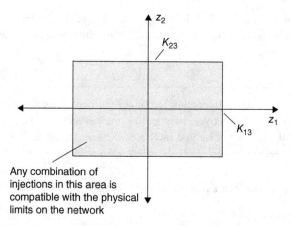

Figure 6.3 The set of feasible injections in a simple two-link network

As noted earlier, this network has the matrix of distribution factors

$$H = \begin{matrix} 1 \rightarrow 2 \\ 2 \rightarrow 3 \\ 1 \rightarrow 3 \end{matrix} \begin{pmatrix} 1/3 & -1/3 \\ 1/3 & 2/3 \\ 2/3 & 1/3 \end{pmatrix}$$

The set of feasible injections is therefore the set of injections $z = (z_1, z_2, z_3)^{\mathrm{T}}$, which satisfy $-K_{1 \rightarrow 3} \leq \frac{2}{3} z_1 + \frac{1}{3} z_2 \leq K_{1 \rightarrow 3}$, $-K_{1 \rightarrow 2} \leq \frac{1}{3} z_1 - \frac{1}{3} z_2 \leq K_{1 \rightarrow 2}$ and $-K_{2 \rightarrow 3} \leq \frac{1}{3} z_1 + \frac{2}{3} z_2 \leq K_{2 \rightarrow 3}$. In addition, $z_1 + z_2 + z_3 = 0$.

This can be represented as the shaded region in Figure 6.4.

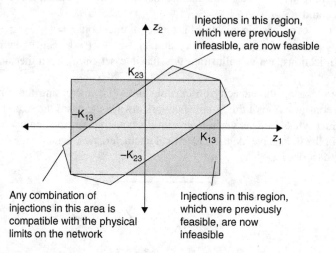

Figure 6.4 Set of feasible injections for a simple three-node network

The size of the new feasible set (the size of the shaded region) depends, amongst other things, on the capacity of the newly constructed link $K_{1\to2}$. If this link is small, the new feasible set will be much smaller than the old feasible set. If this new link is large, the new feasible set may be larger than the old feasible set. However, even if the new link has a large capacity, there may be some injections that were previously feasible and that are now infeasible. At the same time, even if the new link is very small, some injections that were previously infeasible are now feasible (Figure 6.4).

For example, suppose that the capacity on the link $1 \to 3$ is 10 and the capacity on the link $2 \to 3$ is 20. With the original network configuration (without the link $1 \to 2$), the injection of 10 MW at node 1, 20 MW at node 2 and a withdrawal of 30 MW at node 3 (i.e. $z = (10, 20, -30)$) is feasible.

Now, suppose that the link $1 \to 2$ is constructed with a capacity of 10. At the same injection $z = (10, 20, -30)$ the flow on the link $1 \to 3$ is now 13.333, exceeding the link capacity of 10, so this injection is now infeasible.

On the other hand, consider the injection $z = (4, 21, -25)$. This injection is infeasible in the original network configuration (the flow on the link $2 \to 3$ exceeds its physical limit) but is feasible in the new network configuration.

While increasing the capacity of an existing link by a small amount has a small impact on the set of feasible injections, the same is not true for the construction of a new link – indeed, the construction of a very small new link, by changing the topology of the network, may bring about a very large change in the set of feasible injections. Consider, for example, the construction of a new link $1 \to 2$ of infinitesimal capacity (but the same electrical characteristics) in the earlier example. This new link severely constrains the set of feasible injections.[1] In fact, in this new network the only feasible injections are those where z_1 is very close to z_2. This example highlights the fact that *small changes to the network do not necessarily have a small effect on the set of feasible injections*.

Result: The addition of a new link to an existing network may significantly change the shape of the feasible set, making some flows that were previously feasible infeasible and vice versa.

6.6 Summary

As a first step in the process of incorporating network constraints in the optimal dispatch computation, we need some mechanism for representing network constraints in a mathematical form. Network constraints can be expressed as limits on the power flows through network elements. However, power flows in AC power networks cannot easily be controlled directly. Instead, the system operator can typically only control the net injections at each node of the network. Therefore, we need some mechanism that can express the flows over the network elements as a function of the net injections at each node.

---

[1] The new link would not severely constrain the set of feasible injections if its impedance was much greater than that of the other links.

The DC Load Flow model shows that, under certain assumptions, it is possible to represent the flows over a network element as a linear function of the net injections at the different nodes of the network. The matrix that relates net injections to power flows is known as the matrix of Power Transfer Distribution Factors.

It is common to draw a distinction between radial and meshed networks. In radial networks there is only a single path between any two points on the network. Ignoring losses, in a radial network the PTDF matrix takes only the values 1, −1, and zero.

Given a set of limits on the flows over each network element, the DC Load Flow model implies a set of limits on the net injections. A net injection that is consistent with all the power flow limits is said to be feasible. The set of all feasible injections is convex. Given a particular network topology the relaxation of one of the limits always increases the size of the feasible set, but the construction of a new line may change the shape of the feasible set, making some previously feasible injections no longer feasible and vice versa.

Questions

6.1 Provide a simple example to show that the addition of a new line to an existing network may result in a strict reduction in the capability of that network – that is, the new set of feasible injections is a strict subset of the set of feasible injections of the original network.

6.2 Find the matrix of PTDFs for the following two networks assuming that all transmission links have identical electrical impedance, and taking node 4 as the reference node:

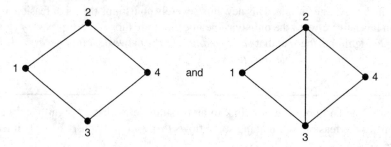

How does the introduction of the new link 2–3 change the set of feasible injections? What conditions must be satisfied for this link to expand the set of feasible injections?

6.3 If the loss of any single transmission circuit is a credible contingency, and if no *ex post* corrective actions are feasible, what condition must be satisfied by any transmission network in order for the transmission network to be able to carry any power flows at all under normal operation (that is, for the set of feasible injections to have any elements other than zero).

6.4 Under what circumstances will the addition of a small (i.e. very low capacity) line to an existing network have a small impact on the set of feasible injections?

7

Efficient Dispatch of Generation and Consumption Resources in the Presence of Network Congestion

In Chapter 6, we showed how we can express the physical limits of an electricity network as a set of linear constraint equations. In this chapter, we introduce these constraint equations into the optimal dispatch task to find the combination of production and consumption that maximises economic welfare subject to the physical constraints of the network.

7.1 Optimal Dispatch with Network Constraints

Previously, we defined the net injection of electric power at a node as the total production of producers at that node less the total consumption of consumers:

$$Z_j = \sum_{i \in N_j} Q_i^{S} - \sum_{i \in N_j} Q_i^{B}$$

Using the benefit function introduced in Section 4.6 and the set of linear constraints derived from the DC Load Flow model introduced in Section 6.2, ignoring losses, the optimal dispatch problem can be written as follows:

$$W(K_l) = \max_{Z_i} \sum_i B_i(Z_i)$$

subject to

a. The energy balance equation $\sum_i Z_i = 0 \leftrightarrow \lambda$ and
b. The generic network constraints $\forall l \in U, \sum_i H_{li} Z_i \leq K_l \leftrightarrow \mu_l$

Recall that node N is chosen to be the reference node ($H_{lN} = 0$ for all links l).

The Economics of Electricity Markets, First Edition. Darryl R. Biggar and Mohammad Reza Hesamzadeh.
© 2014 John Wiley & Sons, Ltd. Published 2014 by John Wiley & Sons, Ltd.

The KKT conditions for this constrained optimisation problem are as follows:

$$B'_i(Z_i) = \text{SMC}_i = \lambda - \sum_i \mu_l H_{li}$$

$$\mu_l \geq 0 \text{ and } \mu_l \left(\sum_i H_{li} Z_i - K_l \right) = 0$$

Here, μ_l is the *constraint marginal value* corresponding to the network element l. The last conditions show that as long as a physical network limit is not binding, the corresponding constraint marginal value is zero. When a physical network limit is binding the constraint marginal value is positive. Using the relationship between the flow on a link and the net injections ($F_l = \sum_i H_{li} Z_i$), the last condition implies that the product of the marginal value and the flow on a link is (at the optimum) equal to the marginal value multiplied by the flow limit on a link:

$$\mu_l F_l = \mu_l K_l$$

The constraint marginal value μ_l is sometimes referred to as the *shadow price of congestion* on link l.

Furthermore, the additional economic value created by relaxing a constraint by a small amount is equal to the constraint marginal value for that constraint:

$$\forall l, \frac{\partial W}{\partial K_l} = \mu_l$$

Result: For any given network constraint, the constraint marginal value is zero unless the constraint is binding, in which case the constraint marginal value is positive. The constraint marginal value is equal to the additional economic value, which is obtained by relaxing the constraint by a small amount under the given market conditions.

Since the Nth node is the reference node, $H_{lN} = 0$ for all links l. As a result, the common marginal cost and marginal utility for generators and consumers at the reference node satisfies

$$\text{SMC}_N = \lambda$$

Hence, we have derived the following result: The common marginal cost and marginal value for all the producers and consumers at any node i is equal to the common marginal cost and marginal value for the producers and consumers at the reference node less, for each binding transmission constraint, the constraint marginal value multiplied by the coefficient of that constraint and that node in the corresponding constraint equation:

$$\text{SMC}_i = \text{SMC}_N - \sum_l H_{li} \mu_l$$

This result shows how marginal cost/marginal utilities vary across geographic locations under optimal dispatch. If there are no transmission constraints binding, then the constraint marginal value for each network element is zero $\mu_l = 0$, and the common marginal cost/ marginal utility is the same across the entire system.

7.1.1 Achieving Optimal Dispatch Using a Smart Market

Furthermore, we can achieve this optimal dispatch outcome using a smart market. We can use the same smart market mechanism introduced in Section 5.1. The only difference with the process set out in that section is that in the second stage of the process the system operator takes into account the physical network constraints in the optimisation process. As before, the output of this process is a set of prices (one for each node) and a rate of production or consumption for each generator or consumer.

As before, provided

a. The market participants pay (or are paid) the correct price that emerges from the smart market process (in this case, the correct nodal price, without price caps or price floors);
b. The market is competitive, so that each producer or consumer is a price-taker; and
c. The network constraints represented in the smart market process are an accurate representation of the true undelrying physical network limits,

Then, the following key results apply:

i. Each producer and consumer has an incentive to truthfully reveal his/her marginal cost and marginal valuation curves;
ii. The resulting dispatch is economically efficient;
iii. Each producer and consumer has an incentive to comply with the dispatch instructions; and
iv. The resulting price at each node is equal to the common marginal cost and marginal value at that node:

$$P_i = \text{SMC}_i$$

The prices at each node are known as *nodal prices* or *locational marginal prices*. From this discussion, we have the following important relationship between nodal prices and constraint marginal values:

$$P_i = P_N - \sum_l H_{li}\mu_l$$

In the case when no constraints are binding, this pricing condition gives $\forall i, P_i = P_N$. In other words, in an uncongested network (ignoring losses) the optimal dispatch involves setting the same price over the entire network. That price is set at a level that clears the market (i.e. a price for which the total supply equals the total demand).

7.2 Optimal Dispatch in a Radial Network

We can gain more insight into these results by applying them to specific simple networks.

In the (very) special case of a radial network, the pricing conditions are even simpler: The price between any two nodes is the same unless there is a binding constraint on the network path between them. The price increases in the direction of the flow.

From the earlier discussion we know that the nodal prices under optimal dispatch will satisfy

$$P_i = P_N - \sum_l H_{li}\mu_l$$

Suppose we take two adjacent nodes i and j at both ends of the link l. Using this expression, the price difference between these nodes is then

$$P_i - P_j = \sum_l \left(H_{lj} - H_{li}\right)\mu_l$$

Earlier (in Section 6.4) we observed that in a radial network

$$H_{li} - H_{lj} = \begin{cases} 1, & \text{if } l \text{ lies on the path from } i \text{ to } j \\ -1, & \text{if } l \text{ lies on the path from } j \text{ to } i \\ 0, & \text{otherwise} \end{cases}$$

Let us assume that link l joins node i directly to node j. It follows that

$$P_j - P_i = \mu_l$$

We can immediately deduce (as before) that a link is uncongested if and only if the prices at each end of the link are the same. Since μ_l is positive when the link is congested in the direction from i to j, we can conclude that the power always flows on a congested link from lower-priced nodes to higher-priced nodes.

Result: On a radial network, ignoring losses

a. Prices on two nodes are equal if and only if the network path between them is uncongested.
b. Electricity always flows across a congested link in the direction from the lower-priced node to the higher-priced node.
c. The number of distinct prices in the network is always one more than the number of congested links. If there is a single congested link, the link divides the nodes of the network into two regions that share a common price.

We will see shortly that these three results do not hold in a meshed network.

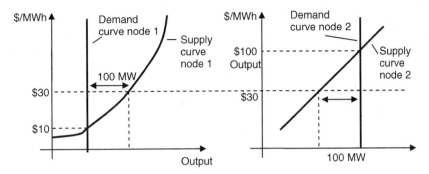

Figure 7.1 Price determination in a simple two-node network without binding network constraints

7.3 Optimal Dispatch in a Two-Node Network

Having developed the theory of optimal dispatch in Section 7.2, let us explore the implications in the context of simple networks. In this section, we will consider a two-node network. In Section 7.4, we will consider a simple three-node network.

Let us consider first the simplest possible radial network – a simple two-node network as illustrated in Figure 7.2. This network has two nodes and a single link, which is potentially congested.

Let us suppose that there is both generation and load located at both of these nodes. The supply and demand curves at each node are illustrated in Figure 7.1.

Figure 7.1 illustrates two extreme cases: where the link has sufficient capacity to be uncongested and where the link has failed (i.e. has no capacity at all). In the event that the link has sufficient capacity to carry all the desired flow, there must be a common price in both regions. As illustrated in Figure 7.1, the common price in this case happens to be $30. At this price node 1 is producing 100 MW more than it consumes (i.e. it is injecting 100 MW into the network). Since there are no losses on the network, this means that precisely 100 MW must be withdrawn from the network at node 2. This means that demand at node 2 must exceed supply by 100 MW as shown in Figure 7.1.

In the case where the link fails (i.e. the capacity of the link goes to zero) the net injection at each node must also be zero. This means that supply must equal demand in each region. In the case of this network this pushes the price up to $100 in node 2 and down to $10 in node 1.

Now consider what happens if the capacity on the link is less than 100 MW (and greater than zero). If the capacity on the link is, say, 50 MW, then a maximum of 50 MW can be injected at node 1. This means that output at node 1 must be backed off precisely 50 MW relative to the unconstrained case. The nodal price at node 1 drops to $18 in this example. This is illustrated in Figure 7.3. Similarly, since only 50 MW can be withdrawn from the network at node 2, the output at node 2 must be increased relative to the unconstrained case in order to balance supply and demand. The nodal price at node 2 increases to $60 in this example.

1   2

Figure 7.2 A simple two-node network

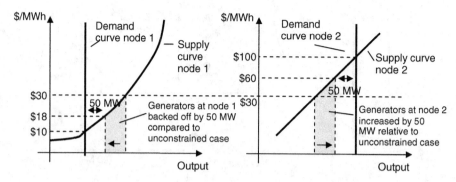

Figure 7.3 A two-node network with a constrained link

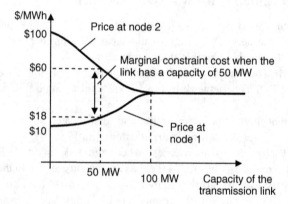

Figure 7.4 Relationship between link capacity and nodal prices

It is clear that, given the demand and supply conditions at both of the nodes of this network, for any level of capacity on this link we can calculate what the optimal dispatch and nodal prices must be. In other words, we can find the function mapping the level of capacity on the transmission link to the output of these generators and the prices at each node. If the capacity on the link is K, we can find a mapping from K to nodal prices $P_1(K)$ and $P_2(K)$. This information can be presented on a graph with the link capacity on the horizontal axis and the prices at each node on the vertical axis, as in Figure 7.4.

Figure 7.3 illustrates the results. When the link has a capacity of 100 MW or larger, the prices at the two nodes are equal, at $30. When the capacity on the link drops to 50 MW, the price at node 2 rises to $60 and the price at node 1 drops to $18. When the capacity of the link drops to zero, the price at node 2 rises to $100 and the price at node 1 reduces to $10.

In Section 7.1, we saw that in a radial network the constraint marginal value for this link is just equal to the price difference between the two nodes ($P_2 - P_1 = \mu_{1\to2}$). Therefore, when the link has a capacity of 100 MW, the constraint marginal value is zero; when the link has a capacity of 50 MW, the constraint marginal value is equal to $42, and when the link has a capacity of 0 MW, the constraint marginal value is equal to $90. The case where the link has a capacity of 50 MW is illustrated in Figure 7.4. We can of course consider the function mapping the link capacity to the constraint marginal value. In addition, we can observe that in this network, as in any radial network, the power flow is always from the lower-priced node to the higher-priced node.

7.4   Optimal Dispatch in a Three-Node Meshed Network

Let us now consider the case of a simple three-node network with three links as discussed earlier.

Recall that, taking node 3 as the reference node, the matrix of distribution factors for this simple network is

$$H = \begin{matrix} 1 \to 2 \\ 2 \to 3 \\ 1 \to 3 \end{matrix} \begin{pmatrix} 1/3 & -1/3 \\ 1/3 & 2/3 \\ 2/3 & 1/3 \end{pmatrix}$$

With this matrix, this simple network has the property that if any line is constrained, the price at the node opposite the constrained line is equal to the average of the prices at the ends of the constrained line.

This is easy to demonstrate. Without loss of generality we can take the constrained line to be the link $1 \to 2$. Using the important result in Section 7.1 that $P_i = P_N - \sum_i H_{li}\mu_l$, we can derive the following two important results:

$$P_1 = P_3 - \tfrac{1}{3}\mu_{1\to2} \quad \text{and} \quad P_2 = P_3 + \tfrac{1}{3}\mu_{1\to2}$$

Adding these two equations together we find that the price at the node opposite the constrained line is equal to the average of the prices at the ends of the constrained line.

$$P_3 = \frac{P_1 + P_2}{2}$$

Note that this result holds under optimal dispatch in this three-node, three-link network whatever the location and nature of the demand and supply curves and therefore whatever the direction and quantity of electricity flow over the network (provided, of course, that just one link is congested). In particular, there is *no* requirement that electricity always flow from a lower-priced node to a higher-priced node.

Result: In a simple three-node, three-link network where each link has identical electrical characteristics, if there is just one constrained line, the least cost feasible dispatch has the property that the efficient nodal price at the node opposite the constrained line is equal to the average of the nodal prices at each end of the constrained line.

This result remains true whatever the location of demand and supply and whatever the direction of electricity flow over the links of the network.

Let us now work through a simple example in which both demand and supply curves are fully specified.

In this example there is only generation at nodes 1 and 2 and only load at node 3. The supply curve at node 1 is $Q_1^S(P) = P$; the supply curve at node 2 is $Q_2^S(P) = (1/2)P$; the demand curve at node 3 is $Q_3^B(P) = 100 - P$. The net injection at each node is $z_1 = Q_1^S$, $z_2 = Q_2^S$ and $z_3 = -Q_3^B$.

In the absence of congestion the socially efficient market price satisfies $P_1 = P_2 = P_3$ and $z_1 + z_2 + z_3 = 0$. Solving the equations gives the unique solution $z = (40, 20, -60)$. The corresponding price is \$40/MWh. At this set of injections the flow on the link 1–2 is 6.666.

Let us now suppose that the link $1 \rightarrow 2$ has a capacity limit of K. Let us suppose this constraint is binding. Using the analysis above we know that under the optimal dispatch the nodal prices must satisfy the condition

$$P_3 = \frac{P_1 + P_2}{2}$$

When combined with the energy balance equation $z_1 + z_2 + z_3 = 0$ and the binding constraint equation

$$\frac{1}{3}z_1 - \frac{1}{3}z_2 = K$$

we can derive the following set of injections: $z_1 = (2/7)(100 + 6K)$, $z_2 = (1/7)(200 - 9K)$ and $z_3 = -(1/7)(400 + 3K)$. Using these injections we can easily compute the corresponding prices.

As before, we can plot the nodal prices as a function of the capacity on the constrained link. This is illustrated in Figure 7.5.

In the case where the transmission link has a capacity of 2, the injections are $Z = (32, 26, -58)$. The corresponding prices are $P = (32, 52, 42)$.

Note that in this example, the price is different at each node even though there is only a single congested link. Furthermore, there is a flow of 28 units from node 2 to node 3, even though the price for electricity is *lower* at node 3 (\$42/MWh) than at node 2 (\$52/MWH).

This example shows that the properties of nodal prices that we noted earlier for a radial network do not carry over to the case of a meshed network:

a. First, earlier we noted that in a radial network, prices on two adjacent nodes are equal if and only if the link between them is uncongested. In the meshed network we see that the prices at nodes 1 and 3 and nodes 2 and 3 are different even though the links $1 \rightarrow 3$ and $2 \rightarrow 3$ are uncongested.

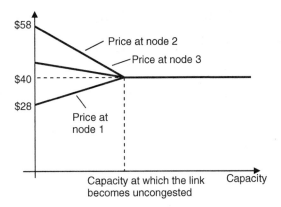

Figure 7.5 Prices in a simple three-node network as a function of 1–2 link capacity

b. Second, earlier we noted that in a radial network, electricity flows in the direction from low-priced nodes to high-priced nodes. In the new example we see that electricity flows from node 2 to node 3 even though the price is higher at node 2 than node 3.
c. Third, earlier we noted that in a radial network a congested link divides the market into two price zones. However, in the new example, a single congested link leads to different prices at all three nodes (or, put another way, a single congested link in this three-node network leads to three price zones).

7.5 Optimal Dispatch in a Four-Node Network

In the case of the simple four-node network with four links illustrated in Figure 7.6 taking node 4 as the reference node, the matrix of distribution factors is

$$
H = \begin{array}{c} 1 \to 2 \\ 3 \to 1 \\ 2 \to 4 \\ 4 \to 3 \end{array} \left(\begin{array}{ccc} 1/2 & -1/4 & 1/4 \\ -1/2 & -1/4 & 1/4 \\ 1/2 & 3/4 & 1/4 \\ -1/2 & -1/4 & -3/4 \end{array} \right)
$$

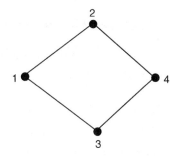

Figure 7.6 A simple four-node network

As in the last example, if any one of these links is congested, the prices around the remainder of the loop follow the 'spring-washer' pattern. For example, if the link $1 \rightarrow 2$ is the only binding transmission constraint, with the constraint marginal value $\mu_{1 \rightarrow 2}$ the pricing relationships are as follows:

$$P_1 = P_4 - \tfrac{1}{2}\mu_{1 \rightarrow 2}, P_2 = P_4 - \tfrac{1}{4}\mu_{1 \rightarrow 2} \text{ and } P_3 = P_4 + \tfrac{1}{4}\mu_{1 \rightarrow 2}$$

We can see immediately that the prices at nodes 3 and 4 are a simple weighted average of the prices at the ends of the congested link:

$$P_3 = \tfrac{2}{3}P_1 + \tfrac{1}{3}P_2 \text{ and } P_4 = \tfrac{1}{3}P_1 + \tfrac{2}{3}P_2$$

For example, if link $1 \rightarrow 2$ is congested and the price at node 1 is \$10 and the price at node 2 is \$40, then it must be that the price at node 3 is \$20/MWh and the price at node 4 is \$30/MWh. As before, this result is completely independent of supply and demand conditions at the different nodes of the network, provided, of course that only link $1 \rightarrow 2$ is congested.

7.6 Properties of Nodal Prices with a Single Binding Constraint

In Section 7.1, we derived the important relationship between nodal prices in a network with binding constraints:

$$P_i = P_N - \sum_l H_{li}\mu_l$$

Let us suppose that there is a single network constraint that is binding on line l for example. The first-order conditions for the problem are now

$$P_i = P_N - H_{li}\mu_l$$

where μ_l is the constraint marginal value for the binding constraint corresponding to l.

Note that if two nodes have a different distribution factor for the link l (i.e. if $H_{li} \neq H_{lj}$), then they must have different prices when link l is binding ($P_i \neq P_j$). In other words, nodal prices can (and usually will) vary between different nodes of a network even though there is only a single congested link.

Let us choose two nodes i and j for which $H_{li} \neq H_{lj}$ (it is always possible to do so since $H_{lN} = 0$ and at least there must be at least one other node for which H_{li} is nonzero. We can rewrite the previous equation to express the price at any node k as a weighted average of the prices at nodes i and j:

$$P_k = \alpha_k P_i + (1 - \alpha_k)P_j \text{ where } \alpha_k = \frac{H_{lj} - H_{lk}}{H_{lj} - H_{li}}$$

In other words, when there is a single constrained link in the network, the prices at all nodes in the network can be expressed as a linear combination of the prices at any two nodes in the network with distinct prices. This is an illustration of a more general property – in a network with a single constrained link there are, in effect, only two independent prices in the entire network (no matter how many distinct nodes there are).

Wu *et al.* (1996) showed that if the only congested link is the line $1 \rightarrow N$ where N is the reference node or swing bus, then the value of H_{li} is positive and $H_{li} < H_{l1}$, so that P_N is the highest-priced node, irrespective of power flows and all the other prices lie between P_1 and P_N.

7.7 How Many Independent Nodal Prices Exist?

At first glance it might appear that nodal pricing leads to a large number of independent prices – indeed, if there are n nodes, it might appear that there are $n-1$ independent nodal prices. However, this is not the case. In fact, under optimal dispatch the prices at every node in the network can usually be calculated given the prices at very few nodes.

As we have seen many times, given the price at the swing bus (node N) and the constraint marginal value μ_l for each congested line, the optimal nodal price at every node can be calculated from the expression

$$P_i = P_N - \sum_l H_{li}\mu_l$$

We showed in Section 7.6 that if there is just one congested line, the price at every node in the network can be determined as a linear combination of the prices at just two distinct nodes with different prices.

More generally, if there are m congested lines, there are only $m+1$ independent prices. Given the prices at $m+1$ nodes with different prices, the prices at all the nodes in the network can be calculated.

Since there are likely to be relatively few network constraints binding at any one time, the number of independent nodal prices is relatively small.

7.8 The Merchandising Surplus, Settlement Residues and the Congestion Rents

Ignoring losses, when no links are congested at the optimal dispatch, prices at all nodes are identical. In this case, and only in this case, the amount the system operator pays for the electricity it purchases from generators is equal to the amount the system operator is paid by consumers for the electricity they consume.

Importantly, this result does not continue to hold when there are binding network constraints.

7.8.1 Merchandising Surplus and Congestion Rents

Let us define the total income received by the system operator in the process of purchasing electricity from generators and selling it to consumers at different nodes as the *merchandising surplus*. This is defined as follows:

$$\text{MS} = -\sum_i P_i Z_i$$

Another possible approach is to value the flow on any link on the network using the constraint marginal value or shadow price. Let us define the *congestion rents* as the sum over all links in the network of the constraint marginal value multiplied by the capacity of that link:

$$\text{CR} = \sum_{l \in U} \mu_l K_l$$

It is straightforward to confirm that, under optimal dispatch, the congestion rents are just equal to the merchandising surplus. Here, we use the result that the flow on any line is related to the net injections by the matrix of PTDFs: $F_l = \sum_i H_{li} Z_i$. In addition, we use the condition derived in Section 8.1, that, under optimal dispatch, the marginal value multiplied by the flow on a link is equal to the marginal value multiplied by the flow limit: $\mu_l F_l = \mu_l K_l$.

$$\text{MS} = -\sum_i P_i Z_i = \sum_{l,i} \mu_l H_{li} Z_i = \sum_l \mu_l F_l = \sum_l \mu_l K_l = \text{CR}$$

In other words, at optimal dispatch the merchandising surplus is equal to the total congestion revenue on the network. This result holds whether or not we take account of losses, as discussed further later. Since the congestion revenue is non-negative, the merchandising surplus at the optimal dispatch is also always non-negative. If there is at least one binding constraint in the network the merchandising surplus is positive.

Note that although we have proved that the merchandising surplus from optimal dispatch is always positive, this does not imply that the merchandising surplus is maximised at optimal dispatch. Later, we will see that the system operator may be able to increase the merchandising surplus by choosing a less efficient dispatch.

There is one more result that turns out to be important. We have proved earlier that the merchandising surplus at the optimal dispatch is equal to the congestion rents. If we have some other feasible dispatch that is not optimal, the merchandising surplus at that dispatch (but valued at the prices corresponding to the optimal dispatch) is always less than the congestion rents. Let us suppose the alternative feasible dispatch is Z_i'. Given this dispatch we can compute the corresponding set of flows $F_l' = \sum_i H_{li} Z_i'$. Since the dispatch is feasible it follows that $\forall l, F_l' \leq K_l$. From this we can conclude that the merchandising surplus at any arbitrary (nonoptimal) dispatch when valued at the optimal dispatch prices is always less than or equal to the congestion rents:

$$\text{MS}' = -\sum_i P_i Z_i' = \sum_{l,i} \mu_l H_{li} Z_i' = \sum_l \mu_l F_l' \leq \sum_l \mu_l K_l = \text{CR}$$

This inequality is strict (meaning that the merchandising surplus is strictly less than the congestion rents) if and only if the flow corresponding to the alternative net injection is less than the flow limit on a link which is constrained in the optimal dispatch.

7.8.2 Settlement Residues

For each link on the network l, we can define an amount of revenue the system operator receives by purchasing power at the originating node, transporting that power, and selling that power at the terminating node. Let us call this amount the *settlement residues*. The settlement residues on link l are defined as follows:

$$SR_l = (P_j - P_i)F_l$$

Here, F_l is the flow on network link l and link l is assumed to connect node i to node j. In other words i is the originating node for link l, and j is the terminating node for link l. Ignoring losses, the total settlement residues are then defined as

$$SR = \sum_{l \in B} SR_l = \sum_{l \in B} (P_{\text{term}(l)} - P_{\text{org}(l)})F_l$$

Here, term(l) is the terminating node for link l and org(l) is the originating node for link l.

It is straightforward to confirm that (ignoring losses) the total settlement residues are equal to the total merchandising surplus. Here, we use the result that (ignoring losses) the net injection at any given node is equal to the net flow out of that node:

$$\forall i, Z_i = \sum_{l \in B:\text{org}(l)=i} F_l - \sum_{l \in B:\text{term}(l)=i} F_l$$

Hence, it follows that

$$MS = -\sum_i P_i Z_i = -\sum_i \sum_{l \in B:\text{org}(l)=i} P_i F_l + \sum_j \sum_{l \in B:\text{term}(l)=j} P_j F_l$$

$$= -\sum_{l \in B} P_{\text{org}(l)} F_l + \sum_{l \in B} P_{\text{term}(l)} F_l = \sum_{l \in B} SR_l = SR$$

7.8.3 Merchandising Surplus in a Three-Node Network

To explore these results further, consider the following example. Suppose that we have a three-node, three-link network with a congested link $1 \to 2$. Suppose that there is production at node 1 and node 2 and consumption at node 3. The supply function at nodes 1 and 2 are given by $z_1(P_1) = 9P_1$ and $z_2(P_2) = P_2$. The demand function at node 3 is given by $z_3(P_3) = 2P_3 - 160$.

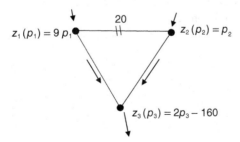

We know that at the optimal dispatch $P_1 + P_2 = 2P_3$ and $z_1 + z_2 + z_3 = 0$. If we assume the link $1 \to 2$ has a capacity of $K_{1 \to 2} = 20$, then, since the flow on this link is equal to the limit, we have that $F_{1 \to 2} = (1/3)z_1 - (1/3)z_2 = 20$. These three equations in three unknowns have a solution at $P_1 = 10$, $P_2 = 30$ and $P_3 = 20$. The corresponding injections are $z = (90, 30, -120)$. Hence, the merchandising surplus is $MS = -10 \times 90 - 30 \times 30 + 120 \times 20 = 600$.

At these prices the marginal value of expanding capacity on the link $1 \to 2$ (the constraint marginal value) is equal to $\mu_{1 \to 2} = (P_3 - P_1)/(H_{1 \to 2,1}) = 30$. The flows over the links are $F_{1 \to 2} = 20$, $F_{1 \to 3} = 70$ and $F_{2 \to 3} = 50$. Hence, the congestion revenue is $CR = \mu_{1 \to 2} K_{1 \to 2} = 30 \times 20 = 600$. The congestion revenue equals the merchandising surplus as expected.

7.9 Network Losses

Up until this point our discussion has completely ignored network losses. Of course, in practice, some electrical energy is consumed in the process of transmission and distribution of electricity (this energy is mostly lost in the form of heat). The amount of energy that is lost in this manner is roughly proportional to the square of the power flow over a line. In the case of lightly loaded lines, the transmission losses are typically small (only a few percent). However, in the case of long, heavily loaded lines, the transmission losses can amount to tens of percent – too large to be ignored in the pricing of electricity networks. Unfortunately, including losses in the simple model of the network that we have been using comes at the expense of the simplicity and elegance of the model we have present earlier.

Let us return to first principles and consider again the basic model set out at the beginning of the chapter. The notation used here will be the same as in the previous sections.

In the previous analysis in which losses were ignored, we had no reason to draw a distinction between the flow on a network element from i to j or from j to i. In the presence of losses, it is important to maintain this distinction. Specifically, we will define the $F_{i \to j}$ to be the flow on a network element as measured at node i. (Similarly, of course, $F_{j \to i}$ is the flow on the same network element measured at node j).

Let us define the link $-l$ to be the opposite direction to link l. The losses on link l are then defined to be the average of the flows in each direction:

$$L_l = \frac{F_l + F_{-l}}{2}$$

For example, suppose that 100 MW leaves node i in the direction of node j, but only 90 MW enters node j from node i. Then the loss on that link from node i to node j is $(100 - 90)/2 = 5$.

By this definition, losses are shared equally between the forwards and backwards direction: $L_l = L_{-l}$. Also, by this definition, although the flow is not the same in each direction, the flow less the losses is the same in each direction (except for the minus sign indicating the change of direction):

$$F_l - L_l = -(F_{-l} - L_{-l})$$

Note that the sum of all the flows on the network is equal to the sum of the losses:

$$L = \sum_{l \in U} L_l = \frac{1}{2} \sum_{l \in U} F_l + \frac{1}{2} \sum_{l \in U} F_{-l} = \sum_{l \in U} F_l$$

From Section 6.2 we know that at each node the total net injection is equal to the flow out from that node:

$$\forall i, Z_i = \sum_{l \in U:\text{org}(l)=i} F_l$$

Let us similarly define the total losses attributed to a particular node as follows:

$$\forall i, L_i = \sum_{l \in U:\text{org}(l)=i} L_l$$

In Section 7.8, we define the merchandising surplus as $MS = -\sum_i P_i Z_i$. In a similar manner, let us define the *loss surplus* as

$$LS = \sum_i P_i L_i$$

7.9.1 Losses, Settlement Residues and Merchandising Surplus

In the presence of losses the settlement residues can be defined as follows (compare with the definition in Section 7.8):

$$SR = \frac{1}{2} \sum_{l \in U} SR_l = \frac{1}{2} \sum_{l \in U} \left(P_{\text{term}(l)} - P_{\text{org}(l)} \right) F_l$$

In Section 7.8, we saw that, ignoring losses, the merchandising surplus is equal to the settlement residues. When losses are present, the merchandising surplus is equal to the sum of the settlement residues less the loss surplus:

$$MS + LS = SR$$

$$MS = -\sum_i P_i Z_i = \sum_i \sum_{l \in U:\text{org}(l)=i} P_i F_l$$

$$= -\frac{1}{2} \sum_i \sum_{l \in U:\text{org}(l)=i} P_i F_l + \frac{1}{2} \sum_i \sum_{l \in U:\text{org}(-l)=i} P_i F_{-l}$$

$$- \sum_i \sum_{l \in U:\text{org}(l)=i} P_i L_l$$

$$= \frac{1}{2} \sum_i \sum_{l \in U:\text{term}(l)=i} P_i F_l - \frac{1}{2} \sum_i \sum_{l \in U:\text{org}(l)=i} P_i F_l - \sum_i P_i L_i$$

$$= \frac{1}{2} \sum_{l \in U} (P_{\text{term}(l)} - P_{\text{org}(l)}) F_l - \sum_i P_i L_i = SR - LS$$

To illustrate this, consider again a simple two-node network. The flow from node 1 to node 2 is 100, and the flow from node 2 to node 1 is −90. Therefore, by the earlier definitions, the loss attributed to the flow from node 1 to node 2 is 5. Similarly, the loss attributed to the flow from node 2 to node 1 is 5. The loss surplus is $LS = 5 \times 10 + 5 \times 100 = 550$. The net injection at node 1 is 100 and the net injection at node 2 is -90. The merchandising surplus is therefore $MS = -10 \times 100 + 90 \times 100 = 8000$. The settlement residue is equal to $SR = \frac{1}{2}(100 - 10) \times 100 + \frac{1}{2}(10 - 100) \times -90 = 8550$. As expected we find that $MS + LS = SR$.

$$P_1 = \$10 \qquad\qquad\qquad P_1 = \$100$$
$$1 \bullet\!\!\!-\!\!\!-\!\!\!-\!\!\!-\!\!\!-\!\!\!-\!\!\!-\!\!\!-\!\!\!-\!\!\!\bullet 2$$
$$F_{1\to2} = 100 \qquad\qquad\qquad F_{2\to1} = -90$$

7.9.2 Losses and Optimal Dispatch

In the case where losses are ignored, the sum of the net injections is zero. Now, when we take losses into account, we find that the sum of the net injections is equal to the sum of the losses attributed to each node:

$$\sum_i Z_i = \sum_i \sum_{l \in U : \text{org}(l) = i} F_l = \sum_{l \in U} F_l = \sum_{l \in U} L_l = \sum_i L_i = L$$

To keep things simple, let us ignore the physical limits on the network. The optimal dispatch problem is therefore to find the set of net injections that solves the following problem:

$$\max \sum_i B_i(Z_i)$$

subject to

$$\sum_i Z_i = L(Z) \leftrightarrow \lambda$$

This problem has the following KKT condition:

$$B_i'(Z_i) + \lambda\left(1 - \frac{\partial L}{\partial Z_i}\right) = 0$$

Alternatively, using the fact that the common marginal cost and marginal value at each node is equal to the negative of the nodal price

$$P_i = \lambda\left(1 - \frac{\partial L}{\partial Z_i}\right)$$

We can distinguish between (a) nodes where a small additional injection increases the total losses on the network $(\partial L/\partial Z_i) > 0$ and (b) nodes where a small additional injection reduces the total losses on the network $(\partial L/\partial Z_i) < 0$. In the optimal dispatch, nodes where a small additional injection increases the total losses on the network are priced at a discount; nodes where a small additional injection increases the total losses are priced at a premium.

7.10 Summary

We can partition the set of producers and consumers by node, and define a net injection and benefit function at each node. The problem of efficient use of a set of production and consumption resources (referred to as optimal dispatch) introduced in Chapter 5 can then easily be extended to incorporate network constraints. In the presence of network constraints, although there is a common marginal cost and marginal value at each node, these values may differ across nodes in a manner that depends on the PTDF matrix and the marginal value of binding constraints.

Furthermore, we can decentralise the process of finding the optimal dispatch using a smart market. Following the fundamental principle of smart markets, if every producer and consumer is a price-taker and provided the network constraints in the optimisation correctly reflect the true network constraints, the resulting dispatch is efficient and the price at each node is equal to the common marginal cost and marginal value at that node. These prices, known as nodal prices or locational marginal prices, bear a strict relationship to each other, which depends only on the PTDF matrix and the marginal value of the binding constraints.

When no network constraints are binding, and ignoring network losses, the prices arising from the optimal dispatch are the same at every node. When constraints arise, the network can be thought of as divided up into separate pricing regions. For a fixed set of production and consumption assets, and a single constrained link, it is possible to draw a diagram that shows how optimal dispatch and prices vary with the limits on the constrained link.

In a radial network, prices at any two nodes are the same unless there is a binding constraint on the path between them, and the price increases across a congested network element in the direction of power flow (power always flows from low-priced regions to high-priced regions). Furthermore, there is always one more pricing region than network constraints (a single constraint divides the network into two pricing regions).

These results do not extend to meshed networks. In a meshed network, prices are the same unless there is a binding constraint on any path between them. Power does not always flow from low-priced nodes to high-priced nodes. In fact, in simple networks it is common for counter-price flows to arise. There is no longer a simple relationship between the number of constraints and the number of pricing regions. A single binding constraint can, in theory, introduce separate prices at every node in the network.

In a meshed network, a single binding network constraint may give rise to as many different prices as there are nodes in the network. However, those prices are not independent. In fact if there is a single binding constraint in the network there are only two independent prices in the network (all prices can be represented as a linear combination of these prices). More generally, if there are n binding network constraints, there are $n + 1$ independent prices.

When prices differ across nodes, the system operator receives a cash-flow stream from transacting in electricity known as the merchandising surplus. For each binding network constraint, we can define the congestion rent for that constraint as equal to the constraint marginal value multiplied by the physical limit for that constraint. If we ignore losses, at the optimal dispatch the merchandising surplus and the total congestion rents are the same. Since the congestion rent is non-negative, the merchandising surplus is also non-negative. Moreover, any other (nonoptimal) dispatch will result in a merchandising surplus that is less than the congestion rent.

For any network link, we can also define the settlement residues as the price difference across the link multiplied by the flow on the link. If we ignore losses, the total settlement residues are equal to the merchandising surplus.

If we take losses into account the flow on a link in one direction is not the same as the flow on the link in the opposite direction. We can define the losses on a link as the average of the flows in each direction. We can also define the losses associated with each node as the sum of losses on the links originating at that node. The loss surplus is the sum of the nodal price multiplied by the losses attributed to that node. We find that the merchandising surplus is now equal to the sum of the settlement residues and the loss surplus.

Questions

7.1 In the case of the simple three-node, three-link network in which the links have identical electrical characteristics, prove the result that when there is a single constraint binding, the price at the node opposite the binding link is equal to the average of the prices at the end of the binding link.

7.2 In much modelling of electricity markets it is common to assume that generators have a marginal cost that is equal to some fixed value (known as the variable cost) up to the capacity of the generator. Suppose that demand is inelastic, that each generator submits an offer curve equal to its marginal cost curve and, further, suppose that each generator has a unique variable cost. In the absence of transmission constraints, in such a market the wholesale spot price is always equal to the variable cost of some generator in the market – known as the 'marginal generator'. Does this remain true when transmission constraints are binding? In other words, is it the case that the wholesale spot price at any given node is always equal to the variable cost of some generator in the market? Can it be the case that the wholesale spot price at a node might be *above* the variable cost of any generator in the market?

7.3 Suppose that for a sequence of nodes in the network, the value of the PTDF matrix H_{li} is decreasing. Prove that, when the link l is congested, the prices on that sequence of nodes are increasing. How does this relate to the 'spring washer' effect?

7.4 Compute the merchandising surplus for the simple three-node network example in Section 7.3. How does the merchandising surplus vary with the capacity of the constrained transmission link?

7.5 Can the merchandising surplus be used to reward socially efficient transmission augmentation decisions? Specifically, if we augment the transmission network by some amount, does the change in the merchandising surplus correspond to the change in the social welfare arising from that augmentation?

7.6 Let us suppose that the nodal prices in a network happen to be negative. Under optimal dispatch should the system operator seek to maximise the network losses? Why or why not?

Further Reading

There is a simple introduction to nodal pricing in Stoft (2000). For the full theory you may like to look at the original work by Bohn *et al.* (1984) and Schweppe *et al.* (1988). It is also worth taking a look at Hogan (1992).

8

Efficient Network Operation

The efficient short-run operation of an electric power system is not just a matter of efficient production and consumption decisions. There may also be decisions to be made about the use of the network itself. In practice, in an AC power network there are relatively few decisions to be made by the network operator. Although there are some devices capable of altering power flows on different lines, electric power generally flows around the network according to the laws of physics. However, there are two important decisions to be made regarding network operation. The first arises in the case of DC interconnectors, where a decision must be made as to how much power to transfer from one location to another. The second relates to the configuration of the network itself. Switching network elements in or out of service can alter power flows in a manner that increases economic welfare.

8.1 Efficient Operation of DC Interconnectors

As we noted in Section 2.6, one of the key features of DC links is that they are controllable. Let us imagine that we have, in addition to the conventional AC network, a network of DC links. How should these links be operated so as to achieve the highest short-run economic welfare?

Let us focus on the DC link from node i to node j. The controllable flow on this link in the direction from i to j is $F_{i \to j}$ (MW) and the capacity of the link is assumed to be $K_{i \to j}$. As before, we will ignore losses on the DC link. There may be many such links in the overall network, but we need only focus on one at a time.

The DC link is like an additional source of consumption at node i and production at node j, so we can add an additional term to the definition of the net injection at node i:

$$Z_i = \sum_{k \in N_i} Q_k^S - \sum_{k \in N_i} Q_k^B - F_{i \to j}$$

Similarly, the net injection at node j satisfies

$$Z_j = \sum_{k \in N_j} Q_k^S - \sum_{k \in N_j} Q_k^B + F_{i \to j}$$

The Economics of Electricity Markets, First Edition. Darryl R. Biggar and Mohammad Reza Hesamzadeh.
© 2014 John Wiley & Sons, Ltd. Published 2014 by John Wiley & Sons, Ltd.

We can now write the benefit function at node i as $B_i(Z_i + F_{i \to j})$ and similarly, the benefit function at node j can be written as $B_j(Z_j - F_{i \to j})$.

Importantly, there are no other changes to the key constraints in the central optimisation task set out in Section 7.1. Specifically, the energy balance constraint remains exactly as before: $\sum_i Z_i = 0$ (since the net injection of the DC link is zero), and the generic AC network constraints remain the same: $\forall l, \sum_i H_{li} Z_i \leq K_l$ (since the DC link does not affect flows on the AC network, except through its impact on the net injection).

The task of the system operator is now to find the combination of net injection at each node Z_i and the DC link flow $F_{i \to j}$ to maximise the short-run economic welfare:

$$W = B_1(Z_1) + \cdots + B_i(Z_i + F_{i \to j}) + \cdots + B_j(Z_j - F_{i \to j}) + \cdots$$

Subject to the following conditions:

a. The energy balance equation $\sum_i Z_i = 0 \leftrightarrow \lambda$ and the generic network constraints $\forall l \in U, \sum_i H_{li} Z_i \leq K_l \leftrightarrow \mu_l$; and
b. The limits on the flows on the DC link $F_{i \to j} \leq K_{i \to j} \leftrightarrow \zeta$ and $F_{i \to j} \geq -K_{i \to j} \leftrightarrow \eta$

From the KKT conditions, we have $B_i'(Z_i + F_{i \to j}) - B_j'(Z_j - F_{i \to j}) + \zeta - \eta = 0$ and $\zeta \geq 0$, $\eta \geq 0$, and $\zeta(K_{i \to j} - F_{i \to j}) = 0$ and $\eta(K_{i \to j} + F_{i \to j}) = 0$. From Section 7.1, we know that $B_i'(\cdot)$ is equal to the nodal price at node i.

Therefore, we can conclude that the optimal dispatch of a DC link (ignoring losses) is as follows: To run the link at its maximum in the forward direction when the price at the terminating node is larger than the price at the originating node; to run the link at its maximum in the backward direction when the price at the terminating node is smaller than the price at the originating node; and indeterminate when the two prices are the same:

$$F_{i \to j} = \begin{cases} K_{i \to j}, & P_j > P_i \\ -K_{i \to j}, & P_j < P_i \\ \text{indeterminate} & P_j = P_i \end{cases}$$

Moreover, let us suppose (without loss of generality) that the DC link is operating at its maximum in the forward direction. In this case the additional economic welfare arising from a small increase in the link capacity is simply equal to the price difference across the DC link:

$$\frac{\partial W}{\partial K_{i \to j}} = \zeta = P_j - P_i$$

We know from previous analysis that the price difference between two nodes in an AC network depends on the presence of congestion on the network: $P_j - P_i = \sum_l (H_{li} - H_{lj})\mu_l$. From this we can conclude that a DC link in an AC network only creates economic value to the extent that the AC network experiences congestion (and vice versa).

> *Result*: The efficient operation of a lossless DC link depends only on the price differences at either end of the link. If the price difference is positive, the link should be run at its maximum in the forward direction. If the price difference is negative, the link should be run at its maximum in the backward direction. When a price difference exists, the additional welfare from increasing the DC link capacity by a small amount is equal to the price difference.

8.1.1 Entrepreneurial DC Network Operation

Can we decentralise decisions regarding the short-run operation of DC links? Let us suppose that we allow the owner of a DC link to capture the settlement residues arising on the link. The profit of the DC-link entrepreneur is then as follows:

$$\pi\left(F_{i\to j}\right) = \left(P_j - P_i\right)F_{i\to j}$$

The task of the DC-link entrepreneur is to maximise this profit subject to the flow limits on the DC-link. Let us assume that this link is so small that it has no impact on the wholesale spot price at each node. In other words, let us assume that the DC-link is a price-taker.

It is straightforward to verify that the profit-maximising choice of flow is (as before) to run the link at its maximum in the forward direction when the price difference is positive and to run the link at its maximum in the backward direction when the price difference is negative.

> *Result*: A profit-maximising, price-taking owner of a DC network link who receives the settlement residues for flows on the link will choose the efficient flow over that DC link.

8.2 Optimal Network Switching

Up until this point we have taken the configuration of the network as given. But in practice, network operators have a degree of control over the configuration of the network. In particular, they can switch certain network elements in and out of service.

We have seen in Section 6.5 how changing the configuration of the network can affect the set of feasible injections. Now let us ask the question: Can we change the configuration of the network in order to increase short-run economic welfare?

Let us suppose that we have a set of network configurations, indexed by θ. Under network configuration θ, the PTDF matrix is assumed to be H_{li}^{θ}. The task of the network operator is to solve the following problem:

$$W = \max_{Z_i, \theta} \sum_i B_i(Z_i)$$

Subject to the following conditions:

a. The energy balance equation $\sum_i Z_i = 0 \leftrightarrow \lambda$ and
b. The generic network constraints $\forall l \in U, \sum_i H_{li}^\theta Z_i \leq K_l \leftrightarrow \mu_l$

It turns out that we may be able to obtain a higher economic welfare by changing the shape of the feasible set by switching network elements in and out of service.

Although the dispatch task presented is reasonably complex in abstract it is possible to make a few limited comments about the nature of optimum. Specifically, let us suppose that we have a set of injections Z_i and a set of prices P_i. Then, if these set of injections and set of prices are the solution to the dispatch task, then it follows (from the result in Section 7.8) that for any other network configuration, the merchandising surplus for any other feasible set of injections Z_i' in that network configuration must be less than or equal to the merchandising surplus at the optimal set of injections. In other words, for any feasible injection Z_i' in network configuration θ:

$$MS(P, Z') = -\sum_i P_i Z_i' = -\sum_i P_i Z_i = MS(P, Z)$$

One implication is that we can test whether or not we have the optimal network configuration: Let us suppose that we have an optimal dispatch in a given network configuration, giving rise to a set of prices P_i and a set of injections Z_i. If there is some other set of injections Z_i', which is feasible in some other network configuration and which yields a higher merchandising surplus (valued at the same prices), then the original dispatch Z_i cannot be optimal overall.

We have demonstrated that at the overall optimum the merchandising surplus is maximised *for a given set of prices*. It is important to be aware that the merchandising surplus is not maximised overall. There could well be other network configurations that yield a higher merchandising surplus. As we will see shortly, this is important when it comes to determining whether we can decentralise the network switching task.

8.2.1 Network Switching and Network Contingencies

In the next part of this text we will discuss the efficient handling of short-term shocks or contingencies to the power system. In practice, many power systems (including the Australian National Electricity Market or NEM) are not able to efficiently respond to network contingencies, such as the outage of a network element. As a consequence (as we will see) the network must be operated at all times as though the contingency has already happened. A consequence of this is that there is substantial redundancy in network capacity and substantially reduced scope for network switching. However, we consider that, in the future, with improved handling of short-run contingencies, there will be substantially greater scope for network switching to form part of day-to-day operations of electric power systems.

8.2.2 A Worked Example

It is easier to gain an insight into how network switching may operate through a simple example.

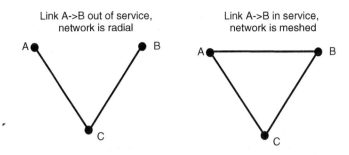

Figure 8.1 Alternative network configurations

Let us suppose that we have a simple three-node network. There are generators at nodes A and B and load at node C. There are network links from A to C and from B to C. There is a third network link, from A to B that can be switched in and out of service. When the link from A to B is out-of-service, the network is radial and takes the form in the diagram on the left of Figure 8.1. When the link from A to B is in service, the network is meshed and takes the form in the diagram on the right of Figure 8.1. As before, to keep things simple we will assume that all the network links have identical electrical characteristics. The capacity of the links from A to C and B to C is 1000 MW each. The link from A to B has a capacity of 100 MW.

The set of feasible injections with and without the link from A to B in service is shown in Figure 8.2.

Let us assume we have a set of generators at nodes A and B. These generators have a constant variable cost and fixed capacity as set out in Table 8.1.

Now let us compute the optimal dispatch for this network with and without the link A–B in place.

For load levels up to 1700 MW, higher economic efficiency can be achieved by leaving link A–B out of service. Intuitively, the reason is that the relatively low capacity link A–B limits the extent to which injections at A and B can depart from each other. Since all of the low-cost generation is at A, for low load levels we would like to draw on generation at A exclusively,

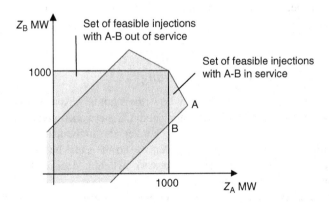

Figure 8.2 Set of feasible injections with and without link A → B in service

Table 8.1 Generator cost data for simple network switching example

Generator	Node	Variable Cost ($/MWh)	Capacity (MW)
G1	A	$10	500
G2	A	$20	500
G3	A	$50	500
G4	B	$100	500
G5	B	$200	500
G6	B	$500	500

leaving generation at B unused. However, as the load levels increase, eventually we reach the capacity limit on flows from A to C. At this point, further increases in load require additional generation from node B. At the load level of 1700 MW, the optimal dispatch is point B in Figure 8.2.

Now suppose we increase the load level to 1900 MW. If we leave link A–B out of service, we must draw 1000 MW from generation at A and 900 MW from generation at B. The total dispatch cost is $145 000/h. But we can do better. By returning link A–B to service we can change the dispatch, drawing more of the cheaper generation (1100 MW) from node A and less of the expensive generation (800 MW) from node B. The dispatch cost drops to $130 000/h. Even though the link A–B is out-of-service for most load levels, at times when the load reaches 1900 MW the link A–B is worth an additional $15 000/h to the system.

Let us suppose the load is 1900 MW. With link A–B out of service, the prices are $P = (50, 200, 200)$ and the net injection is $Z = (1000, 900, -1900)$, so the merchandising surplus is $150 000/h. But with the link A–B in service there is another dispatch $Z' = (1050, 850, -1900)$, which is also feasible. At the same prices this yields a merchandising surplus of $157 500/h. Since this merchandising surplus is higher, the original dispatch choice could not be optimal.

Result: Network switching can result in materially higher economic welfare. A given network configuration is not optimal if there is some other set of net injections that is feasible in some network configuration that yields a higher merchandising surplus valued at the current prices.

8.2.3 Entrepreneurial Network Switching?

In the previous sections, we have shown how the optimal dispatch task can, in part, be decentralised to profit-maximising generators and DC network operators. In each case we saw that provided the entrepreneur was a price-taker, the individual decisions can be made consistent with the overall social optimum. Can we do the same here? Is there some reward mechanism that allows us to decentralise the network switching decision to a profit-maximising network operator?

The answer seems to be no. The reason is simple. The impact of a generator or a DC link on the overall market outcomes can be quite small. For small generators or small consumers, it is not unreasonable to assume that these parties are price takers. However, the same is not true for a network switching decision. A network switching decision may have a very profound impact on prices.

For example, in the earlier simple network, with the load at 1900 MW, switching A–B into service results in prices $P = (50, 200, 125)$. The new optimal net injection is $Z = (1100, 800, -1900)$. However, with these prices, the merchandising surplus is only \$22 500/h. Compared to the case where A–B is out of service the overall economic welfare has improved but the merchandising surplus is much lower. A profit-maximising network switcher, which is rewarded with the merchandising surplus, would not make this choice.

Having said that, there still remains scope for the design of incentive schemes on a regulated network service provider. For example, let us suppose that we know the set of prices and dispatch that would arise in a given network scenario under normal operations. We may want to create incentives for the network service provider to discover new ways of operating the network that yield more economic value. We could reward the service provider for discovering new network switching possibilities by rewarding them the difference in the merchandising surplus valued at the previous optimal prices.

For example, let us suppose that in the previous example network, the possibility of switching link A–B into service was not a normal operational process, and let us suppose that the regulator sought to reward the network service provider for discovering new such possibilities. As we saw earlier, at the optimal dispatch with link A–B out of service, the prices are $P = (50, 200, 200)$ and the merchandising surplus is \$150 000/h. Now suppose that the network service provider discovers the switching possibility. The new optimal dispatch is $Z = (1100, 800, -1900)$, which yields a merchandising surplus (when valued at the old prices) of \$165 000/h. The network service provider could be rewarded with this change in the merchandising surplus which, in this case, is \$15 000/h – exactly equal to the social benefit from network switching.

Result: A network service provider that receives the merchandising surplus will not have an incentive to choose an efficient network configuration. However, we could envisage an incentive scheme in which the network service provider is rewarded for discovering new network configurations by receiving the change in the merchandising surplus valued at the current prices.

8.3 Summary

Efficient short-term operation of an electricity network is not merely a matter of efficient production and consumption decisions. There can also be decisions regarding the operation of the network.

The simplest network operation decisions involve the operation of a DC link. In the case of a lossless DC link, the efficient operation involves operating the link at its capacity in the forward direction when the price at the terminating node is above the price at the originating node and vice versa when the price at the terminating node is below the price at the originating node. Furthermore, we can decentralise this decision to a profit-maximising DC network operator, by allowing the DC network operator to keep the settlement residues on the DC link. In this case, provided the network operator is a price-taker, the network operator will have an incentive to operate the link efficiently.

In the case of an AC network, a change in the network configuration can have a substantial effect on the shape of the set of feasible injections. In certain circumstances it may make economic sense to switch network elements in or out of service in order to increase economic welfare. In the text we discussed an example that involved switching between a radial and a meshed network configuration.

A set of injections and prices that are optimal in a particular network configuration are only optimal overall if there is no other injection that is feasible in some network configuration that yields a higher value for the merchandising surplus when valued at the original prices.

It is not possible to simply reward a network operator on the basis of the merchandising surplus. The reason is that such a network operator would not be a price-taker. However, we could envisage an incentive scheme for an AC network operator that provides rewards based on changes in the merchandising surplus holding the prices constant.

Specifically, provided we know the prices and dispatch at the current optimal dispatch outcome, we could reward the network operator for finding new network configurations for which there is a set of injections that yields a higher merchandising surplus when valued at the original prices.

Questions

8.1 If a DC-link entrepreneur is considering creating a new for-profit DC link in an AC network, which two nodes should the entrepreneur be seeking to connect with the DC link? To what extent does the entrepreneur's profit depend on investment decisions in the AC network?

8.2 Let us suppose we have a given network configuration and a corresponding set of prices and optimal dispatch. How do we know if there is some other network configuration that might yield a higher economic welfare?

8.3 Is it possible to design a mechanism that might reward a network service provider for changing the network configuration when it is socially beneficial to do so?

Further Reading

See Ventosa *et al.* (2005).

Part IV

Efficient Investment in Generation and Consumption Assets

Interior of Freeman's Bay power station, New Zealand, ca 1908 (Source: Neil Rennie)

The previous part of this book dealt with the efficient use or operation of a stock of generation, consumption and network assets. But as we emphasised in Part I, efficient use of the existing stock of assets is only part of what we require for economic efficiency. It is just as important to achieve efficient investment in new assets – efficient investment in generation assets and efficient investment in new consumption assets. Do decentralised smart markets yield efficient signals for long-term investment? Can markets deliver the right amount of investment in the right location, at the right time?

9

Efficient Investment in Generation and Consumption Assets

In this chapter, we focus on the characteristics of investment in generation assets and, to an extent, in buyer-side consumption assets. In Chapter 10, we go on to ask whether or not this outcome can be achieved in a competitive market.

9.1 The Optimal Generation Investment Problem

We will start by setting out the general problem of optimal generation investment. In the following sections we will clarify aspects of this model by looking at special cases.

The problem of optimal generation investment is a problem of choosing the right amount of investment, of the right type, in the right location, at the right time. We will assume that we have a range of different generation types available at each location. The optimal generation investment problem is then a question of choosing the optimal capacity of each generation type at each location.

Importantly, this is inherently a *long-term* decision. The capacity of a generator is assumed to be fixed in the short-run. We therefore need a model that makes a distinction between short-run and long-run.

We will model short-run effects as follows. We have seen that the willingness of generators to produce can be represented in the cost function $C_i(Q_i^S)$ and the willingness of customers to consume can be represented in the demand function $U_i(Q_i^B)$. In the previous chapters these were taken as fixed. Now we will imagine that there are short-term factors that affect these demand and supply conditions. Specifically, we will assume that the utility functions are given by $U_i(Q_i^B, \varepsilon_i^B)$, where ε_i^B are short-term utility-shifting factors (such as changes in demand, perhaps due to changes in the weather). Similarly, the cost functions will be given by $C_i(Q_i^S, \varepsilon_i^S)$, where ε_i^S are short-term cost-shifting factors (such as outages or changes in input fuel costs). (Later we will also allow for the possibility of short-term factors that affect network capacity).

We will model the utility-shifting factors and cost-shifting factors in the short term as random variables. Specifically, let us define a state of the system as a specific realisation of these random variables, $s = (\varepsilon^S, \varepsilon^B)$. A given short-run state of the network s is assumed to

The Economics of Electricity Markets, First Edition. Darryl R. Biggar and Mohammad Reza Hesamzadeh.
© 2014 John Wiley & Sons, Ltd. Published 2014 by John Wiley & Sons, Ltd.

occur with probability p_s. We will use the notation $U_{is}(Q_i^B)$ and $C_{is}(Q_i^S)$ to refer to the utility function and cost functions in state s, respectively.

For the moment let us make the important assumption that there are no sunk costs or lumpiness in generation investment. This greatly simplifies the analysis by allowing us to ignore the past and the future. We can focus on finding the optimal mix of generation investment in any time period, independent of investment in the past or the path of investment in the future.

Let us assume that at each location i, we have a range of generation types indexed by t. The capacity of generation type t at location i is assumed to be K_{it} and the rate at which expenditure is incurred when producing at the rate Q_{it}^S is $C_{it}(Q_{it}^S, K_{it})$. The capacity of a generator is assumed to be a hard limit on the rate of production in the short-run so that $Q_{it}^S \leq K_{it}$.

Let us focus first on the short-run optimal dispatch decision. Given the capacity of each generation type at location i, K_{it}, in state of the world, s, the benefit function at location i is defined as follows:

$$B_{is}(Z_{is}, K_{it}) = \max_{Q_{its}^B, Q_{its}^S} \sum_t U_{is}(Q_{its}^B) - \sum_t C_{its}(Q_{its}^S, K_{it})$$

$$\text{subject to } Z_{is} = \sum_t Q_{its}^S - \sum_t Q_{its}^B \leftrightarrow \alpha_{is}$$

$$\forall t, 0 \leq Q_{its}^S \leftrightarrow \sigma_{its}$$

$$\forall t, Q_{its}^S \leq K_{it} \leftrightarrow \tau_{its}$$

From the KKT conditions, we have that

$$\frac{\partial B_{is}}{\partial K_{it}} = \tau_{its} - \frac{\partial C_{its}}{\partial K_{it}}$$

In addition, $\partial C_{its}/\partial Q_{its}^S = \alpha_{is} - \tau_{its} + \sigma_{its}$ and $\tau_{its}(K_{it} - Q_{its}^S) = 0$, so we can write the constraint marginal value on the generator production constraints as follows:

$$\tau_{its} = \left(\alpha_{is} - \frac{\partial C_{its}}{\partial Q_{its}^S} \right) I(K_{it} = Q_{its}^S)$$

The overall problem of finding the optimal dispatch and the optimal mix and level of investment is as follows:

$$W(K_l) = \max_{Z_{is}, K_{it}} E\left[\sum_i B_{is}(Z_{is}, K_{it}) \right] = \max_{Z_{is}, K_{it}} \sum_{i,s} p_s B_{is}(Z_{is}, K_{it})$$

subject to the following:

$$\forall s, \sum_i Z_{is} = 0 \leftrightarrow p_s \lambda_s$$

$$\forall l, s, \sum_i H_{li} Z_{is} \leq K_l \leftrightarrow p_s \mu_s$$

From the KKT conditions, the condition for the optimal capacity of generation is as follows:

$$\sum_s p_s \left(\alpha_{is} - \frac{\partial C_{its}}{\partial Q^S_{its}} \right) I\left(K_{it} = Q^S_{its} \right) = \sum_s p_s \frac{\partial C_{its}}{\partial K_{it}}$$

Result: The level of investment in the generation of a given type at a given location is optimal when the expected gap between the system marginal cost at that location and the marginal cost of generation (at times when the generator is capacity-constrained) is equal to the expected cost of adding generation capacity of that type at that location.

9.2 The Optimal Level of Generation Capacity with Downward Sloping Demand

The result derived in the following section is quite general. To gain more insight into what this means, let us assume that generators have a simple cost function, with a constant variable cost up to the total capacity and a constant cost of adding capacity.

In addition, just for this section, we will also assume that there is only one type of generation (and all generators of this type have the same variable cost). This allows us to set aside the task of choosing the optimal mix of different types of generation to focus on the task of choosing the optimal total capacity of generation.

As before, in order to draw a distinction between the long-run and the short-run, we must define what is fixed and what varies in the short-run and long-run. As in Section 1.6 we will assume that demand is uncertain in the short-run (i.e. demand varies faster than it is possible to vary the capacity of generators). In the state of the world, s, consumers are assumed to consume at the rate Q_s and the utility of consumers is assumed to be $U_s(Q_s)$.

By our assumptions, the cost of production of the generators varies linearly with the rate of production, up to the total capacity, which we will label K (MW). We will suppose that an additional unit of capacity costs f ($/MW). In other words, the cost function of the generators is assumed to take the following form:

$$C(Q, K) = cQ + fK \quad \text{for} \quad 0 \le Q \le K$$

For simplicity we will assume that this cost function is fixed and certain and therefore does not vary with the state of the world.

What is the socially efficient level of capacity K? We can answer this question by solving an optimisation problem, which is a simplification of the problem in Section 9.1:

$$\max \sum_s p_s [U_s(Q_s) - C(Q_s, K)]$$

$$\text{subject to } Q_s \le K \leftrightarrow p_s$$

Here, p_s is the probability of state s arising. (We have ignored the case where the demand is so low that it is desirable to not produce or consume anything at all). The KKT conditions for a maximum of this problem are as follows:

$$\mu_s = U'_s(Q_s) - c \text{ and } \mu_s(K - Q_s) = 0 \text{ and } \sum_s p_s \mu_s = f$$

In other words, the optimal dispatch has the property that, in the short-run, the rate of production is where the demand curve intersects the variable cost (i.e. $U'_s(Q_s) = c$) provided that the rate of production is less than the total capacity (i.e. provided $Q_s < K$). When the rate of production is equal to the total capacity (i.e. when $Q_s = K$) the marginal valuation can be above the variable cost. This can be expressed as follows:

$$\sum_s p_s \left(U'_s(Q_s) - c \right) I(Q_s = K) = f$$

In other words, the optimal level of capacity has the property that the average distance between the marginal valuation and the variable cost (at times when the generator is operating at its capacity) is equal to the cost of adding a unit of capacity. This is illustrated in Figure 9.1.

In this chapter, we are focusing on the determination of optimal investment outcomes rather than whether this can be achieved through decentralised decisions in a competitive market. But we know from previous chapters that in a competitive market, the market price is equal to the common marginal cost and marginal value at the optimum dispatch. In other words, the price in state s satisfies $P_s = U'_s(Q_s)$. It therefore follows that at the optimal level of generation capacity, the average difference between the price and the variable cost is equal to the per-unit cost of adding capacity:

$$\sum_s p_s (P_s - c) I(P_s > c) = f$$

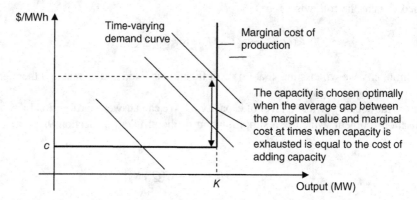

Figure 9.1   Determination of the optimal level of capacity of generation of a particular type

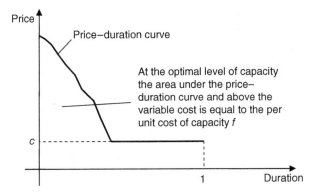

Figure 9.2 Determination of the optimal level of capacity in a system with a single type of generation

We can illustrate this using a price–duration curve. In this market, when demand is low the market price is equal to the variable cost of generation, c. When demand is high the price can increase above the variable cost of generation. When the generation capacity is optimal the area under the price duration curve and above the variable cost is equal to the per-unit cost of expanding capacity. This is illustrated in Figure 9.2.

9.2.1 The Case of Inelastic Demand

Let us focus for a moment on the special case in which the demand curve takes the special shape set out in Figure 5.4 discussed in Section 5.5 – that is, where the demand is vertical up to some threshold (denoted V) at which point the demand drops to zero.

With a demand curve of this form, and a single generator type with variable cost, c, the market price is either equal to the variable cost, c, when the total volume demanded is less than or equal to the total generation capacity, or is equal to V when the total volume demanded exceeds the generation capacity (in which case load is voluntarily reduced).

In this simple case, the earlier condition states simply that, at the optimal level of capacity, the difference between V and the variable cost, c, multiplied by the probability that load shedding is required, is equal to the unit cost of adding capacity:

$$(V - c)\Pr(Q_s \geq K) = f$$

We can define the loss of load probability (LOLP) as the probability that voluntary load shedding will be required. This clearly depends on the volume of installed capacity K. Using the equation at the optimal level of capacity with this simple demand curve, the loss of load probability satisfies the following:

$$\text{LOLP}(K) = \Pr(Q_s \geq K) = \frac{f}{V - c}$$

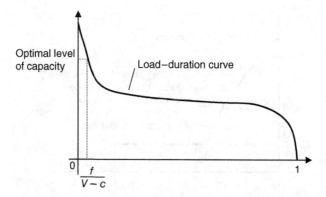

Figure 9.3 Deriving the optimal level of generation capacity from the load–duration curve

Note that:

- The higher the value of consumption to customers (the higher V is), the lower the optimal probability of load shedding;
- An increase in either the cost of adding generation capacity, f, or the variable cost of generation, c, reduces the optimal level of capacity;
- As long as there is some fixed cost of holding extra capacity (i.e. as long as $f > 0$), it will not be efficient to choose a level of capacity that guarantees 100% reliability (that is, the probability that load will be left unserved is not zero). In other words, it will, in general, not be efficient to choose a level of generation capacity sufficient to serve all load under all circumstances.

> *Result*: In an optimal dispatch with inelastic demand up to some valuation V, at the optimal level of generation capacity, there will remain a finite probability that some load will be left unserved.

Note also that in this simple case the optimal level of generation capacity can be obtained from the load–duration curve. Given a loss of load probability of $f/(V - c)$, the optimal level of generation capacity is the corresponding point on the load–duration curve as illustrated Figure 9.3.

9.3 The Optimal Mix of Generation Capacity with Downward Sloping Demand

Section 9.2 focused on a very strong simplification in which it was assumed that generation was all of one type, with a single variable cost, c. In reality, of course, there is a wide range of generation types with different fixed and variable costs. A key problem for the hypothetical system planner is to choose the efficient mix of different generation types. Let us now generalise the previous results to the case of many different types of generation.

Specifically, let us assume that when there are K_t units of capacity of generation of type t, and when that type of generation is producing at the rate Q_t, costs are incurred at the rate

$$C_t(Q_t, K_t) = c_t Q_t + f_t K_t \quad \text{for} \quad 0 \le Q_t \le K_t$$

The social optimisation problem is as follows:

$$\max \sum_s p_s \left(U_s \left(\sum_t Q_{ts} \right) - \sum_t C_t(Q_{ts}, K_t) \right)$$

subject to $Q_{ts} \le K_t \leftrightarrow p_s \mu_{ts}$ and $Q_{ts} \ge 0 \leftrightarrow p_s \nu_{ts}$

The KKT conditions for a maximum of this problem are as follows:

$$\forall t, s, \mu_{ts} - \nu_{ts} = U_s'(Q_s) - c_t$$

$$\mu_{ts} \ge 0 \text{ and } \nu_{ts} \ge 0$$

$$\mu_{ts}(K_t - Q_{ts}) = 0 \text{ and } \nu_{ts} Q_{ts} = 0 \text{ and}$$

$$\sum_s p_s \mu_{ts} = f_t$$

We know from the previous chapters what the optimal dispatch looks like in this situation. For any given total level of production, the generators with the lowest marginal cost are chosen in order from lowest variable cost to highest (the 'merit order'). Each generator with a lower variable is used up to its capacity before a generator with a higher variable cost is called on.

From the KKT conditions, we can see that the level of capacity of each generator type is optimal when the expected gap between the marginal value of the customer and the variable cost of that generator (which is only positive when that generator is operating at capacity) is equal to the unit cost of adding capacity to that generator type. Furthermore, as we know, in a competitive market, the market price will be equal to the marginal value of each consumer. By the earlier formulae we can derive an expression for the relationship between the average price above a threshold and the fixed and variable costs of a generator.

From the KKT condition we see that, as before, the average gap between the price and the variable cost when the price is above the variable cost must be equal to the unit cost of adding capacity:

$$\sum_s p_s \mu_{ts} = \sum_{s:P_s \ge c_t} p_s(P_s - c_t) I(P_s > c_t) = \sum_{s:P_s \ge c_t} p_s(P_s - c_t)$$

$$= E[(P - c_t) I(P \ge c_t)] = f_t$$

Furthermore, the average gap between the price and the variable cost when the price is above the variable cost can be rewritten as the difference between the expected price given the price is above variable cost and the variable cost multiplied by the probability that the price is above the variable cost:

$$E[(P - c_t) I(P \ge c_t)] = \left(E(P|P \ge c_t) - c_t \right) \Pr(P \ge c_t)$$

Hence, we have the result that, at the efficient level of capacity, the expected price given that the price is above variable cost is equal to that variable cost plus the fixed cost of capacity discounted by the probability the price is above the variable cost:

$$E(P|P \geq c_t) = c_t + \frac{f_t}{\Pr(P \geq c_t)}$$

The expression on the right-hand side is often known as the long-run marginal cost (LRMC) of a generator of type t. The LRMC is equal to the variable cost plus the fixed cost (per unit of capacity) discounted by the probability that the generator is operating.

As before, we can relate the problem of determining the optimal level of capacity of each generation type to the price–duration curve. From the expression, we know that for each generation type, the area under the price duration curve and above the variable cost of that generation type must be equal to the fixed cost of that type.

In other words, starting with the highest-variable-cost generator, the area under the price–duration curve and above that variable cost must be equal to the fixed cost of the highest-variable-cost generator. Then, taking the second-highest-cost generator, the area under the price–duration curve and above the variable cost of that generator must be equal to the fixed cost of that second-highest-variable-cost generator, and so on. This is illustrated in Figure 9.4.

Collectively, the set of fixed costs and variable costs of all the generators determine the shape of the price–duration curve when the optimal level of capacity of each generator is chosen.

Note, in particular, that the price must, on occasions, be higher than the marginal cost of the highest-variable-cost generator. In fact the price must be higher than this threshold sufficiently often that the area under the price–duration curve and above this threshold is equal to the fixed cost of the highest-variable-cost generator.

We have not yet shown that this investment is privately profitable. We return to that question in a later chapter.

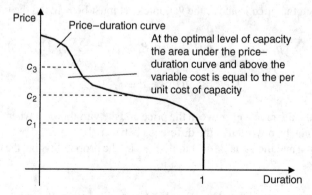

Figure 9.4   Determination of the optimal capacity of each generation type from the price–duration curve

> *Result*: In an electricity industry with a finite number of generation types, with a constant variable cost of production and a constant cost of adding generation capacity, and with no limit on the amount of capacity of each generation type that can be added, the optimal mix of generation is where, for each generation type, the area under the price–duration curve and above the variable cost of that generation type is equal to the fixed cost of adding capacity of that generation type.
>
> Another way of saying the same thing is that in the optimal mix of generation, for each generation type, the expected price given that the price is above the variable cost of that generation type is equal to the LRMC of that generation type.

9.4 The Optimal Mix of Generation with Inelastic Demand

Let us once again focus our attention on the special case in which demand is perfectly inelastic and equal to Q, up to some threshold, which we will denote V. If we assume that demand is represented by a continuous random variable, then with positive probability the marginal value and the price can only be equal to a variable cost of one of the generators $c_1, c_2, c_3, \ldots, c_N$ or V. As before, we will assume that these variable costs are placed in order from the lowest variable cost to the highest variable cost:

$$c_1 < c_2 < \cdots < c_N$$

Furthermore, if we let S^n denote the sum of the capacity of the first n generator types, $(S^n = \sum_{j=1}^{n} K_j)$ then (using the result in Section 9.3) we can state that under optimal dispatch the marginal value and the price takes the value V when the demand exceeds S^N, and the value c_i when the demand lies between S^{i-1} and S^i.

$$P(Q) = \begin{cases} V, & \text{if } Q > S^N \\ c_i, & \text{if } S^{i-1} < Q \leq S^i, i = 1, \ldots, N \end{cases}$$

We know from Section 9.3 that at the optimal level of capacity of each generation type the area under the price–duration curve and above the variable cost of each generation type is just equal to the fixed cost of that generation type. Since the quantity is assumed to be a continuous random variable, we have that

$$\int_{S^i} (P(Q) - c_i) f(Q) dQ = f_i$$

Substituting in the price as a function of the quantity, we have the following expression:

$$(V - c_i)\Pr\left(Q > S^N\right) + \sum_{j=i+1}^{N} (c_j - c_i)\Pr(S^{j-1} < Q \leq S^j) = f_i$$

From this expression, solving recursively, we have that

$$f_N = (V - c_N)\Pr(Q > S^N) \text{ and}$$

$$\forall i, f_{i-1} - f_i = (c_{i-1} - c_i)\Pr(Q > S^{i-1})$$

Let us define $\theta^i = \Pr(Q > S^i)$. θ^i represents the proportion of the time that the generator of type i is operating at capacity. It therefore represents the *capacity factor* of the ith generator. Then we can write that, at the optimal level of capacity for each generation type

$$f_N + \theta^N c_N = V \theta^N \text{ and}$$

$$f_i + \theta^i c_i = f_{i+1} + \theta^i c_{i+1}, \ i = 1, \dots, N-1$$

More specifically, these equations state that in the optimal dispatch, the probability that there will be voluntary load shedding is given by the intersection of the line $y = c_N x + f_N$ and the line $y = Vx$. The optimal capacity factor of the highest-variable-cost generator is where the line $y = c_{N-1} x + f_{N-1}$ intercepts the line $y = c_N x + f_N$. For every other generator, the optimal capacity factor of the generator is where the line $y = c_i x + f_i$ intercepts the line $y = c_{i+1} x + f_{i+1}$.

These equations are known as *screening curves* and have a simple graphical interpretation. Consider the graph of a line $y = c_i x + f_i$. This line has a y-intercept equal to the fixed cost of the generator, and a slope equal to the variable cost of the generator. The equations state that the x-value of the point of intercept of two of these lines gives the optimal *capacity factor* of each type of generator.

Figure 9.5 illustrates the case where there are two generator types: A peaking generator (with a higher variable cost and lower fixed cost) and a baseload generator (with a lower variable cost and a higher fixed cost). The optimal level of load shedding is where the line $y = c_2 x + f_2$ intercepts the line $y = Vx$. The optimal level of capacity of the peaking generator is where the line $y = c_1 x + f_1$ intercepts the line $y = c_2 x + f_2$. The remainder is the optimal level of time during which the baseload generator is the marginal generator.

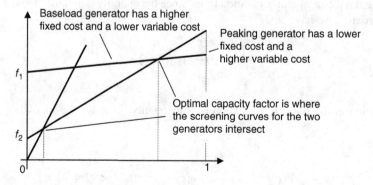

Figure 9.5 Determining optimal mix of generation using screening curves

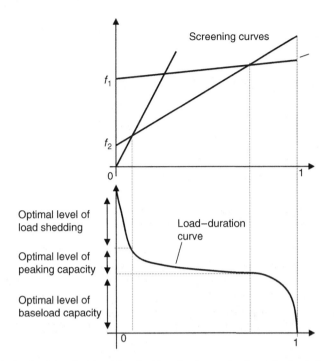

Figure 9.6 Derivation of optimal level of baseload capacity from the load–duration curve

Given the load–duration curve and the optimal capacity factor of each generation type, we can directly read the optimal capacity of each generation type off the load–duration curve as shown in Figure 9.6.

Result: In an electricity industry with inelastic demand and a finite number of generation types, the optimal mix of generation can be determined using screening curve analysis. The lower bound of the screening curves determines the proportion of time that the corresponding generation type is the marginal generator.

9.5 Screening Curve Analysis

Despite the fact that screening curves are only valid under a number of strict assumptions (such as inelastic demand, all generators have a well-defined fixed and variable cost, no lumpiness in investment), screening curves are a useful tool for gaining back-of-the-envelope style intuition about the effect of a change in one of the key parameters on the optimal mix of generation.

For example, what effect does an increase in the fixed cost of a particular type of generation have on the capacity of that generation type in the optimal mix? What effect does an increase in the variable cost have on the capacity of that generation type in the optimal mix?

These questions are easy to answer using screening curve analysis. An increase in the fixed cost of a particular type of generation always results in a reduction in the period of time in

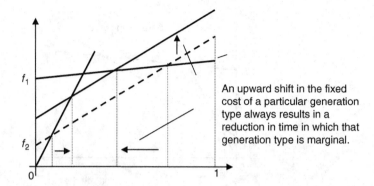

An upward shift in the fixed cost of a particular generation type always results in a reduction in time in which that generation type is marginal.

Figure 9.7   Impact of an increase in the fixed cost of a generation type on the optimal mix of generation

which that generation type is the marginal generator, and a reduction of the capacity of that generation type in the optimal mix. This is illustrated in Figure 9.7.

Similarly, an increase in the variable cost of a particular type of generation always results in a reduction in the period of time in which that generation type is the marginal generation, but may result in a larger or smaller actual capacity of that generation type in the optimal mix.

We can derive these results mathematically. The capacity factor of a generator was defined earlier as follows:

$$\theta_i = \frac{f_{i-1} - f_i}{c_i - c_{i-1}}$$

We can define the proportion of time that a generator is the marginal generator as follows:

$$\mu_i = \theta_i - \theta_{i-1}$$

It is easy to check that μ_i is decreasing in both f_i and c_i (this is one of the exercises set out later).

9.5.1 Using Screening Curves to Assess the Impact of Increased Renewable Penetration

One of the interesting questions that we might ask is the following: What is the impact of a substantial penetration of renewable generation resources on the optimal mix of conventional generation?

Let us assume that the renewable generation has a very low marginal cost and that it is uncontrollable (the term that is normally used is *intermittent*). Can we incorporate the impact of such renewable generation into the screening curve analysis?

Since the renewable generation is assumed to have a low marginal cost, it will always be dispatched for the full amount of its production (assuming that the marginal cost of all other generation types is positive). The conventional generation will service the rest of the load – the load that remains after the renewable generation has produced as much as it is able to produce. Therefore, one way to incorporate uncontrollable, low-marginal-cost generation into the

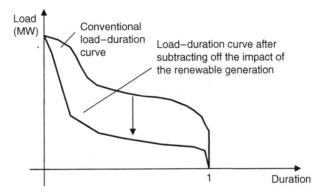

Figure 9.8 Derivation of the 'net load' duration curve

screening-curve analysis is to subtract the production of the renewable generation from the total demand for electricity. In particular, we can replace the load–duration curve in our analysis with a 'net load'–duration curve where the net load is the load after subtracting the renewable production.

The net-load–duration curve is stretched downwards relative to the conventional load duration curve. As long as there is a positive probability that the renewable generation will not be producing at all at peak times, the peak demand (the highest point on the load duration curve) remains the same. Otherwise, the remainder of the load–duration curve is stretched downwards as illustrated in Figure 9.8.

What is the impact of such renewable generation on the optimal mix of generation? Importantly, the proportion of time that each generator is the marginal generator remains exactly the same. However, the optimal amount of capacity of each type of generation changes. Importantly, although the proportion of time in which load shedding occurs does not change, the total volume of potential load shedding is much larger. This is illustrated in Figure 9.9.

9.5.2 Generation Investment in the Presence of Network Constraints

The previous sections ignored the possibility of network constraints. Does this analysis change in the presence of network constraints?

The simple answer is no. If we assume that there is a finite set of generation types at each node, and each type has a constant variable cost and a fixed per-unit cost of adding a unit of capacity, then the previous analysis still applies, although the analysis applies at the level of each node.

Specifically, there is the optimal mix of generation at each node if and only if, at each node, for each generation type at that node, the area under the price–duration curve for that node and above the variable cost of that generation type is equal to the fixed cost of adding capacity of that generation type.

9.6 Buyer-Side Investment

This chapter has focused primarily on generation investment. However, as emphasised in the first part of this text, buyers must also make investments that increase their value from

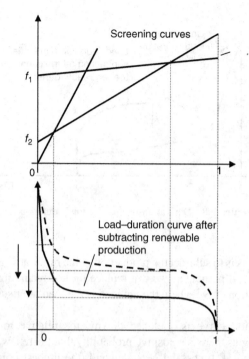

Figure 9.9   Impact of intermittent baseload generation on optimal mix of generation

electricity – this might include investment in electricity-consuming appliances and devices. Can we say anything about the optimal level of buyer-side investments?

It is much harder to say anything in the abstract about buyer-side investment. The optimal decision in any case will depend heavily on the manner in which the buyer-side investment affects the utility or value received from consuming electricity.

In the case where the buyer has the potential to install a local generator, so that the utility function of the buyer takes the form set out in Section 4.5, the previous analysis set out in this chapter applies, provided that the generator is effectively paid the local nodal price for its production. As we have seen previously, the optimal capacity of that generation type is where the area under the local price–duration curve and above the variable cost of the on-site generation is equal to the fixed cost of the on-site generation. As we noted in Section 4.5, there is complete symmetry in the treatment of generation integrated with load and stand-alone generation.

However, there are other ways in which the buyer's investment may affect his/her own valuation for electricity. For example, let us suppose that a buyer can make an investment that shifts his/her demand curve vertically upwards. This might be the case, for example, if the investment is in some form of energy efficiency, which allows the user to extract more value per unit of electricity consumed. Under this assumption, the utility function of the buyer might take the following form:

$$U(Q, K) = V(Q) + f(K)Q + c$$

Here, $V(Q)$ is an 'underlying' utility function, $f(K)$ is some function that shows how much extra value per unit of consumption arises from an investment of K units in energy efficiency, and c is a constant. Let us assume that this function is increasing ($f'(\cdot) > 0$) but at a decreasing rate ($f''(\cdot) < 0$) to reflect diminishing returns from investment. Let us suppose that each additional unit of investment costs r. The social optimisation problem is then as follows:

$$\max_{Q_s, K} \sum p_s(U_s(Q_s, K) - C(Q_s)) - rK$$

From the KKT conditions for the optimisation, the efficient level of production and consumption in the state of the world, s, satisfies the following:

$$\frac{\partial U}{\partial Q}(Q_s, K) = V'_s(Q_s) + f(K) = C'(Q_s)$$

This shows that an investment shifts the demand curve vertically. In addition, the condition for the optimal choice of investment is as follows:

$$\sum p_s \frac{\partial U_s(Q_s, K)}{\partial K} - r = \sum_s p_s Q_s f'(K) - r = 0$$

In other words, the optimal level of investment in assets that shift the demand curve upwards is given by

$$f'(K) = \frac{r}{\sum_s p_s Q_s}$$

By these assumptions, an increase in the cost of these investments reduces the optimal level. More importantly, an increase in the average consumption increases the value of such investments, increasing the optimal level.

Of course, this example only illustrates one potential type of buyer-side investment. The precise conditions for efficient investment in a more general case will depend on how the buyer-side investment affects the utility of the buyer.

9.7 Summary

In the previous chapters we looked at the characteristics of optimal *use* of a set of generation and consumption assets. In this chapter we explored the features of optimal *investment* in generation and consumption assets. In order to keep things simple we focused on the case where there are a finite number of types of conventional controllable generation. Each generation type was assumed to have a constant variable cost up to some total capacity. The question of the optimal mix of generation is the question of the optimal choice of capacity of each generation type.

This problem has a particularly straightforward solution. At the optimum choice of generation capacity, for each generation type, the expected market price when the market price is above the variable cost of that generation type is equal to the long-run marginal cost of that generation type. The long-run marginal cost is equal to the variable cost, plus the fixed cost divided by the probability that the generation type is producing.

Another way of saying the same thing is that, for each generation type, the area under the price–duration curve and above the variable cost of that generation type must be equal to the fixed cost of that generation type.

In the case where demand is inelastic, the optimal mix of generation will require load shedding with some positive probability. More generally, the market price must, on occasion, be higher than the generator with the highest variable cost (so that, even for the highest-variable-cost generation, the area under the price–duration curve and above that variable cost is equal to the fixed cost of that generation type).

When demand is inelastic and continuously distributed, we can use screening curves to graphically determine the optimal mix of each different generation type. The screening curve for a generation type has a y-intercept equal to the fixed cost of that generation type and a slope equal to the variable cost of that generation type. The intersection of the screening curve of a generation type and the screening curve of the next generation type in the merit order gives the optimal capacity factor for that generation type in the optimal mix of generation. The distance between two capacity factors gives the proportion of time that generation type is marginal. We can apply the capacity factors to the load duration curve to read off the actual capacity of each generation type that is required in the optimal mix.

Uncontrollable (intermittent) renewable generation can be incorporated in the screening curve analysis by considering a net load–duration curve, after subtracting the production of the renewables. Importantly, the capacity factor of each type of conventional generation remains unchanged (although the actual amount of capacity of each generation type may be changed significantly). In particular, the amount of load that may need to be shed increases.

This analysis extends to the case where there are network constraints, but the optimal mix of generation will vary from node to node and the relevant price–duration curve is the price–duration curve for each node separately.

Buyers must also make investments to increase their value from the use of electricity. Overall economic efficiency requires that these investments are also carried out efficiently. The precise conditions for efficiency will vary from case to case depending on how the investment affects the utility of the buyer. In the special case where the buyer is investing in local generation, which is paid the local nodal price, the same analysis applies.

Questions

9.1 Under the assumption that demand is inelastic and is continuously distributed, show that the proportion of time a particular generation type is the marginal generator is decreasing in that generation type's variable cost and fixed cost.

9.2 Assume that we have a set of N generation types ranked in order of increasing variable cost. Can you provide conditions on the fixed and variable costs under which generation type i is not used at all in the efficient mix of generation?

9.3 What impact do ramp rate constraints have on the optimal mix of generation?

9.4 Let us suppose that demand is inelastic. The demand–duration curve is given by $Q = 1000 - 1000z$. Suppose that there are three different types of generation with a variable cost of \$10, \$20 and \$50, together with load-shedding at \$1000/MWh. The fixed costs of these generation types are $15, $5 and $1 MW^{-1} h^{-1}, respectively. Find the optimal mix of generation in this industry.

9.5 How does increasing the capacity on a network link affect the optimal mix of generation at either end of that network link?

Further Reading

There is a useful discussion of screening curves in Stoft (2002).

10

Market-Based Investment in Electricity Generation

Chapter 9 addressed the question of the computation of the socially efficient level of investment in generation and consumption assets. However, if we are to exploit the benefits of competitive markets, we would like to know if we can decentralise these investment decisions to private entrepreneurs. Is it the case that decentralised investment decisions by private entrepreneurs will lead to an efficient level of investment?

10.1 Decentralised Generation Investment Decisions

In Chapter 9, we derived the conditions that an omniscient central planner would follow to make optimal investment decisions. Now let us ask the question whether those outcomes can be achieved through decentralised decisions by private generation entrepreneurs.

As in Chapter 9, we will maintain the following assumptions:

- There are a number of distinct generator technologies, each with a constant variable cost;
- There are constant returns to scale in each technology, with a distinct fixed cost of adding capacity for each generation type;
- Capacity can be added in arbitrarily small increments (i.e. there is no 'lumpiness' in generation investment) and there is no limit on the volume of capacity that can be added of any one generation type; and
- There are no sunk costs – capacity can be added or withdrawn at will.

We will assume that there are a large number of independent generation entrepreneurs. Since each can make an arbitrarily small investment in the capacity of a particular type of generation, we will focus on the case where the investment by each entrepreneur is so small as to have a negligible effect on the market price (or, more precisely, the price–duration curve).

Also, as before, let us assume that all of the generators are price-takers.

Let us focus on the decision facing an entrepreneur considering an investment of additional capacity in a generation type with constant variable cost c ($/MWh) and fixed cost per unit of capacity f ($/MW/h).

The Economics of Electricity Markets, First Edition. Darryl R. Biggar and Mohammad Reza Hesamzadeh.
© 2014 John Wiley & Sons, Ltd. Published 2014 by John Wiley & Sons, Ltd.

Since there are no sunk costs, we can focus on the investment decision in a single period (we do not need to consider the impact over multiple periods). We will assume that the generation entrepreneur chooses to invest in a period if and only if doing so is profitable.

Since the generator is a price-taker, the generator will choose to produce if and only if the local market price exceeds the variable cost. Furthermore, when the local market price exceeds the variable cost, it will produce at its capacity. Let us suppose that in state of the world s, the forecast market price is P_s. The profit of the generator in this state of the world ($/h) is

$$\pi = (P_s - c)KI(P_s \geq c) - fK$$

The expected profit of the generator per unit of capacity is therefore

$$E(\pi) = \sum_{s:P_s \geq c} p_s(P_s - c) - f$$

The generation entrepreneur will add a small amount of capacity of this generation type if and only if this expression is positive. Similarly, the generation entrepreneur will withdraw a small amount of capacity of this generation type if this expression is negative. Therefore, we can conclude that in a free-entry-and-exit equilibrium, this expression will be zero.

We can write this condition in two equivalent ways: First, assuming that the state of the world is a continuous random variable, we find that in the free-entry-and-exit equilibrium capacity will be adjusted to the point where the area under the price–duration curve and above the variable cost of a generation type is just equal to the fixed cost of adding capacity of that generation type.

$$E[(P - c)I(P \geq c)] = (E(P|P \geq c) - c)\Pr(P \geq c) = f$$

This is illustrated in Figure 10.1.

Another way of writing this expression is that the expected price given that the price is above the variable cost of a generation type must be equal to the long-run marginal cost of that

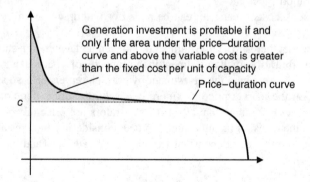

Figure 10.1 The profitability of a generation investment depends on the area under the price–duration curve and above the variable cost

generation type:

$$E(P|P \geq c) = c + \frac{f}{\Pr(P \geq c)}$$

However, these are just the same conditions we derived earlier for the optimal level of capacity and the optimal mix of generation. Therefore, we can conclude that decentralised investment decisions will, under the assumptions we have set out earlier, lead to an optimal level of generation investment and an optimal mix of generation investment.

> *Result*: Under the assumptions we just set out, in a free-entry-and-exit equilibrium, with a high level of competition between generators, the equilibrium level of capacity chosen for each type of generator is optimal.

This is an important result. It suggests that, at least under these fairly restrictive conditions, decentralised competitive markets yield an efficient level of generation investment.

Of course we can extend this result to the case where there are network constraints. In this case the result applies to investment at each node in the network separately.

10.2   Can We Trust Competitive Markets to Deliver an Efficient Level of Investment in Generation?

In the context of electricity markets, a market that provides compensation to generators solely for the provision of electrical energy is known as an *energy-only market*. This analysis can be summarised as asserting that (at least under certain assumptions, such as the assumption that the market is competitive) an energy-only market will deliver an efficient level of investment.

This result is not at all controversial in most markets in a modern economy. Nearly all modern economies rely on conventional market forces to deliver an efficient level of hotel rooms, aircraft flights, hairdressers and so on. In all of these markets we expect entrepreneurs to earn sufficient revenue from the sale of their services alone, whether those services are hotel rooms, flights or haircuts. We do not refer to these markets as service-only markets. We do not expect that other mechanisms will be developed to compensate investors in hotels, airlines or hairdressing salons.

However, this result remains somewhat controversial in the context of electricity markets. There is a concern that competitive electricity markets will not deliver an efficient level of reliability. Specifically, there is a concern that if every generator offers its output to the market in a way that reflects its marginal cost curve, these generators will not earn sufficient revenues to cover their total costs.

10.2.1   Episodes of High Prices as an Essential Part of an Energy-Only Market

Let us suppose that every generator offers its output to the market in a manner that reflects its marginal cost curve. How is it, exactly, that each generation type is compensated for its fixed costs?

The simple answer is that generators earn a contribution to their fixed costs whenever the market price rises above their variable cost of operation, as normally occurs when generators further up the merit order are being dispatched.

As we saw in Section 9.3, under certain simple assumptions, for each generation type, the area under the price–duration curve and above the variable cost for that generation type will, in equilibrium, be equal to the fixed cost per unit of capacity of that generation type.

This holds for all generation types. In particular, it must hold for even the highest-cost generation type. This means, of course, that the wholesale spot price must, at least at certain times, increase above the level of the variable cost of the most expensive generating unit in the market. Indeed the wholesale spot price must be above the variable cost of the highest-cost unit in the market by far enough and for long enough so that the area under the price–duration curve is equal to the fixed cost per unit of capacity of that high-cost generator.

In other words, in a conventional electricity market, some episodes of high prices (perhaps price spikes) are necessary if all generators are to be able to cover their fixed costs. Episodes of high prices are not, in themselves, evidence of market power. Indeed, as we have just observed, episodes of high prices are essential to allow all generators (including the highest-cost generators) to cover their fixed costs. In this respect the electricity market is just like any other market with time-varying prices.

Result: In an energy-only market (in which generators are only compensated for the energy they produce), the wholesale spot price must at times be higher than the variable cost of the highest-variable-cost generating unit in the market. Episodes of high prices and/ or price spikes are not in themselves evidence of market power or evidence of market failure.

10.2.2 The 'Missing Money' Problem

However, what if there are political or administrative restrictions on prices going to very high levels? If prices cannot go high enough for long enough the market cannot yield an efficient mix of generation. This is sometimes known as the 'missing money' problem.

In the United States it is common for wholesale market prices to be capped around $1000/ MWh. This is probably not high enough for peaking generators to receive adequate compensation for their fixed costs. Or, more accurately, a price cap of $1000/MWh is probably not high enough unless prices are able to rise to this level for many hours in the year. Episodes of persistent high prices may not be politically sustainable.

In short, although in theory a competitive market will yield an efficient mix of generation types, in practice, market distortions, such as price caps, may prevent prices rising to a level that is sufficient to provide adequate compensation to some generation types. In our view, the preferable solution is to remove the price caps. In practice, the common solution in the United States is to introduce separate secondary markets for 'capacity'. These are discussed further in Part VIII.

> *Result*: Political or administrative constraints on wholesale energy prices may prevent prices from rising high enough for long enough to justify generation investment. Such political or administrative constraints should be avoided wherever possible.

10.2.3 Energy-Only Markets and the Investment Boom–Bust Cycle

Sometimes, the concern is expressed that competitive energy-only electricity markets will result in a boom–bust cycle with alternating periods of under-capacity (and very high or volatile electricity prices) followed by periods of over-capacity (with very low and stable electricity prices).

We can distinguish these criticisms into two categories: those that challenge the assumptions in the earlier simple model and those that consider that the market outcomes needed to drive the results will not be politically sustainable.

The simple model relied on a number of extreme assumptions. In practice many generators have a degree of market power (see Part VII). Generation investment (especially commercial-scale generation investment) is very lumpy. A decision to build a larger generator will almost certainly have an impact on prices. Generation investment also involves very substantial sunk costs. Under these conditions it is likely that the previous conditions will not always hold. There will likely be episodes of over-capacity and under-capacity and periodic bouts of high prices. However, experience to date from liberalised electricity markets (not to mention the many other markets around the world) has not yet provided evidence that electricity markets will systematically fail to deliver needed investment in practice.

Another category of criticism is that an energy-only market will result in outcomes that are not politically sustainable, such as episodes of high prices, and that politicians will be forced to intervene.

This is a harder criticism to assess. Tolerance for market outcomes varies from country to country. It takes time for market participants to adjust to the demands of a market environment. However, we note that in a well-functioning electricity market, participants purchase risk management instruments (see Part VI) that largely insulate them from spot market price risk. Episodes of high prices need not correspond with episodes of hardship and need not test the tolerance of politicians.

10.3 Price Caps, Reserve Margins and Capacity Payments

We have seen how a decentralised energy-only market can lead to an optimal level of investment in generation capacity and an optimal mix of generation types. However, in order for this to hold, it must be that the market price is equal to the common marginal valuation and marginal cost of all generators at all times. At times of high demand and/or low supply, this price could be very high indeed (on the order of tens of thousands of dollars per MWh).

In practice, most liberalised electricity markets around the world have chosen to impose a price cap – that is, an artificial limit on the level at which prices can rise. As we saw in Section 5.5, the presence of a price cap distorts short-run production and consumption decisions resulting in the need for involuntary rationing (known as load shedding).

A price-cap will also have a serious impact on incentives for investment. Specifically, a price-cap will tend to reduce the revenue that generators receive at high-price times, reducing

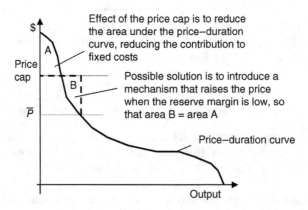

Figure 10.2 Administered price arrangements to restore incentives for investment in the presence of a price cap

the incentive for investment in generation and reducing the incentive for the buyer side of the market to invest in reducing electricity consumption at peak times.

Outside of the electricity industry, price controls (such as controls on the rents that landlords can charge) are well known for causing many undesirable effects.

However, given that price caps are in place, is it possible to design some mechanism to offset the harmful effects of the price cap?

Let us suppose that we start from a situation with an efficient mix and level of generation. Now suppose that a price cap is imposed.

The effect of a price cap is to 'truncate' the price–duration curve, limiting the high prices in the manner shown in the following diagram (Figure 10.2). If the price cap is below the variable cost of any generation type, then (as we saw in Section 5.5) the operational decision of that generation type is distorted – that generation type is not dispatched. The same applies if the price cap is below the marginal value of any load curtailment. In addition, as we have seen previously, a price cap will affect other operational and maintenance decisions of generators, reducing their reliability.

In regard to investment, the effect of the price cap is to reduce the revenue received by all generators. At the optimal mix of generation, generators would not receive sufficient revenue to earn a positive profit. If no further action is taken, there would be an adjustment in the mix and volume of generation investment, to the point where, despite the price cap, the area under the price–duration curve and above the variable cost of each generation type is equal to the fixed cost of that generation type. This could result in a substantial reduction in the volume of generation and an increase in the requirement for involuntary load shedding.

10.3.1 Reserve Requirements

One possible policy response to the missing money problem brought about by the price cap is to restore the area under the price–duration curve by increasing the price at times when the cap is not binding to offset the reduction in price brought about by the cap when the price cap is binding. This is illustrated in Figure 10.2.

One possible solution to this problem is to increase the price at certain times to the price cap. This increases the area under the price duration curve. For example, at all times when the price exceeds a threshold $\overline{P}$, the price could be set at the price cap. If $\overline{P}$ is chosen correctly, the additional area under the price–duration curve (area B in Figure 10.2) would equal the area lost due to the price cap (area A in Figure 10.2).

These administered pricing arrangements distort incentives for both production and consumption at such times – the market participants have the incentive to produce too much (and consume too little) when the price would (but for the intervention) fall between the threshold and the price cap, and (as we have already seen) have an incentive to consume too much (and produce too little) when the price would otherwise exceed the price cap.

10.3.2 Capacity Markets

A price cap results in a loss of revenue to all generators equal to the area 'A" in Figure 10.2. If we are to sustain the efficient mix of generation, generators must somehow be compensated for this loss in revenue.

One way to achieve this is to make a payment to each generator equal to the area 'A" multiplied by the capacity of that generator.

If the system operator happens to know the optimal level of total capacity, it can determine the size of the subsidy required (i.e. the size of the area A) by establishing a separate *capacity market*. In this market the system operator would effectively tender for a certain level of capacity. Generators would offer their capacity to that capacity market. The equilibrium in that market is the minimum subsidy that generators require to install that level of capacity. This is equal to the area A in Figure 10.2.

In the view of the authors, capacity markets represent a band-aid solution – attempting to fix up a problem created by the unwillingness to allow the core market to operate effectively. Capacity markets can, at best, offset the investment decisions faced by generators. They do not solve the investment decisions faced by buyers and they do not solve the operational problems created by the use of a price cap.

The underlying problem here is the price cap. The price cap induces inefficient usage decisions (as we have seen, there is too much incentive to consume and too little incentive to produce at times when the price exceeds the price cap) as well as distorting investment decisions. We do not compensate investors for providing production capacity in any other market in the economy. It is not clear that there is a need for capacity markets in the electricity industry either.

Result: A cap on the price in the wholesale market that is binding at times reduces the revenue that generators can earn from the market thereby reducing their incentives to invest. This is known as the 'missing money' problem and results in an inefficient mix of generation. The incentives for investment can be restored by making additional payments to generators based on their available capacity. These payments are often determined through a market process known as a 'capacity market'. Capacity markets represent a response to an existing market defect (the price cap) and are not necessary where the price cap has been removed.

10.4   Time-Averaging of Network Charges and Generation Investment

In practice, many electricity market participants do not receive or pay the short-run wholesale market price for their production or consumption of electricity. This is particularly the case for smaller electricity market participants – particularly customers connect to electricity distribution networks. Such customers typically pay charges that are averaged over time and across locations.

Time-averaging of network charges results in inefficient usage decisions – customers have an incentive to consume too much (and produce too little from local generation) at times of network congestion. At the same time, customers have an incentive to consume too little (and produce too much from local generation) at times when the network is not congested.

In addition to these inefficient usage decisions, the time-averaging of network charges has a significant impact on the incentives to make investment in local generation. This is discussed further in Section 20.1. Specifically, as we will see, under time-averaging of network charges, customers have no incentive to invest in generation with a high variable cost (since that generation will never be called on to produce). Conversely, customers have too much incentive to invest in generation with a low variable cost.

To see this, consider the following: We know from the previous analysis that a customer facing the correct local nodal price has an incentive to invest in a small amount of controllable generation with a constant variable cost and a constant cost of adding capacity if and only if

$$E[(P - c)I(P \geq c)] = (E(P|P \geq c) - c)\Pr(P \geq c) > f$$

Now consider a customer facing a fixed time-averaged price P^B. This customer will earn a profit of $(P^B - c)I(P^B \geq c)$ per unit of capacity for each period. Therefore, the customer has an incentive to invest in a small amount of controllable generation if and only if

$$(P^B - c)I(P^B \geq c) > f$$

Figure 10.3 illustrates these two functions in the case where the time-averaged price is above the expected spot price: $P^B > E(P)$. In this case, for generators with high variable cost, the

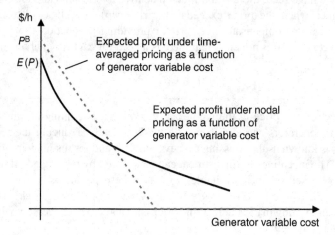

Figure 10.3   Impact of time-averaging of prices on generation investment incentives

expected profit is higher under nodal pricing. For generators with a low variable cost, the expected profit may be higher under time-averaged pricing.

In other words, time-averaging of pricing results in too much incentive to invest in low-variable-cost 'baseload' local generation and too little incentive to invest in high-variable-cost 'peaking' generation.

Result: Time-averaging of network charges results in both inefficient usage decisions and inefficient investment decisions in local generation. Specifically, customers have too much incentive to invest in local baseload generation and too little incentive to invest in local peaking generation.

10.5 Summary

Chapter 9 looked at the conditions for optimal investment under a set of simplifying assumptions. In this chapter, we explored whether or not generation entrepreneurs making decentralised commercial investment decisions would achieve the efficient mix and level of generation investment. In the case of a price-taking generator with constant variable cost, the expected profit of a generation entrepreneur is the difference between the market price and the variable cost whenever the generator is producing. The entrepreneur will invest when this value is positive (and will reduce investment when this value is negative). In equilibrium, the expected profit of the generation entrepreneur will be zero. This is the same condition that we derived for optimal investment by an omniscient central planner. In other words, under the simplifying assumptions assumed here, a decentralised competitive electricity market will yield the efficient level and mix of generation investment.

Electricity markets that compensate generators solely for the electrical energy they produce are called energy-only markets. There is some debate whether or not energy-only markets will deliver an efficient level and mix of generation investment in the long run. Experience with energy-only markets has not yet given rise to cause for concern, but there is the potential that some future outcomes may not prove politically acceptable.

In practice, many markets distort price outcomes in various ways. One such price distortion is the presence of a price cap. A price cap reduces the average revenue that all generators receive, distorting both operational decisions and giving rise to a 'missing money' problem, reducing generation investment. The effects of a price cap can be partially offset through mechanisms, such as a capacity market, which separately compensate generators for providing generation capacity.

Many smaller electricity customers pay a time-averaged network charge. The time-averaged price distorts both usage and investment decisions. Such customers have too little incentive to invest in high-variable-cost 'peaking' local generation and too much incentive to invest in low-variable-cost 'baseload' local generation.

Questions

10.1 True or False: In a market with inelastic demand the set of generation technologies (the set of variable costs and fixed costs of adding capacity for each generation type) completely determines the shape of the price–duration curve.

10.2 Not all liberalised electricity markets have price caps. Why do wholesale market price caps exist? What is their primary role?

10.3 The analysis in this chapter has focused on the private incentives for investment in controllable generation. What are the conditions for private investment to be profitable in, say, hydro-generation? Do these correspond to the socially optimal investment conditions?

Part V

Handling Contingencies: Efficient Dispatch in the Very Short Run

Construction of the tower to support the power cable across the Tory Channel in the Marlborough Sounds, New Zealand (Source: Neil Rennie)

The previous sections focused on the question of efficient use of a given stock of generation, consumption and network assets at a point in time, and efficient investment in generation and consumption assets. We saw in Part III that when there is variability in the supply, demand and network conditions, that variability is of no consequence for optimal short-run dispatch unless there are intertemporal constraints binding such as ramp-rate constraints or energy limits. When ramp rates or energy limits may be binding *ex post*, generators may adjust their dispatch *ex ante*. We expand on these ideas in this part. In reality, a major component of the real-time minute-by-minute operation of any commercial power system is handling short-run contingencies – preparing for the contingencies that may happen and ensuring the power system remains in balance after they do. All real power systems choose to alter the dispatch of some generators from the short-run optimum in order to reduce the cost of handling contingencies *ex post*. Furthermore, many power systems do not seek to achieve efficient dispatch in the very short run. Instead, short-run dispatch of generation is often achieved through heuristic procedures, such as frequency balancing, which ignore the impact of network constraints. When network constraints are ignored in short-run dispatch, the physical limits of the power system must be defined in such a way that the network is able to handle the power flows following any network contingency. This can significantly limit the precontingent power flows relative to the underlying physical limit.

11

Efficient Operation of the Power System in the Very Short-Run

In practice, for reasons that will become clear shortly, all real power system operators spend a considerable amount of time and effort anticipating and protecting the power system against the next change in supply, demand or network conditions. But why is this necessary? And how does this fit with the concept of optimal dispatch discussed in previous chapters?

11.1 Introduction to Contingencies

In any electricity industry, no matter how it is organised, supply and demand must be kept in balance and any physical network constraints must be satisfied at all points in time.

In the real world, power system operators spend a considerable amount of time and energy ensuring that the power system remains in balance and that it will remain in balance following any contingency.

A *contingency* can be defined as any short-run change to the supply, demand or network conditions in the market, such as the following:

- Any change in the production conditions facing a generator so that it is unwilling or unable to produce according to its previously submitted marginal cost curve, such as the physical outage of all or part of the plant, a change in its input prices, or a change in the physical availability of a resource (such as solar or wind availability);
- Any change in consumption decisions so that a consumer is unwilling or unable to consume according to its previously submitted demand curve, such as the physical outage of a local plant, a change in downstream production conditions affecting the demand for electricity or a change in the weather;
- Any change in network conditions, resulting in a change in the set of feasible power flows: the outage of a network element, a change in a line rating (perhaps due to a change in weather) and so on.

The Economics of Electricity Markets, First Edition. Darryl R. Biggar and Mohammad Reza Hesamzadeh.
© 2014 John Wiley & Sons, Ltd. Published 2014 by John Wiley & Sons, Ltd.

For example, the following are typical contingencies:

- The sudden loss of a single unit of a generating station;
- The sudden loss of one circuit of a double-circuit transmission line;
- The loss of a distribution transformer;
- A large drop in the production from wind generators;

As set out in Section 2.2, short-run changes in the supply and demand conditions in an interconnected AC power system are typically reflected (in the very short run) in changes in the common system frequency. Therefore, short-run supply/demand balancing of the power system is closely related to the task of *frequency control*.

However, efficient short-run operation of the power system is more than just frequency control. The frequency of an interconnected power system depends only on the *overall* (network-wide) supply/demand balance. However, as emphasised in Part III, the efficient operation of a power system involves more than just ensuring an overall supply/demand balance. Efficient operation of a power system also involves ensuring that the power flows remain within their physical limits at all points in time. Although efficient short-run operation of a power system involves a frequency control aspect, maintaining the system frequency within narrow bands is not the only task required for efficient operation of a power system.

Recall that we have defined a contingency as any short-run change in the supply, demand or network conditions. Contingencies differ in their significance and how likely they are to arise. We will see later that it is efficient for power systems to take certain actions *ex ante*, before a contingency arises, to reduce the economic harm that occurs once a contingency has occurred. However, it is neither desirable nor feasible to take such actions for all possible contingencies, especially contingencies that are extremely remote, or that are likely to cause little harm. We will define *credible contingencies* as those contingencies for which the economic benefits of taking action *ex ante* exceeds the cost. We will see later how to define these benefits and costs.

11.2   Efficient Handling of Contingencies

In Part III we explored how to achieve efficient use of a given set of generation, consumption and network assets. Do we need to change the principles we derived in Part III in order to accommodate the potential for contingencies?

As we have seen, every power system must have mechanisms for ensuring that the power system remains in balance and within the physical limits of the network at all points in time, both before and after a contingency occurs.

The optimal dispatch process set out earlier achieves these objectives at a fixed point in time. It ensures that supply matches demand overall and any physical network limits are satisfied. The optimal dispatch process should therefore be all that is required to achieve these objectives on a continuous basis provided we could operate that dispatch process sufficiently frequently to accommodate the (potentially rapid) changes to the power system required following a contingency.

In practice, the optimal dispatch process is typically carried out at intervals of 5 min or longer. But why not carry out the dispatch process much more frequently, say, every 5 s? Is there a theoretical limit to how rapidly we could (at least in theory) perform the optimal dispatch computation set out in Part III? Perhaps the optimal dispatch calculation could be performed every 20 ms (one complete cycle of a 50 Hz waveform)? (Of course, in reality there may be practical issues about information flows and computational time that may limit the frequency with which the dispatch process can be operated, but let us put these issues aside for a moment).

The key idea here is that if the optimal dispatch computation were carried out sufficiently frequently, there would never be any need for separate handling of contingencies. Contingencies would be efficiently addressed through the postcontingency optimal dispatch adjustment of the power system.

Result: If the optimal dispatch process set out in Part III could be performed sufficiently frequently so as to accommodate the changes to the power system following a contingency, there would be no need for any separate handling of contingencies.

This result is striking. In practice, as we have emphasised many times, much time and effort of power system operators goes into anticipating and managing contingencies as they arise. This line of reasoning suggests that this effort is only required as a consequence of the inability or unwillingness to operate the dispatch process more frequently.

11.3   Preventive and Corrective Actions

In practice, power system operators do not just simply wait until a contingency arises and then take whatever actions are necessary to restore the supply/demand balance following the contingency. Instead, it is conventional for power system operators to take into account the potential for contingencies and to alter the dispatch away from the theoretical optimal level, *even before a contingency has happened.*

This raises a fundamental question. When and/or under what circumstances does it make sense to distort the dispatch of the power system away from its optimal (no contingency) level *ex ante* (before a contingency arises), rather than rely on adjustments to the power system *ex post*?

As we saw in Section 4.9, it is useful to make a distinction between preventive and corrective actions. Once a contingency has occurred, *corrective actions* are those actions that are taken by the power system to adjust to a new equilibrium. This may involve the temporary and/or rapid change in the output of some generators or loads. In contrast, *preventive actions* are those actions taken by the power system to distort the dispatch of the power system away from its optimal (no contingency) level before a contingency has occurred.

Why would we ever distort the dispatch of the power system away from its optimal level *ex ante*? Put another way, why does it ever make sense to take preventive actions?

We have already seen the answer to this question in Part III (Sections 4.8 and 4.9). The answer depends on how quickly and easily the power system adjusts to the new equilibrium. If, following the contingency, no intertemporal constraints will be binding, the power system can adjust immediately to the new optimal dispatch. In effect there is no need for any corrective actions. In this case, there is no need for the power system to make any adjustment or allowance for the contingency *ex ante* – in other words, no need for any preventive actions. Each time period can be considered as independent. The optimal dispatch process need only consider the one-off short-run optimal dispatch in that period. Where the corrective actions are costless, there is no benefit in taking preventive actions.

However, if, following the contingency, intertemporal constraints will be binding, the power system will not be able to adjust immediately to the new optimal dispatch. Now, as we saw in Section 4.8, there may be circumstances where the optimal dispatch precontingency is different ('distorted away from') the optimal one-off short-run dispatch. There is a trade-off between the cost of corrective actions and the *ex ante* dispatch. By altering the *ex ante* dispatch (by taking preventive actions) we can reduce the cost of adjustment to the new equilibrium (i.e. reduce the cost of corrective actions). If the probability of the contingency is high enough, it may make economic sense to take preventive actions to reduce the cost of corrective actions *ex post*.

We can be somewhat more precise about what we mean. We can imagine that, prior to the contingency happening, the power system is in some form of stable equilibrium state without any binding intertemporal constraints. Once the contingency happens (such as the loss of a generator, a large load, or a network element), perhaps after some period of transition, there will be some new long-run stable equilibrium without any binding intertemporal constraints. In that new equilibrium the output of some generators may be higher or lower and the consumption of some loads may be higher or lower. As we will see, the question for us is how quickly and easily the power system can adjust to that new stable equilibrium.

If the power system can move immediately to the new equilibrium, there is no need to take any actions *ex ante* – the power system can be operated up to its full physical limits *ex ante* and, in the event that a contingency happens, will instantaneously adjust to its new physical limits.

However, in reality, there are limits to how quickly a power system can adjust to a large shock. There are limits to how quickly generators can ramp up or down (known as ramp rate constraints, as discussed in Sections 4.8 and 4.9). There are limits to how quickly loads are willing or able to change their consumption. In addition, even if some facilities can respond very quickly to a contingency (such as a battery storage facility), that facility may have a limited amount of energy that it can produce in restoring the supply/demand balance.

In the presence of ramp rate constraints or energy limits, the power system cannot adjust immediately to its new equilibrium. Instead the power system must transition to its new equilibrium over time. This may involve temporarily and rapidly increasing the output of some generators and/or some loads. There may be substantial costs incurred in the short-run as the power system adjusts to its new equilibrium.

Where the system can transition immediately to its new long-run equilibrium, the cost of taking corrective actions is zero. In this context, as we have seen, it makes no sense to take any actions *ex ante* in anticipation of a contingency arising. However, where there are costs of taking corrective actions it may be possible to take actions *ex ante* which, even though they reduce the economic surplus relative to the optimal dispatch without contingencies, reduces the cost of taking contingencies when they do occur.

Efficient power system operation involves not just efficient dispatch, but efficient dispatch following a contingency (minimising the cost of taking corrective actions) and efficient trade-off of the cost of altering the dispatch precontingency (the cost of preventive actions) against the cost of adjusting to the new equilibrium *ex post* (the cost of corrective actions).

> *Result*: Efficient power system operation requires not just efficient dispatch before a contingency occurs, and efficient dispatch following a contingency, but, when inter-temporal constraints are binding *ex post*, efficient anticipation of future contingencies and trading off of the cost of preventive actions against the cost of corrective actions.

11.4   Satisfactory and Secure Operating States

In Section 11.3, we saw that the need to take preventive actions arises from the inability of the power system to adjust instantaneously to a new equilibrium – instead, the power system must transition to the new equilibrium over time.

However, the length of that transition phase matters. If the transition is particularly long, a second contingency may arise, which significantly complicates the task of finding the optimal dispatch. In order to keep things simple, let us make the assumption that the transition phase is sufficiently short (say, on the order of minutes) that the probability of a second contingency happening during the transition phase is so low that the possibility of a second contingency can be ignored.

We can now make a distinction between the state of the power system during the initial and final equilibria and the state of the power system during the transition phase. During the transition phase all of the power system physical constraints must be satisfied, but (as just noted) the possibility of a further contingency can be ignored. However, during the initial and final equilibrium states, we must take into account the possibility that further contingencies may arise and therefore there is a need for preventive actions to minimise the cost of handling those contingencies.

We can distinguish between a *satisfactory operating state* and a *secure operating state*. A power system is in a satisfactory operating state if supply and demand is in balance and all physical network constraints are satisfied. A power system is in a secure operating state if it is in a satisfactory operating state and the power system will remain in a satisfactory operating state following any credible contingency and will return to a secure operating state within a short period of time (say 30 min).

All real power systems must be maintained in a satisfactory operating state at all times. Under normal operation (that is, at all times except in the short period of time immediately following a contingency), it is typical to require that the power system be in a secure operating state. In a secure operating state the power system is capable of accommodating any credible contingency while remaining within a satisfactory operating state and returning to a secure operating state within 30 min.

The official definitions of a satisfactory and a secure operating state in the Australian electricity market are set out next. Should the definition of a secure operating state include the requirement that no intertemporal constraints (such as ramp rate constraints) are binding?

> *Definition*: In the Australian electricity market, the power system is said to be in a *satisfactory operating state* when all the following conditions are met:
>
> a. The frequency of the system is within a predefined range (49.9–50.1 Hz with some exceptions);

  b. The voltage at each node of the system is within a predefined range;
  c. The current flows on all transmission lines of the power system are within the applicable operating limits (including short-term limits);
  d. All other plants forming the power system are operating within their normal ratings (including short-term ratings);
  e. The circuit breakers on the system are capable of disconnecting any potential fault; and
  f. The power system is stable in the sense that any oscillations caused by a small disturbance will tend to die down within 10 s.

The power system is said to be in a *secure operating state* if it is in a satisfactory operating state and will remain in a satisfactory operating state following any credible contingency, and will return to a secure operating state within 30 min.

11.5 Optimal Dispatch in the Very Short Run

How can we go about determining the optimal operation of the power system in the face of credible contingencies?

In the previous sections, we saw that even if there are no intertemporal constraints binding *ex ante*, if intertemporal constraints will be binding following a contingency, that possibility must be taken into account in the precontingency dispatch. Therefore, we need to model the postcontingency outcomes in the optimal dispatch process. One way of doing this is set out here.

Let us suppose that we have an initial state of the power system (describing the supply, demand and network conditions) s. Let us suppose that no intertemporal constraints are binding in this state. However, a contingency may arise at any point in time. Let us suppose that the kth contingency occurs with probability p_k. Immediately following that contingency the power system transitions to a new state s_{k1}, and then follows a path of adjustment $s_{k1}, s_{k2}, \ldots, s_{kT}$, before reaching a new long-run state with no intertemporal constraints binding, s'_k. This is illustrated in the following diagram (Figure 11.1).

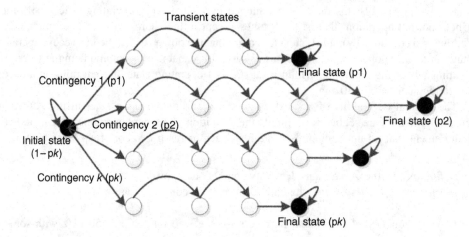

Figure 11.1 Illustration of initial state, transient states and final states

Let TB(s) be the total economic surplus received each time period when the power system is in state s (the total economic surplus is the sum of the benefit function introduced in Section 4.6, or the sum of the utility function less the cost function, in Section 1.4).

The economic problem is to find the path of dispatch (i.e. the initial dispatch, the transition path and final the steady state) that yields the highest surplus. Since this is an intertemporal problem, we will seek the highest *present value* of the surplus. Let us define the present value of the surplus in steady-state s to be $V(s)$. The present value of the surplus in the initial steady state is equal to the sum of (a) the surplus in the initial state plus the present value of the surplus in that initial state, discounted by the probability that no contingency arises, and (b) for each contingency, the surplus during the transition to the new steady state and the present value of the surplus in that new steady state, discounted by the probability that the contingency will arise:

$$V(s) = (1 - P)\left(\frac{\text{TB}(s)}{1+r} + \frac{V(s)}{1+r}\right) + \sum_k p_k \left(\sum_{t=1}^{T} \frac{\text{TB}(s_{kt})}{(1+r)^t} + \frac{V(s'_k)}{(1+r)^T}\right)$$

Here, $P = \sum_k p_k$ is the probability of any contingency occurring and r is the discount rate.

In principle, once the new steady state, s'_k, is reached, a further contingency could occur. The possibility for multiple contingencies significantly complicates the analysis. In order to keep things simple, let us assume that after a period of time in the new steady state, s'_k, the power system is restored to its original state, s. With this assumption, we can write the present value of the power system in the original state as a linear combination of (a) the per-period economic welfare in the original state; (b) the adjustment cost during the transition (i.e. the difference in the per-period economic welfare during the transition from the final steady state); and (c) the per-period economic welfare in the final state:

$$V(s) = \alpha(1 - P)\text{TB}(s) + \alpha \sum_k p_k \left(\text{AC}\left(s \rightarrow s'_k\right) + \gamma \text{TB}\left(s'_k\right)\right)$$

Here, α and γ are constants and

$$\text{AC}\left(s \rightarrow s'_k\right) = \sum_{t=1}^{T} \frac{\text{TB}(s_{kt}) - \text{TB}\left(s'_k\right)}{(1+r)^{t-1}}$$

The final steady state of the power system will depend on the contingency that has occurred, but not on the initial steady state.

Let us consider the decision of the system operator who is choosing whether or not to modify the initial steady state from its one-off efficient level, in order to account for contingencies. In other words, let us consider under what circumstances it is efficient to take preventive actions in order to reduce the cost of taking corrective actions. It is clear from the preceding expression that the task of the system operator is to choose the initial state and the postcontingency transition path in order to maximise the following:

$$(1 - P)\text{TB}(s) + \sum_k p_k \text{AC}\left(s \rightarrow s'_k\right)$$

We can immediately make a few observations. First, if either the probability of a contingency is zero ($p_k = 0$), or if there is no cost of adjusting to the new steady state equilibrium ($AC = 0$), then the system operator should make no adjustment to the initial steady state – the system operator should choose the dispatch that maximises the economic welfare in each period. Put another way, when taking corrective actions is costless, the system operator should take no preventive actions. In the absence of ramp rate constraints, energy limits and so on, it is never efficient to take preventive measures – that is, to alter the dispatch in advance of a contingency occurring. Preventive measures are only required to the extent that there are limitations on the ability of the power system to respond *ex post*.

Of course, we would expect that under normal circumstances, in the event of a contingency, ramp rate constraints and so on will prevent instantaneous adjustment to the new equilibrium. Instead, optimal dispatch may require the sequential dispatch of a range of plant, including some highly flexible plant, at varying levels until the new steady-state equilibrium is reached.

Suppose that s_1 and s_2 are two different initial steady states. Let us suppose that s_1 involves more significant preventive actions than s_2, so that the per-period economic welfare in state s_1 is less than s_2. When is it efficient for the system operator to choose s_1 over s_2? The answer is that it is efficient to choose more preventive actions when the increased economic welfare during the transition (discounted by the probability that a contingency will arise) is larger than the per-period loss in welfare. In other words, it makes sense to take more significant preventive actions if and only if

$$\sum_k \frac{p_k}{(1-P)} \left(AC(s_1 \rightarrow s'_k) - AC(s_2 \rightarrow s'_k) \right) > TB(s_2) - TB(s_1)$$

For example, it may be efficient to reduce the output of a unit in the steady-state dispatch relative to the efficient merit-order dispatch where either (a) the loss of that unit is a credible contingency, and reducing its output reduces the requirement for expensive postcontingency balancing services; or (b) the inefficiency from reducing the output of that unit *ex ante* is outweighed by the ability to use this unit in the transition to the new steady state in the event of a contingency.

11.6   Operating the Power System Ex Ante as though Certain Contingencies have Already Happened

Historically, there has been a significant asymmetry in the costs of handling certain contingencies. Also, it has been very difficult to respond to contingencies by adjusting the consumption of electricity consumers. In the absence of effective price signals (to most consumers), a reduction in the consumption of consumers is almost entirely involuntary and is seen as unreliability. Power system operators have historically been loath to ration electricity through involuntary load shedding. At the same time it has been practically impossible to get electricity customers to increase their consumption at short notice.

Given the unwillingness and inability to alter the consumption of electricity consumers in response to contingencies, the burden of responding to contingencies has fallen to other generators. In practice, it has been significantly easier to rapidly reduce the production of generators than to rapidly increase production.

As a consequence, the costs of adjustment to the power system following a sudden increase in supply (i.e. the loss of a large load) were historically much smaller than the potential costs of adjustment following a sudden reduction in supply (i.e. the loss of a larger generator). Similarly, the loss of a network connection is of more significance in the importing region (requiring an increase in local supply) than in the exporting region (where a local reduction in supply is required).

Given a set of contingencies, therefore, the contingencies involving the loss of a large generator were historically likely to be more significant than the loss of a small generator which, in turn, is more significant than the loss of a load.

From this analysis, if the adjustment cost is large enough, it outweighs the distortion to the dispatch cost *ex ante*. In fact, if the adjustment cost is large enough, the optimal dispatch minimises the adjustment cost or, more specifically, minimises the worst-case adjustment cost. The dispatch that minimises the worst-case adjustment cost is the dispatch corresponding to the event that the contingency has already happened. In other words, if the adjustment cost is large enough, the optimal precontingency dispatch is the worst-case postcontingency outcome. The power system must be operated as though the worst-case contingency has already occurred.

Result: If the adjustment costs are high enough, it is efficient to operate the power system *ex ante* as though the worst-case credible contingency has already occurred.

This result is important. In practice, power systems are routinely operated in a way that in effect assumes that the worst case scenario has already happened. We will come back to this issue in Chapter 12.

11.7 Examples of Optimal Short-Run Dispatch

To illustrate the determination of optimal dispatch, consider the following examples: Let us start with a simple example in which we have 1000 MW of $10/MWh generation with a ramp rate of 40 MW/interval, and a large amount of $50/MWh generation, with a ramp rate of 20 MW/interval, and a large amount of $1000/MWh with a large enough ramp rate that it is never ramp rate constrained.

Let us suppose that demand is initially 2000 MW, and that the loss of 500 MW of $50/MWh generation is a credible contingency.

Let us start by considering the optimal 'myopic' dispatch ignoring the contingency. Clearly, the lowest-cost (merit order) dispatch involves dispatching the $10/MWh generation for 1000 MW, and the $50/MWh generation for the rest (1000 MW). This yields a per-period dispatch cost of $1000 \times 10 + 1000 \times 50 = \$60,000$.

Now consider what happens when the contingency occurs. This is illustrated in Figure 11.2. The output of the $50/MWh generation immediately drops to 500 MW. The shortfall is made up by the expensive $1000/MWh generation. The adjustment takes 25 periods (as the $50/MWh generation slowly ramps back up to 1000 MW). After 25 periods the per-period dispatch cost returns to the original level of $60 000/interval. However, the transition process is costly. The additional costs incurred (above the normal per-period dispatch cost) is, on average, $249 400 per period.

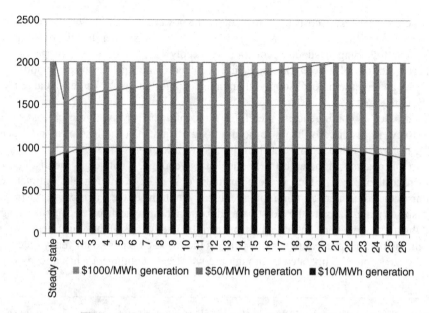

Figure 11.2 Optimal dispatch ignoring the contingency

However, we can do better than this. Let us suppose we reduce the output of the cheap $10/ MWh generation down to 800 MW in the original steady state. This provides scope to ramp this generation up to 1000 MW in the event of a contingency. This raises the per-period dispatch cost in the steady state to $68 000/interval, but it reduces the average cost of dispatch during the transition period to $103 200 per period. The new dispatch is illustrated in Figure 11.3.

Now suppose that the contingency occurs with a probability of 10%. We can, in principle, determine the average dispatch cost (pre- and postcontingency) for all possible combinations of

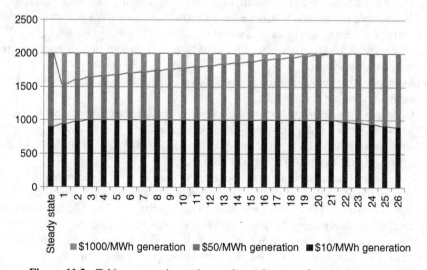

Figure 11.3 Taking preventive actions reduces the cost of corrective actions

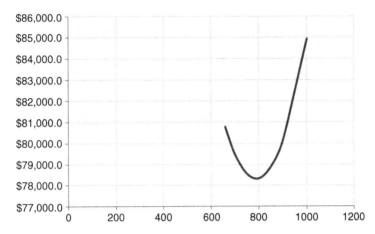

Figure 11.4 Average dispatch cost as a function of the dispatch of $10/MWh generation in the steady state

dispatch in the initial steady state. It turns out (see Figure 11.4) that as we reduce the dispatch of the $10/MWh generation from the myopic optimal dispatch (1000 MW), the average dispatch cost drops rapidly. However, beyond a certain point, further reductions in the dispatch of the cheap generation do not reduce the adjustment costs and continue to increase the initial steady state dispatch, so the total average dispatch cost starts to rise. Figure 11.4 shows that, in this example, the optimal initial dispatch is around 800 MW for the $10/MWh generation and around 1200 MW for the $50/MWh generation.

This example clearly illustrates the general point that, when intertemporal constraints are binding *ex post*, it may be preferable to alter the dispatch *ex ante*. Or, put another way, when the cost of taking corrective actions is high, it may make sense to take additional preventive actions.

11.7.1 A Second Example, Ignoring Network Constraints

Now let us consider a slightly more complicated example. The generating units available and their capabilities are set out in Table 11.1.

There is a single credible contingency: with probability of 0.01, each period, generator 2 will trip and become unavailable. The interest rate is 2% per interval. As can be seen, only unit 1 has a production (capacity) limit and only unit 3 has a ramp rate limit. In this first example, the physical location of each generating unit is irrelevant since we are ignoring transmission constraints and losses.

Table 11.1 Generating unit data for examples of short-run dispatch

Unit	Node	Variable Cost ($/MWh)	Initial Capacity (MW)	Ramp Up and Down Rates (MW/h)
1	2	$10	750	n/a
2	3	$100	n/a	n/a
3	1	$200	n/a	100
4	3	$2000	n/a	n/a

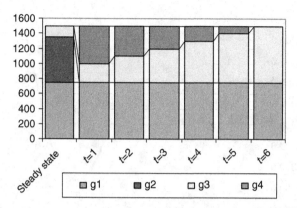

Figure 11.5 Optimal pre- and postcontingency dispatch in a simple network with no binding transmission constraints

Let us suppose the load is 1500 MW. In the absence of any contingencies, the optimal dispatch follows the merit order; unit 1 would be dispatched up to 750 MW and the remainder of the load (750 MW) would be met by unit 2.

Now let us explore the optimal dispatch allowing for the contingency of the loss of unit 2. As illustrated in Figure 11.5, in the initial steady state, it is not efficient to dispatch each unit according to the simple merit order. Instead, under the contingency-constrained dispatch, unit 2's output is reduced (relative to the theoretical merit order) to 600 MW, with unit 3 making up the remainder. As Figure 11.5 shows, following the contingency, unit 4 initially ramps up rapidly to 500 MW, and then ramps down steadily as unit 3 ramps up.

This example illustrates the same points we saw in the previous example. First, in general it will be efficient to make use of preventive measures to reduce the cost of corrective measures *ex post*. Second, in general, optimal dispatch *ex post* involves an evolution of different generating units ramping up and down at different rates over time. With a wide range of different types of generating units with different ramp rate constraints and/or energy limits, the transition path to the new steady-state equilibrium could be complex.

11.7.2 A Further Example with Network Constraints

Now let us consider the optimal dispatch problem in the context of the simple three-node, three-link transmission network illustrated in Figure 11.6. In this network all the transmission lines have identical electrical characteristics. There is a single flow constraint between nodes 2 and 3 of 500 MW. The total system load at node 3 is 1000 MW; as before, the loss of unit 2 is the only credible contingency. In this example we will assume the variable cost of unit 4 is now $10 000/ MWh (otherwise the characteristics of each generating unit are as given in Table 11.1.

The optimal dispatch in the absence of contingencies is to dispatch 750 MW at unit 1 and 250 at unit 2. As shown in Figure 11.7, computation of the optimal pre- and postcontingent dispatch shows that in the initial steady state, unit 1 is backed off to 700 MW and unit 2 is backed off to 200 MW, leaving unit 3 to take up the remainder of 100 MW.

The interesting outcome in this case is what happens in the postcontingent dispatch following the loss of unit 2. In this case, the optimal dispatch involves unit 1 ramping *down* (rather than up) and units 3 and 4 ramping up to substitute for both the loss of unit 2 and

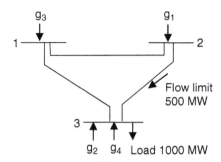

Figure 11.6   Simple three-node network diagram

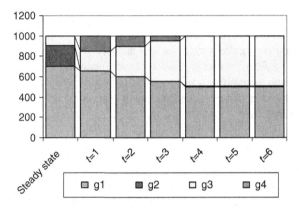

Figure 11.7   Optimal pre- and postcontingency dispatch in a simple meshed network with one binding transmission constraint

the reduction in output at unit 1. As before, since unit 3 is constrained in how fast it can ramp up, unit 4 increases in output first and then ramps down as its output is replaced by unit 3. This is shown in Figure 11.7.

This example illustrates the perhaps counter-intuitive result that, in meshed networks, in order to efficiently address the contingency of the loss of a generating unit, the optimal procurement and dispatch of balancing services may require the procurement and dispatch of lower services from some generators (and the offsetting procurement and dispatch of even larger quantities of raise service from some other generators). This is, however, consistent with optimal dispatch.

11.8   Optimal Short-Run Dispatch Using a Competitive Market

This chapter has so far focussed on the question of the computation of optimal dispatch by a hypothetical omniscient system operator with perfect information about supply, demand and network conditions (as well as all contingencies). However, as we have noted several times, one of the objectives of electricity market reform is to decentralise generation and consumption decisions, allowing private agents to make their own operational and investment decisions under the incentives brought about by market prices.

Table 11.2   Generating data for short-run dispatch example

Type	Variable Cost ($/MWh)	Initial Capacity (MW)	Ramp Up and Down Rates (MW/h)
1	$10	1000	n/a
2	$50	n/a	50
3	$120	n/a	n/a

Is it the case that the efficient economic outcomes discussed in this chapter could be achieved through a decentralised market process? Specifically, is it the case that a private for-profit generating entrepreneur would voluntarily reproduce the optimal short-run dispatch outcomes discussed earlier?

The answer seems to be yes. This is an illustration of the general principle set out in Section 1.5 that, provided the prices are set correctly, and provided the producers and consumers are price-takers, competitive market processes will achieve welfare-maximising outcomes.

As we will see, a profit-maximising generator that expects to face an intertemporal constraint (such as a ramp rate constraint or an energy limit) following a contingency will voluntarily choose to 'distort' its offer curve precontingency in order to be able to better exploit the prices that arise postcontingency.

11.8.1   A Simple Example

This can be seen in the following simple example. This example illustrates how a profit-maximising generator will voluntarily choose to increase its output precontingency, even though its variable cost exceeds the market price, in order to increase its ability to exploit the high prices that arise postcontingency.

This example features three types of generators. The variable cost, capacity and ramp rate limits of the generators are set out in Table 11.2. Note that only the type 1 generators are capacity constrained and only the type 2 generators are ramp-rate constrained.

Let us suppose that the system is in an initial steady state where demand is 2000 MW. However, each period there is a probability of, say, 10% that a contingency will occur and demand will jump immediately to 2500 MW. Following that contingency, the system has a finite number of periods to transition to a new steady state equilibrium. Then, for simplicity, we will assume the system returns to its original state without further ramp rate constraints (in effect, this assumes that the ramp rate limits are ramp up limits only).

The first observation we can make is that the simple steady-state least-cost dispatch ignoring contingencies involves type 1 generators dispatched to 1000 MW, and type 2 generators dispatched for the remainder (1000 MW). The dispatch cost is $60 000 per period.

Now let us solve the welfare-maximisation problem to find the optimal dispatch taking into account the contingency. From the analysis in Section 11.2, we can write the present value of the economic welfare as a weighted sum of the welfare with no contingency and the welfare following the contingency:

$$V_0 = (1 - P)\left(\frac{TB_0}{1 + r} + \frac{V_0}{1 + r}\right) + P\left(\sum_{t=1}^{T} \frac{TB_t}{(1 + r)^t} + \frac{V_0}{(1 + r)^T}\right)$$

Here, V_0 is the present value of the economic welfare for the power system in the initial steady state, $P = 0.1$ is the probability of the contingency, r is the discount rate, TB_0 is the single period total benefit or total surplus in the initial steady state, TB_t is the single period total surplus during the transition to the new steady state, and T is the total number of periods that elapses before the system reverts back to the initial state.

It turns out that the economic welfare can be expressed as a linear combination of the welfare in the initial steady state and the welfare during the transition

$$V_0 = \alpha_0 TB_0 + \sum_{t=1}^{T} \alpha_t TB_t$$

where

$$\alpha_0 = (1 - P)\left(P + r - \frac{P}{(1+r)^{T-1}}\right)^{-1} \text{ and } \alpha_t = \frac{P}{(1+r)^{t-1}}\left(P + r - \frac{P}{(1+r)^{T-1}}\right)^{-1}$$

Since each generator has a constant variable cost, the total economic cost each period is just the variable cost of each generator multiplied by the rate of production of each generator:

$$TB_t = c_1 Q_{1t} + c_2 Q_{2t} + c_3 Q_{3t}$$

The welfare-maximisation task is as follows:

$$\min V_0 = \alpha_0 TB_0 + \sum_{t=1}^{T} \alpha_t TB_t$$

subject to the following conditions:

a. $\sum_i Q_{i0} = 2000 \leftrightarrow \alpha_0 \lambda_0$ and
b. $t = 1, \ldots, T, \sum_i Q_{it} = 2500 \leftrightarrow \alpha_t \lambda_t$ and
c. $t = 0, \ldots, T, Q_{1t} \leq 1000 \leftrightarrow \mu_t$
d. $t = 1, \ldots, T, Q_{2t} \leq Q_{2t-1} + 50 \leftrightarrow \gamma_t$.

The KKT conditions for this problem include the following:

i. For the generator of type 1: $t = 0, \ldots, T, \alpha_t c_1 - \alpha_t \lambda_t - \mu_t = 0$
ii. For the generator of type 2:

$$\alpha_1 c_2 - \alpha_1 \lambda_1 + \gamma_1 = 0$$

$$t = 1, \ldots, T, \alpha_t c_2 - \alpha_t \lambda_t - \gamma_t + \gamma_{t+1} = 0$$

$$\alpha_T c_2 - \alpha_T \lambda_T - \gamma_T = 0$$

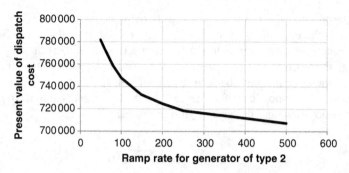

Figure 11.8 Impact of ramp rate limits on economic welfare

The solution to this problem is set out in Table 11.3. As can be seen, in the initial steady state, it is welfare-maximising for generation of type 1 to withhold some generation capacity compared to the no-contingency optimal dispatch. Generation of type 2 produces more in the steady state relative to the no-contingency optimal dispatch. In other words, in this simple example it is efficient to take some preventive actions to reduce the cost of taking corrective actions.

The presence of the ramp rate constraint reduces the overall economic welfare. Or, put another way, the ramp rate constraint raises the cost of handling this contingency. The higher the ramp rate, the lower the overall cost. This can be seen in Figure 11.8. This graph shows how the present value of the dispatch cost changes as the ramp rate limit on generator of type 2 increases. As the ramp rate increases, a point is eventually reached (at the ramp rate of 80 MW/ interval) where it is no longer efficient to take preventive actions (instead the steady state dispatch is the same as the no-contingency optimal dispatch). Further increasing the ramp rate limit reduces the need to use the most expensive generation after a contingency occurs. In the limit when the ramp rate limit is 500 MW/interval, the power system adjusts immediately to the new equilibrium and so the cost of corrective action is zero.

Table 11.3 Illustration of efficient dispatch in the presence of ramp rate limits

Time Interval	Generation			Dispatch Cost ($/h)	Prices ($/MWh)
	Type 1 ($10/MWh)	Type 2 ($50/MWh)	Type 3 ($120/MWh)		
Steady State	850	1150	0	66 000	$10.00
1	1000	1200	300	106 000	$120.00
2	1000	1250	250	102 500	$120.00
3	1000	1300	200	99 000	$120.00
4	1000	1350	150	95 500	$120.00
5	1000	1400	100	92 000	$120.00
6	1000	1450	50	88 500	$120.00
7	1000	1500	0	85 000	$93.66
8	1000	1500	0	85 000	$50.00
9	1000	1500	0	85 000	$50.00
10	1000	1500	0	85 000	$50.00

11.8.2 Optimal Short-Run Dispatch through Prices

Now consider the profit-maximising decisions of each generator type, taking the preceding prices as given. Each generator is assumed to maximise the present value of its future stream of profits. The future stream of profit is a weighted average of the profit in the steady state (discounted by the probability the steady state will recur) and the profit in the transition (discounted by the probability the contingency will occur):

$$\Pi_{i0} = (1 - p)\left(\frac{\pi_{i0}}{1 + r} + \frac{\Pi_{i0}}{1 + r}\right) + p\left(\sum_{t=1}^{T} \frac{\pi_{it}}{(1 + r)^t} + \frac{\Pi_{i0}}{(1 + r)^T}\right)$$

where $\pi_{it} = (P_t - c_i)Q_{it}$ for $t = 0, 1, \ldots, T$. As before, this reduces to

$$\Pi_{i0} = \alpha_0 \pi_{i0} + \sum_{t=1}^{T} \alpha_t \pi_{it}$$

Let us consider first the position of the generator of type 1. This generator's profit maximisation problem is as follows:

$$\max \Pi_{10} = \alpha_0 \pi_{10} + \sum_{t=1}^{T} \alpha_t \pi_{1t}$$

subject to $t = 0, \ldots, T, Q_{1t} \leq 1000 \leftrightarrow \mu_t$
The key KKT condition for this optimisation is as follows:

$$\alpha_t(P_t - c_1) - \mu_t = 0$$

By inspection we can see that this is the same condition as in the earlier welfare-maximisation problem provided that the price in each period is equal to the marginal value of the energy-balance constraint (i.e. provided $P_t = \lambda_t$). Therefore, provided this generator faces the correct prices, it will choose the efficient level of production in each period.

Intuitively, since this generator faces no intertemporal constraints, it just seeks to produce to maximise its profit each period. In the first period, the price is equal to the variable cost of this generator ($10/MWh), so the generator is indifferent as to how much it produces. In all subsequent periods the price is above the variable cost of this generator, so the generator produces at its capacity limit (1000 MW).

Now consider the position of the generator of type 2. This generator's profit maximisation problem is as follows:

$$\max \Pi_{20} = \alpha_0 \pi_{20} + \sum_{t=1}^{T} \alpha_t \pi_{2t}$$

subject to $t = 1, \ldots, T, Q_{2t} \leq Q_{2t-1} + 50 \leftrightarrow \gamma_t$

The key KKT conditions for this optimisation are as follows:

$$\alpha_0(P_0 - c_2) + \gamma_1 = 0, t = 1, \ldots, T, \alpha_t(P_t - c_2) - \gamma_t + \gamma_{t+1} = 0 \text{ and } \alpha_T(P_T - c_2) - \gamma_T = 0.$$

Again, by inspection we can see that provided the prices are set correctly (i.e. provided $P_t = \lambda_t$) this condition is the same as the welfare-maximisation problem above. Provided the generator faces the correct prices it will choose the optimal level of dispatch in each period.

It is interesting to note that, in the initial steady state the market price is \$10/MWh. Yet this generator, with a variable cost of \$50/MWh, chooses to be dispatched to 1150 MW. In other words, in the initial steady state this generator is making economic losses (of \$−46 000/period). Yet, this generator chooses to be dispatched to this level so that it can take advantage of the high prices that arise when a contingency arises. Specifically, when a contingency arises the price goes to \$120/MWh for several time intervals. The generator expects to make a profit overall despite the fact that it makes a loss in normal operation.

> *Result*: Provided the market prices are set correctly, efficient handling of contingencies (that is, efficient short-run dispatch) can be achieved through the standard optimal dispatch process alone.

This result is important as we have emphasised many times. A large part of the day-to-day operation of a power system involves anticipating and handling contingencies. This analysis shows that this focus on contingencies is a consequence of the decision to not provide the correct price signals to market participants at very high frequencies.

11.8.3 Investment Incentives

As we have just emphasised, achieving economic efficiency is not just a matter of efficient use of the existing stock of assets, but efficient investment.

In the simple problem earlier, we could consider both investment in the capacity of each generation type and investment in improving the ramping capability of the generation of type 2.

Let us focus here on the latter question. We saw in Figure 11.8 that economic welfare increases as the ramp rate limits on the type 2 generator increase. However, does this generator have an economic incentive to increase its ramp rate limit? Put another way, does this generator increase its profit by increasing its ramp rate?

The simple answer is yes. From the earlier profit-maximisation problem we can see that the increase in the profit from a small change in the ramp rate is equal to the sum of the marginal values on the ramp-rate constraint:

$$\frac{\partial \Pi_{20}}{\partial RR} = \sum_{t=1}^{T} \gamma_t$$

Since the marginal value on the ramp rate constraint is non-negative, the generator cannot reduce its profits by increasing its ramping capability (as we would expect). Furthermore, the greater the price rise (over the variable cost of this generator) and the longer the duration that this generator is ramp-rate constrained, the greater the profit incentive to increase the ramp-rate capability.

Finally, it is useful to observe that the social benefit from a small change in the ramp rate on generator 2 is exactly the same as the change in the profit on generator 2 from a small change in the ramp rate. In other words, the generator has a socially efficient incentive to expand its capability.

$$-\frac{\partial V_0}{\partial \text{RR}} = \sum_{t=1}^{T} \gamma_t = \frac{\partial \Pi_{20}}{\partial \text{RR}}$$

11.9 Summary

In any real power system, supply and demand must be kept in balance at all points in time. Contingencies are shocks to the power system – due to changes in supply conditions, demand conditions or changes in network conditions. Examples include the sudden loss of a generating unit, the loss of a large electricity customer, or the loss of a major network asset. Power system operators spend a considerable amount of day-to-day effort in anticipating and managing contingencies.

In a real power system, supply–demand imbalances result in changes to the system frequency. In practice, managing contingencies is linked to maintaining system frequency within a narrow band. However, since the frequency is common across the power system, the maintenance of system frequency only relates to the overall supply–demand balance and ignores physical limits on individual elements of the power system. Therefore, maintaining system frequency within a narrow band is only a small part of the task of efficient operation of the power system in the short-run.

Efficient handling of contingencies is an extension of the optimal dispatch task discussed in detail in Part III. In principle, if the optimal dispatch task could be carried out sufficiently frequently, there would be no need for any separate handling of very short run contingencies.

In practice it is very common for power systems to take actions in advance of a contingency arising to mitigate the consequences of the contingency once it arises. We can distinguish preventive and corrective actions. Corrective actions are the actions taken postcontingency to restore the power system to a new equilibrium. Preventive actions are the actions taken precontingency in anticipation of the potential for a contingency to occur.

The need for preventive actions is closely linked to the concept of intertemporal constraints. If there are no intertemporal constraints (i.e. no ramp rate constraints or energy limits) binding postcontingency, the power system can adjust immediately to its new equilibrium. In this case there is no need for any corrective actions and no need for any preventive actions.

At the other extreme, if the adjustment of the power system to the new equilibrium is very costly it may be prudent to operate the power system *ex ante* as though the contingency has already occurred. Improving the efficiency of short-run dispatch reduces the cost of adjusting to a new equilibrium postcontingency and therefore reduces the need for preventive actions.

Since preventive actions are closely associated with intertemporal constraints, if intertemporal constraints may be binding postcontingency, the optimal dispatch task must be modified to take into account those contingencies that might lead to binding intertemporal constraints *ex post*. This requires (of course) an intertemporal optimisation.

Importantly, the efficient contingency-constrained dispatch can be achieved through a decentralised market process. Market participants, responding to the market prices, voluntarily choose the efficient level of preventive and corrective actions. Furthermore, market participants face efficient incentives to invest in expanding their physical capabilities (e.g. increasing their ramping capability) when it is efficient to do so.

Questions

11.1 Why, in practice, is it costly to take corrective actions? What are the primary sources of the cost of taking corrective actions? (Hint: the most significant sources of the cost of taking corrective actions are where the power system does not have the ability to make the necessary changes *ex post* – what sorts of short-run changes in dispatch may be outside the control of the power system operator?)

11.2 True or False: The optimal level of preventive actions depends on a balancing of the need to reduce the cost of corrective actions; if the preventive actions for different contingencies balance out, there may be no need to take preventive actions at all? Can you provide an example?

11.3 In an example in the text, generation of type 2 increased its output voluntarily. This decreased the market price. Is it possible to find an example where the market price is increased? What are the conditions required?

11.4 Find an expression for the γ_t as a function of $P_t - c_2$ (Section 11.8) and hence show that the incentive to increase the ramp rate depends on both the duration of the ramp rate constraint and the price path during that constraint.

Further Reading

The question of how to achieve optimal dispatch outcomes in the very short run is relatively new. You might like to look at Biggar and Hesamzadeh (2011) and Hemsazadeh, Galland and Biggar (forthcoming, International Journal of Electrical Power and Energy Systems).

12

Frequency-Based Dispatch of Balancing Services

Chapter 11 focused on the question of the efficient handling of contingencies. We saw that, at least in theory, contingencies can be efficiently handled through the normal optimal dispatch process operating over very short timeframes, taking into account the ramp rate and energy limit constraints that arise on those short timeframes. However, this is not how contingencies are handled in real power systems. Instead, it is typically the case that during the dispatch interval, balancing services are dispatched to handle frequency variation. This chapter focuses on the implications of this approach.

12.1 The Intradispatch Interval Dispatch Mechanism

In principle, as we have seen in Chapter 11, the efficient short-run dispatch can be achieved through the conventional optimal dispatch process. However, this might require running the dispatch process very frequently – perhaps even at time intervals much less than 1 s. There could be genuine practical difficulties in doing so – not least of which is handling the large information flows and computational requirements. In addition, frequent operation of the dispatch process gives rise to frequent determination of prices which, while efficient, results in a very substantial information processing and storage task by market participants and which complicates the process of defining risk-management instruments.

Let us suppose that it is simply not feasible to carry out the optimal dispatch process on a very short timeframe. What can we do instead? We still need some mechanism for maintaining the supply/demand balance in the very short-run. At the same time, we would like to make use of the economic efficiency of the optimal dispatch process as much as possible.

The only practical solution is to run the optimal dispatch process as frequently as is reasonably feasible, and then to rely on some other mechanism to approximate optimal dispatch in the interval between subsequent runs of the dispatch process. Let us suppose that the dispatch process is operated over a time period known as the *dispatch interval* or scheduling interval. This could be anywhere from a few seconds to a few minutes. Let us refer to the mechanism by

The Economics of Electricity Markets, First Edition. Darryl R. Biggar and Mohammad Reza Hesamzadeh.
© 2014 John Wiley & Sons, Ltd. Published 2014 by John Wiley & Sons, Ltd.

which the rate of production or consumption of market participants is adjusted during the dispatch interval as the *intradispatch interval dispatch mechanism* (IDIDM).

The key problem here is that the intradispatch interval dispatch mechanism is (almost by definition) imperfect and inefficient. It cannot (except perhaps by chance) replicate the outcome of an efficient dispatch process. In particular, the intradispatch interval dispatch mechanism either results in inefficient use of resources for a given set of physical network limits and/or does not ensure that all physical network limits are respected. There are several consequences:

- The cost of corrective actions are higher than they could be. In turn, the need for preventive actions is higher than it could be.
- The Power system must be operated more conservatively during the dispatch process to ensure that physical constraints are not violated during the intradispatch interval period.

12.2 Frequency-Based Dispatch of Balancing Services

In practice, it is typically the case that the dispatch of generators during each dispatch interval occurs in response to changes in the system frequency.

The situation in the Australian National Electricity Market (NEM) is typical: In each dispatch interval, a certain set of market participants is 'enabled' to provide a certain volume (MW) of these frequency control services (also known as frequency control ancillary services or FCAS). There are two types of frequency control services:

- *Regulation service,* which is intended to address small supply–demand imbalances. Market participants enabled to provide regulation service respond to signals from the system operator at a relatively high frequency (every few seconds);
- *Contingency services*, which address larger supply–demand imbalances. Market participants that are enabled to provide contingency services act autonomously in response to the observed local frequency (without waiting for a signal from the system operator).

The contingency services are further divided into three types: fast, medium and slow. Market participants enabled for the fast FCAS services must be able to offer up to the full promised MW increase or decrease in net injection within 6 s and must be able to sustain this increase or decrease for up to 60 s. Participants enabled for the medium FCAS must be able to offer the full MW increase or decrease in net injection within 60 s and must be able to sustain this increase or decrease for up to 5 min. Participants enabled for the slow FCAS are allowed to ramp up or down to the promised volume over 5 min.

The combined effect of these frequency control services is to create an *ad hoc* short-term dispatch process that operates during each 5 min dispatch interval. During each dispatch interval, market participants that are not enabled for these frequency control services are expected to change their production or consumption in a smooth, linear manner to their new dispatch target, independent of system contingencies. On the other hand, market participants enabled for frequency control services are expected to, in addition, either (a) respond to increase/decrease signals from the system operator (in the case of market participants providing the regulation service); or (b) respond to variations in the system frequency by ramping up or down at the fast, medium or slow rate promised (in the case of market participants providing the contingency service).

This frequency-based intradispatch interval dispatch mechanism is adequate for maintaining supply/demand balance in the very short run. However, it is of course not economically efficient, for several reasons as follows:

- All relative cost information is ignored. In principle, dispatch should be according to merit order, with lower-cost generators dispatched ahead of higher-cost generators. Instead, with this mechanism, all market participants enabled for, say, fast raise service, respond at the same time independent of their cost.
- The fixed fast/medium/slow time classifications may not reflect the physical capability of market participants. For example, a market participant that is capable of responding in, say, 1 s, can offer a fast frequency raise service, but has no incentive to offer this service over 1 s, since the fast service only requires the service be offered over 6 s. A generator that is capable of ramping in, say, 2 min can offer the slow raise service, but has no incentive to offer the increase in output over 2 min, since it is only technically required to provide the service over 5 min.
- Most important of all, all physical network constraints are ignored. No account is taken of the geographic location of the market participants that provide the frequency control services. Network flow limits may be violated postcontingency. The implications of this are quite significant and are discussed in more detail in Section 12.3.

12.3 Implications of Ignoring Network Constraints when Handling Contingencies

As we have just noted, the typical intradispatch interval dispatch mechanism is based entirely on correcting variation in the system frequency. However, as noted earlier, the loss of a network element will not, in general, have much (if any) impact on system frequency. After all, putting aside losses, the amount of power injected into a network element is equal to the amount of power withdrawn. Immediately following the loss of the network element, provided the loss of the network element does not result in the physical disconnection of different parts of the power system (which is very unusual in a meshed electricity network), the total power injected and withdrawn remains the same.

In other words, an IDIDM that focuses entirely on system frequency cannot efficiently handle network outages. Even more importantly, physical network limits may be violated *ex post*, following a contingency.

However, many network constraints cannot be violated, even in the very short run. As a consequence, if the IDIDM focuses entirely on system frequency, the power system must be operated in each dispatch process in such a way that the next credible contingency results in power flows that are within the physical network limits.

In effect, since the adjustment costs are arbitrarily large (no *ex post* adjustment is feasible), the power system must be operated *ex ante* as though the next credible contingency has already occurred. This results in substantial unutilised capacity on electricity networks (especially on transmission networks).

For example, Figure 12.1 illustrates a simple two-circuit transmission line. If each circuit of the two-circuit transmission line is capable of carrying 1000 MW of power (and if the two circuits are electrically identical), the combined transmission link is physically capable of carrying 2000 MW of power.

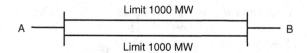

Figure 12.1 A simple two-circuit transmission line

However, with a frequency-based IDIDM, what is the maximum power flow that can be permitted to be carried from A to B? If one of the circuits should fail, the power flow would immediately be transferred to the remaining circuit (no *ex post* corrective actions are feasible). Since each circuit has a limit of 1000 MW, no more than 1000 MW of power can be allowed to flow in the normal dispatch process. In other words, 1000 MW of physical capacity is kept idle at all times.

The key result here is that the need to ensure that physical limits are not violated *ex post* significantly reduces the effective operating limits of the power system *ex ante*.

In practice, the result is not quite as bad as we have presented here. Some network elements can tolerate short periods of operation above their physical steady-state long-term limits. This might allow the power system long enough for the conventional dispatch process to bring about the corrective actions needed. For example, let us suppose that the circuits in the network could tolerate a power flow of 1100 MW for 5 min. Also, let us suppose that there is sufficient adjustment capability at both A and B to reduce the flow on the network element back to 1000 MW within a 5 min interval, then the *ex ante* power flow from A to B could be allowed up to 1100 MW. This is an improvement, but still (of course) very much less than the theoretical physical maximum of 2000 MW.

In the example above, under normal operation, each network element was only operated to 50% of its physical capability. However, the situation could be significantly worse. For example, Figure 12.2 illustrates a simple four-node, four-link network carrying power from A to B. All the links in the loop are assumed to have identical electrical characteristics. As a consequence, of the 2000 MW injected at A, 1500 MW flows via the short route and 500 MW flows via the longer route. Now suppose the loss of the short path is a credible contingency. Immediately following the loss of this network element, the 2000 MW of power would flow around the longer path. The power flow around the longer path would increase from 500 to 2000 MW. If this network is to be capable of delivering 2000 MW of power from A to B, the network elements on the longer path must be capable of carrying 2000 MW. In other words, under normal operation, the network elements on the longer path are only operated to 25% of their physical capability.

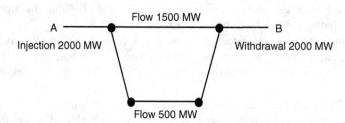

Figure 12.2 The loss of a network element may substantially increase the flows on the remaining network elements

> *Result*: In the event that corrective actions only focus on maintaining the system frequency within certain bounds, the power system must be operated in such a way that the power system remains in a satisfactory operating state following the loss of any network element. This can result in substantial excess capacity on power networks under normal operation.

12.3.1 The Feasible Set of Injections with a Frequency-Based IDIDM

In Chapter 7, we introduced the notion of the set of feasible injections. However, as we have seen, if we ignore network limits in the intradispatch interval dispatch mechanism, then we must operate the power system in the dispatch process in such a way that it can handle power flows immediately postcontingency. Let us explore the implications of this for the set of feasible injections in a simple three-node network.

Consider the simple three-node network illustrated in Figure 12.3. The loss of any of these network links is considered to be a credible contingency. The failure of one of the elements of a network will change the network's configuration and (therefore) its matrix of distribution factors.

The loss of any one of the three links leads to a new network with a new matrix of distribution factors. For example, the loss of the link $1 \to 2$ yields the network illustrated in Figure 12.4. As we noted in Section 12.2, taking node 3 as the reference node, this network has the following matrix of power transfer distribution factors:

$$H = \begin{matrix} 1 \to 3 \\ 2 \to 3 \end{matrix} \begin{pmatrix} 1 & 0 \\ 0 & 1 \end{pmatrix}$$

Similarly, the loss of the link $2 \to 3$ yields the network illustrated in Figure 12.5, which has the following matrix of power transfer distribution factors:

$$H = \begin{matrix} 1 \to 3 \\ 1 \to 2 \end{matrix} \begin{pmatrix} 1 & 1 \\ 0 & -1 \end{pmatrix}$$

Figure 12.3 A simple three-node network

Figure 12.4 The network in Figure 12.3 following the loss of the link 1-2

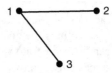

Figure 12.5 The network in Figure 12.3 following the loss of the link 2-3

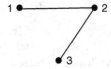

Figure 12.6 The network in Figure 12.3 following the loss of the link 1-3

Finally, the loss of the link $1 \rightarrow 3$ yields the network illustrated in FIgure 12.6, which has the following matrix of power transfer distribution factors:

$$H = \begin{matrix} 1 \rightarrow 2 \\ 2 \rightarrow 3 \end{matrix} \begin{pmatrix} 1 & 0 \\ 1 & 1 \end{pmatrix}$$

The set of injections that are feasible must not only be feasible in the original network, but feasible under each of these contingencies. This implies that the set of feasible injections must simultaneously satisfy all of the following conditions:

(a) For the original network

$$-K_{13} \le \tfrac{2}{3} z_1 + \tfrac{1}{3} z_2 \le K_{13}$$

$$-K_{12} \le \tfrac{1}{3} z_1 - \tfrac{1}{3} z_2 \le K_{12} \text{ and}$$

$$-K_{23} \le \tfrac{1}{3} z_1 + \tfrac{2}{3} z_2 \le K_{23}$$

(b) For the network with the loss of link 1–2

$$-K_{13} \le z_1 \le K_{13} \text{ and } -K_{23} \le z_2 \le K_{23}$$

(c) For the network with the loss of link 2–3

$$-K_{13} \le z_1 + z_2 \le K_{13} \text{ and } -K_{12} \le z_2 \le K_{12}$$

(d) For the network with the loss of link 1–3

$$-K_{12} \le z_1 \le K_{12} \text{ and } -K_{23} \le z_1 + z_2 \le K_{23}$$

In fact the conditions (a) are redundant since any injection that satisfies the conditions (b), (c) and (d) also satisfies the conditions (a). The effect of these additional constraints on the set of

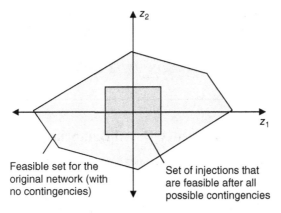

Figure 12.7   The set of feasible injections after contingencies with small capacity on link 1-2

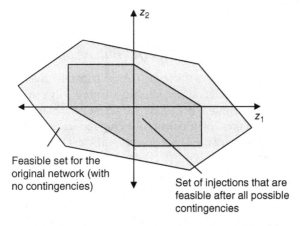

Figure 12.8   The set of feasible injections after contingencies with large capacity on link 1-2

feasible injections is illustrated in Figure 12.7 and Figure 12.8. As before, the shape of the feasible set depends on the size of the link $1 \rightarrow 2$. If the link $1 \rightarrow 2$ is small relative to the capacity of the other two links, the feasible set is as illustrated in Figure 12.7.

If the link $1 \rightarrow 2$ is larger than, or equal to, the capacity of the other two links, the feasible set is as illustrated in Figure 12.8.

In either case, as we have observed earlier, the set of injections that remains feasible after all possible contingencies is much smaller than the set of injections that is feasible in the absence of any contingencies. This is the typical case.

Result: Suppose that the system operator must take preventive action against a contingency corresponding to the loss of transmission link l. Let H^{-m} be the matrix of distribution factors for the network in the event of the contingency corresponding to the loss of the link m. Let us assume that the thermal and stability limits on the remaining lines are independent of the contingency. Then, a set of injections, Z, is feasible under all contingencies if and only if the power flows under each possible contingency are feasible, that is if and only if

$$\forall m, l, \sum_i H_{li}^{-m} Z_i \leq K_l$$

12.4 Procurement of Frequency-Based Balancing Services

In the previous sections, we have assumed that there are a set of market participants that are 'enabled' to provide balancing services – that is, market participants that stand ready to increase or reduce their rate of production or consumption either in response to signals from the system operator (in the case of regulation service) or in response to observed variation in the system frequency (in the case of contingency services).

However, several questions remain: Which market participants should be enabled to provide these services, and by how much? What overall volume of frequency control services is required? And how should the costs be allocated?

12.4.1 The Volume of Frequency Control Balancing Services Required

If we assume that contingencies are sufficiently rare such that the chance of two contingencies happening in any one dispatch interval is negligible, we need only cater for one contingency in each dispatch interval. Furthermore, if we have sufficient balancing services to handle, say, a 100 MW contingency, then there is sufficient balancing services to handle any contingency of less than 100 MW. As a result, we need only procure sufficient balancing services to handle the largest credible contingency.

It is not normally the case that demand increases suddenly by a very large amount. The largest electricity consumers may have some impact on the system frequency, but this can usually be managed through requirements, such as advanced notice before commencing operations or scaling-up operations. In practice, the largest credible contingency leading to shortage of supply relative to demand (and therefore leading to a reduction in system frequency) is almost always the loss of the single largest generating unit. For example, if the largest generating unit is producing 500 MW, then 500 MW of frequency raise service is required.

Similarly, it is not normally the case that generation increases suddenly by a very large amount. As a consequence, the largest credible contingency leading to an increase in the system frequency is the loss of the largest single load.

Therefore, the volume of raise service purchased is typically equal to the output of the largest single generating unit and the volume of lower service purchased is equal to the consumption of the largest consuming unit.

Let us suppose the largest generating unit is producing 500 MW. Should the system operator always procure 500 MW of raise service no matter what the price?

The answer depends (amongst other things) on the effectiveness of the integration of the demand side into the wholesale market. As we saw earlier, if the demand side cannot participate effectively in the wholesale market, it may make sense to impose a price cap on the energy market. The same principle applies here. If the demand side of the market cannot provide balancing services, it may not make sense to continue to procure a given volume of raise service, no matter what the cost. Instead, if the price is high enough, it may make sense to, in effect, provide balancing services through involuntary load shedding. In this case, the system operator can reduce the amount of raise service it procures through the market and rely instead on shedding load in the event of a contingency. This is undesirable, but may be better than the alternative.

Result: In the event that a large number of customers are unresponsive to wholesale market conditions, it may not make sense to always procure sufficient balancing services to cater for the loss of the largest load. Instead, if the price for balancing services is large enough, it may be preferable to reduce the volume procured and, instead, allow for a volume of involuntary load-shedding.

12.4.2 Procurement of Balancing Services

How should the system operator determine which market participants should be enabled to provide balancing services? One approach is to select the cheapest provider through a tendering process. That tendering process might take place periodically – say, every month or every week.

In the Australian NEM, the market participants that are enabled to provide frequency control services are selected through a tendering or procurement process, which operates on the same time frequency as the dispatch process – in other words, the market participants enabled to provide balancing service are chosen at each dispatch interval.

Procuring balancing services on the same time schedule as the dispatch process has the advantage that it allows the provision of balancing services by a market participant to be traded-off against the provision of electrical energy.

The volume of balancing services that a market participant can provide depends on (a) the rate at which that market participant is producing or consuming energy, and (b) the physical limits on the rate at which that market participant can produce or consume energy.

For example, a generator that is physically capable of producing 500 MW, and which is given a dispatch target of 490 MW, cannot offer more than 10 MW of frequency raise service. Conversely, suppose the same generator has a minimum stable generation level of, say, 300 MW. If it is dispatched for 350 MW it cannot offer more than 50 MW of frequency lower service.

In general, if a generator has a physical maximum rate of production $\overline{K}$, the sum of the current rate of production Q and the amount of raise service the generator can offer Q^R must be less than the physical maximum:

$$Q + Q^R \leq \overline{K}$$

In addition, there is usually an independent maximum on the total amount of raise service a generator can offer: $Q^R \leq \overline{K}^R$. The set of all feasible combinations of energy output and raise service (Q, Q^R), therefore, has a shape like that illustrated in Figure 12.9.

The optimal-dispatch task of the system operator is now generalised to include the procurement of balancing services. Let us suppose that a system operator seeks to procure a volume Q^E of electrical energy, and amounts $Q^{FR}, Q^{FL}, Q^{SR}, Q^{SL}$ of fast-raise, fast-lower, slow-raise and

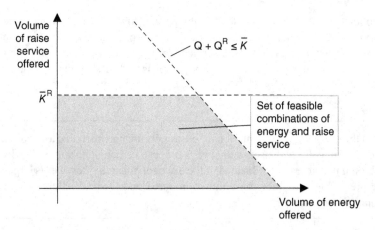

Figure 12.9   The set of feasible combinations of energy and raise service for a typical generator

slow-lower balancing services, respectively. Let us suppose that generator i submits bids into each of these markets, offering to produce energy at the variable cost of $\hat{c}_i^{E}$ and offering to the other markets at the costs of $\hat{c}_i^{FR}, \hat{c}_i^{FL}, \hat{c}_i^{SR}, \hat{c}_i^{SL}$, respectively. Let us suppose that this generator has a maximum and minimum rate of production $\overline{Q}_i^{E}$ and $\underline{Q}_i^{E}$ and maximum amounts of raise and lower service $\overline{Q}_i^{FR}, \overline{Q}_i^{FL}, \overline{Q}_i^{SR}, \overline{Q}_i^{SL}$, respectively.

The task of the system operator is as follows:

$$\min \sum_i \hat{c}_i^{E} Q_i^{E} + \sum_i \hat{c}_i^{FR} Q_i^{FR} + \sum_i \hat{c}_i^{FL} Q_i^{FL} + \sum_i \hat{c}_i^{SR} Q_i^{SR} + \sum_i \hat{c}_i^{SL} Q_i^{SL}$$

subject to the following:

$$\sum_i Q_i^{E} = Q^{E}, \sum_i Q_i^{FR} = Q^{FR}, \sum_i Q_i^{FL} = Q^{FL}, \sum_i Q_i^{SR} = Q^{SR}, \sum_i Q_i^{SL} = Q^{SL},$$

$$\forall i, \ Q_i^{E} + Q_i^{FR} + Q_i^{SR} \le \overline{Q}_i^{E}$$

$$\forall i, \ Q_i^{E} - Q_i^{FL} - Q_i^{SL} \ge \underline{Q}_i^{E}$$

$$Q_i^{FR} \le \overline{Q}_i^{FR}, Q_i^{FL} \le \overline{Q}_i^{FL}, Q_i^{SR} \le \overline{Q}_i^{SR}, Q_i^{SL} \le \overline{Q}_i^{SL},$$

This dispatch process will not, in general, achieve the efficient dispatch outcome.

12.4.3   Allocating the Costs of Balancing Services

The remaining question to address is how to recover the costs of balancing services. (Cost recovery is not an issue in optimal dispatch since, in that case, the prices ensure that each market participant is adequately compensated in equilibrium).

It seems sensible to attempt to allocate the costs of balancing services to those market participants that have deviated from their target rate of production or consumption and which therefore gives rise to the need for balancing services. One approach would operate as follows:

- No costs would be recovered in dispatch intervals in which there is no need for balancing services. In other dispatch intervals, a price would be set for either raise services (in the case where there is a need for increased injection) or for lower services (in the case where there is a need for reduced injection). This price would presumably vary with the volume of raise services required (or the volume of lower services).
- Market participants that have departed from their target in a way that increases the need for raise or lower services would be charged the raise or lower price; market participants that have departed from their target in a way that reduces the need for raise or lower services would be paid the raise or lower price.
- The difference between the amount raised in revenue and the amount paid out reflects the charge on the imbalance that should be used to offset the costs of procuring balancing services.

In practice, in the Australian NEM, the cost of raise services are charged to all generators on the basis of their registered capacity. The cost of lower services are charged to loads.

As an aside, it is worth noting that, as in any market, there is the potential for market power to arise in these markets for frequency control balancing services. Market power is discussed further in Chapter 15.

12.5 Summary

Chapter 11 showed that, at least in theory, the efficient handling of contingencies can be achieved through high-frequency application of the optimal dispatch process. However, this may not be feasible in practice. In any case, it has not been tried. In practice, the dispatch process is only operated at best every few minutes over a period of time known as the dispatch interval or scheduling interval. However, the system operating constraints, such as the supply–demand balance constraint, must be maintained over much shorter timeframes. Therefore, in practice, power systems that make use of an optimal dispatch process also make use of another process to dispatch generators on a timescale much shorter than the dispatch interval. This process is referred to here as the intradispatch interval dispatch mechanism or IDIDM.

The most common form of IDIDM dispatches generation during the dispatch interval on the basis of the system frequency alone. Commonly, some market participants are chosen to respond to high-frequency balancing signals from the system operator (known as AGC or regulation). Other market participants are chosen to respond autonomously to system frequency variation, standing ready to increase or decrease their rate of production or consumption in response to frequency changes.

This frequency-based IDIDM ignores network constraints. The outage of a network element will not normally have an impact on the overall system frequency. A frequency-based dispatch mechanism therefore cannot achieve efficient dispatch of the power system following a network contingency. As a consequence, since the physical network limits of the power system must be maintained at all times, even following a contingency, the power system must be operated (*ex ante*, in the normal dispatch process) in such a way that no system limits will be violated *ex post*. In effect this means that the power system must be operated as though a network outage has already happened. This results in operating limits on network elements that are potentially much less than the physical capability of the network element – requiring substantial idle capacity. Improvements in the IDIDM has the potential to significantly improve network utilisation.

In many markets, the market participants that provide frequency-control balancing services are chosen through a market tendering process. Typically, the volume of raise service procured is equal to the production of the largest single generating unit and the volume of lower service procured is equal to the consumption of the largest single load. In markets such as the Australian NEM, this procurement occurs every 5 min on the same time cycle as the dispatch process. This allows the procurement of balancing services to be traded-off with the provision of electrical energy. Consideration must be given to how the costs of balancing services procured in this way will be allocated across market participants.

Questions

12.1 Should the output of the largest generator be traded off with the price of balancing services?

12.2 Should the volume of FCAS procured be price-dependent?

Further Reading

You might want to take a look at the *AEMO Guide to Ancillary Services in the National Electricity Market*.

Part VI
Managing Risk

Twin hardwood poles carrying 110 kV transmission lines in the Horowhenua area, New Zealand
(Source: Neil Rennie)

We have seen that economic efficiency requires that market participants be exposed to the wholesale spot price for electricity. However, that price can be highly volatile. Electricity sellers face the risk of prolonged periods of low prices. On the other hand, electricity buyers face the risk of episodes of very high prices. Both parties may be reluctant to make investments without some way of reducing the risk they face. This is the role played by risk-management instruments. Risk-management tools bridge the gap between the volatile spot prices that provide the correct signal for efficient short-run use and operation of a set of assets, and the long-term price signals needed for efficient investment.

13

Managing Intertemporal Price Risks

Participants in wholesale electricity markets face volatile wholesale prices. This chapter discusses the standard tools for managing those risks, and some of the ways those tools can be combined to obtain the insurance that participants in the market desire. In this chapter, we focus on the case where there is a single wholesale spot price (network constraints are ignored). This allows us to focus on the question of managing intertemporal price risk. In Chapter 14, we focus on the question of managing differences in prices across different locations on the network (interlocational price risk).

13.1 Introduction to Forward Markets and Standard Hedge Contracts

Wholesale electricity markets are known to be highly volatile. The wholesale spot price may vary by multiples of hundreds or even thousands over the course of the day.

This volatility exposes market participants to risk. Where market participants are risk averse, exposure to risk reduces short-run economic welfare. Exposure to risk can also dampen incentives to make valuable sunk investments, reducing long-run economic welfare.

Participants in wholesale electricity markets will typically seek to minimise their exposure to price volatility. Market participants can reduce their exposure to this volatility by participating in the *forward markets*.

A trade in the forward market can be an agreement to take physical delivery of the good or service at a given time in the future at a fixed price. Alternatively (and more likely) a trade in the forward market may represent a financial agreement to pay the difference between the agreed forward price and the wholesale spot price at the designated time in the future.

Trading in the forward market has two substantial advantages for market participants: First, it allows market participants to shift risk. Second, it provides important signals about future expected market prices, allowing market participants to make investment decisions in the light of collective information, pooled by the market, about future average price levels.

For example, a baseload generator thinking about making a substantial investment in new generation capacity may seek to reduce the risk it faces by entering into a fixed-price sale agreement in advance. This agreement may cover the bulk of the output of the new generation capacity over many years of the life of the asset.

The Economics of Electricity Markets, First Edition. Darryl R. Biggar and Mohammad Reza Hesamzadeh.
© 2014 John Wiley & Sons, Ltd. Published 2014 by John Wiley & Sons, Ltd.

Many forward contracts, especially long-term forward contracts, are specially tailored arrangements, negotiated between a buyer and seller. Prior to the establishment of formal wholesale markets in electricity, a large proportion of electricity was purchased through long-term power purchase agreements.

Although specially tailored arrangements persist, in an active market with many buyers and sellers, the forward contracts often take certain standardised forms. This facilitates trade in the forward contracts themselves. This trade can be between buyers and sellers directly (which is known as *over-the-counter* trade), or it can be facilitated by an *exchange*.

One of potential problems with forward transactions is that their value depends on the financial capacity of the trading partner to complete the transaction. A promise to make a financial payment in the future is of little value if, at the point in time when the payment must be made, the trading partner is insolvent. Therefore, in the case of direct or over-the-counter trade, the parties will pay particular attention to the credit-worthiness of their counterparties. Many firms have rules that prevent them from dealing with counterparties with a credit rating lower than a threshold level. On the other hand, in the case of exchange-based trade, the exchange itself usually acts as the counterparty, taking on the counterparty risk for each side of the transaction, and reducing one of the transaction costs associated with trade.

13.1.1 Instruments for Managing Risk: Swaps, Caps, Collars and Floors

Financial contracts whose primary purpose is the management of risk are known as *hedge contracts*. Hedge contracts can take a wide range of forms. As noted earlier, hedge contracts may be specially tailored arrangements. However, we will focus here on the standardised hedge contracts known as swaps and caps (and their relatives – collars and floors). Contracts of this form make up the bulk of the contracts traded in the over the counter hedge market.

In order for a hedge contract to reduce the risk faced by a firm, the hedge contract must require the firm to make a financial payment (or to accept a financial payment) that is based on some observable outcome correlated with the firm's profit. The most common form of hedge contracts base the financial payment on the observed spot market price – which is usually highly correlated with the profit of the market participants.

Other risk-management instruments are sometimes available, such as contracts based on the weather (weather is highly correlated with total demand which is, in turn, a primary driver of profit). It is also possible to conceive of risk-management contracts based on other outcomes, such as generator outages or outages on transmission lines. There are also a number of other risk-management instruments, collectively known as 'exotics'. These include 'swaptions', 'weather derivatives', 'Asian options' and many others.

For the moment, let us focus on the most common form of hedge contracts, for which the financial payment is based exclusively on the market price – these include the well-known swaps and caps, as described next.

13.1.2 Swaps

The simplest forward contracts involve fixed volumes. For example, a farmer might agree to sell 100 tonnes of grain, for delivery in June 2020, at a fixed price today. A generator might agree to deliver 100 MWh of output between 5 and 6 p.m. on 3 January 2018. In the electricity industry, simple fixed-volume forward contracts are known as *contracts-for-differences* or *swaps*.

Swaps are a financial agreement under which one party (known as the seller) agrees to pay another party (known as the buyer) an amount equal to the difference between the spot price at a predetermined half-hour period and a predetermined fixed price multiplied by a predetermined quantity. Contracts of this form are called swaps because the effect of the contract is to swap a floating revenue stream for a fixed revenue stream.

In mathematical notation, if S is the fixed price and P is the (uncertain) future spot price at a given date and time, then a swap for a volume of X MW is a financial contract under which the buyer agrees to pay the seller the following amount at a given future date and time:

$$\text{Swap}(P, X, S) = (S - P)X$$

For example, when the spot price is \$100/MWh and the swap price is \$50/MWh, the amount $50X$ is paid by the buyer to the seller at the given date and time.

Of course, swaps may involve two-way flows of payments. If the spot price is higher than the fixed price, the swap seller pays the buyer the difference. On the other hand, if the spot price is lower than the fixed price, the swap buyer pays the seller the difference.

Although it is technically feasible to trade swap contracts applying only to a given date and time in the future, in practice it is common to group swaps with the same volume and similar time periods together. For example, in the Australian National Electricity Market (NEM), so-called peak swaps cover the hours from 7:30 a.m. to 10:00 p.m. on weekdays, over a month, a quarter, a calendar year or longer. So-called flat swaps cover all 48 half-hour periods in a day for (again) a month, a quarter, a calendar year or longer. Off-peak swaps can be fabricated by purchasing a flat swap and selling a peak swap for the same time volume and period.

Note that a peak swap can be viewed as a swap with a volume component that varies in a prespecified manner (the volume is zero at off-peak times and a fixed given volume at peak-times). More generally, it is technically feasible to construct a swap that has a volume that varies in a prespecified manner over the course of a day or year. This is known as a *sculptured swap*.

13.1.3 Caps

Caps (also known as 'call options' or 'one-sided obligations') are a financial agreement under which one party agrees to pay the difference between the spot price at a predetermined half-hour period and a given strike price multiplied by a predetermined quantity, but only in the event that the spot price exceeds the strike price. As compensation for this future payment stream the buyer pays the seller a fixed up-front sum.

In mathematical notation, if S is the strike price and P is the (uncertain) future spot price at a given date and time, then a cap for X units is a financial contract under which the seller agrees to pay the buyer $(P - S)X$ dollars if $P \geq S$ and nothing otherwise. In exchange the buyer pays the seller the fixed price P_C per unit of cap. The total payment from the buyer to the seller is as follows:

$$\text{Cap}(P, X, S, P_C) = (S - P)XI(P \geq S) + P_C X$$

Recall that $I(.)$ is the indicator function that takes the value 1 when the expression in brackets is true and 0 otherwise. Comparing the equations for the cap and a swap, we can see that the payoffs for a swap and a cap take the same form when the spot price is above the strike price for the cap. As a consequence we can see that a swap can be viewed as a special case of a cap contract – specifically, a cap contract with a strike price less than or equal to all possible realisations of the spot price. Suppose that S is set at such a low level that $P \geq S$ is always true then

$$\text{Cap}(P, X, S, P_C) = (S - P)XI(P \geq S) + P_C X = (S - P + P_C)X = \text{Swap}(P, X, S + P_C)$$

> *Result*: A swap can be viewed as a special form of a cap with a strike price lower than any possible realisation of the spot price.

As with swaps, caps are usually combined into groups covering several dates and times. For example, a cap might be sold covering a one-month period. The seller must then pay the buyer the difference between the spot price and the strike price times the cap volume in any half-hour period in that one-month period. (Technically, this form of cap is equivalent to a European call option with multiple exercise dates within the contract period).

13.1.4 Floors

Although swaps and caps are the most common standardised hedge products, it is worth mentioning a couple of variations, such as floors and collars.

A *floor* (or 'put option') is just the opposite of a cap. A floor is a financial agreement under which one party agrees to pay the difference between the spot price at a predetermined half-hour period and a given strike price multiplied by a predetermined quantity, but only in the event that the spot price falls below the strike price. As compensation for this future payment stream the buyer pays the seller a fixed up-front sum.

In mathematical notation, if S is the strike price and P is the (random) future spot price at a given date and time, then a floor for X units is a financial contract under which the seller agrees to pay the buyer $-(P - S)X$ dollars if $S > P$ and nothing otherwise. In exchange the buyer pays the seller the fixed price P_F per unit of floor. The total payment from the buyer to the seller is as follows:

$$\text{Floor}(P, X, S, P_F) = -(S - P)XI(P < S) + P_F X$$

It is worth observing that selling a swap of a given quantity is identical to selling a cap and buying a floor at the same strike price and quantity. This is an illustration of a result in financial mathematics known as put-call parity. To see this, observe that

$$\begin{aligned}
\text{Cap}(P, X, S, P_C) &- \text{Floor}(P, X, S, P_F) \\
&= (S - P)XI(P \geq S) + P_C X + (S - P)XI(P < S) - P_F X \\
&= (S - P + P_C - P_F)X = \text{Swap}(P, X, S + P_C - P_F)
\end{aligned}$$

Note that this implies that given any two of a swap, a cap and a floor (of the same quantity and strike price), it is possible to synthetically construct the third. For example, selling a floor can be constructed by selling a cap and buying a swap. Purchasing a financial asset is equivalent to taking a 'long' position in that asset. Conversely, selling a financial asset is equivalent to taking a 'short' position in that asset. We can therefore conclude that holding a floor (i.e. taking a long position in a floor) is equivalent to a long position in a cap plus a short position in a swap.

> *Result*: A swap can be 'created' from a long position in a cap combined with a short position in a floor with the same strike price and volume. A floor can be created from a long position in a cap and a short position in a floor of the same strike price and volume.

13.1.5 Collars (and Related Instruments)

A collar is an instrument under which the seller pays the buyer the difference between the spot price and a ceiling price when the spot price exceeds the ceiling and the buyer pays the seller the difference between the spot price and a floor price when the spot price drops below the floor price.

A collar can be constructed using a portfolio of caps and floors: Selling a collar with a ceiling price of S_C and a floor price of S_F and a quantity of X is identical to selling a cap with a strike price of S_C and buying a floor with a strike price of S_F. In other words

$$\text{Collar}(P, X, S_C, S_F, P_{\text{Coll}}) = Cap(P, X, S_C, P_C) - \text{Floor}(P, X, S_F, P_F)$$

Alternatively, using the result just proved above (that a floor is equivalent to buying a cap and selling a swap) a collar could be constructed by buying a swap, buying a cap at the ceiling price and selling a cap at the floor price.

Other combinations of caps and floors are, of course, possible. For example, a 'straddle' is constructed by selling a cap and a floor with the same strike price and volume. A 'strangle' is constructed by selling a cap and a floor with different strike prices and the same volume.

13.2 The Construction of a Perfect Hedge: The Theory

Let us explore how a price-taking generator might go about hedging the risks that it faces.

As we have throughout this text, let us assume that the generator can be represented in the short-run by a cost function. The generator's cost function $C(Q, \varepsilon)$ is assumed to depend on its rate of production Q and a set of uncertain cost-shifting factors ε, which affect the total cost of production (such as the level of input costs or generator availability).

Let us assume that this generator is a price-taker and faces an uncertain wholesale spot price for its output equal to P. Finally, let us assume that the generator sells a portfolio of hedge contracts. In the most general case this portfolio has a payoff that depends only on spot price P and the factors that affect the cost function ε. The total payout on the hedge portfolio is given by $H(P, \varepsilon)$.

The total hedged profit of the generator when it faces a price P and chooses a level of output Q is therefore given by

$$\pi(P, Q, \varepsilon) = PQ - C(Q, \varepsilon) - H(P, \varepsilon)$$

As before, we will assume that the generator chooses its rate of production in such a way as to maximise its profits, given the market price. The profit-maximising level of production for a price-taking generator is where the marginal cost is equal to the price:

$$\frac{\partial C}{\partial Q}(Q(P, \varepsilon), \varepsilon) = P$$

Substituting the profit-maximising rate of production back into the profit function, we find that the profit of the generating firm is a function of spot price and the cost-shifting factors as follows:

$$\pi(P, \varepsilon) = PQ(P, \varepsilon) - C(Q(P, \varepsilon), \varepsilon) - H(P, \varepsilon)$$

From this expression we can see that a generating firm faces the following sources of uncertainty in its profit function (before taking into account the impact of hedging):

a. Uncertainty in the generator cost function arising from (i) changes in input prices, (ii) outages, (iii) other factors, such as industrial action. These factors are reflected in the random variable ε.
b. Uncertainty in production volumes (due to changes in the supply, demand and network conditions, reflected in the variable rate of production $Q(P, \varepsilon)$);
c. Uncertainty in the spot price, which is reflected in the random variable P.

13.2.1 The Design of a Perfect Hedge

Let us define the concept of a perfect hedge as follows:

- A *perfect price hedge* eliminates the variability of the hedged profit of the generator with respect to the spot price.
- A *perfect cost-shifting hedge* eliminates the variability of the hedged profit of the generator with respect to the cost-shifting factors.

An overall *perfect hedge* completely eliminates the variability of the hedged profit of the generator.

Differentiating the hedged profit function and using the envelope theorem, we have the following results:

$$\frac{\partial \pi}{\partial P}(P, \varepsilon) = Q(P, \varepsilon) - \frac{\partial H}{\partial P}(P, \varepsilon) = 0$$

and

$$\frac{\partial \pi}{\partial \varepsilon}(P, \varepsilon) = -\frac{\partial C}{\partial \varepsilon}(Q(P, \varepsilon), \varepsilon) - \frac{\partial H}{\partial \varepsilon}(P, \varepsilon) = 0$$

Hence, we have the following conditions: A generator obtains a *perfect price hedge* if the rate of change of the hedge payoff with respect to the price is equal to the rate of production of the generator:

$$\frac{\partial H}{\partial P}(P, \varepsilon) = Q(P, \varepsilon)$$

We can define the *volume associated with a hedge portfolio* as the rate of change of the payout with respect to a change in the price:

$$V(P, \varepsilon) = \frac{\partial H}{\partial P}(P, \varepsilon)$$

This analysis shows that, a generator with a certain cost function (i.e. no cost-shifting factors) can obtain a perfect price hedge by obtaining a portfolio that has a volume that precisely matches the profit-maximising rate of production of the generator:

$$V(P, \varepsilon) = Q(P, \varepsilon)$$

Result: A perfect price hedge for a generator can be obtained by creating a portfolio of hedge products with a volume equal to the output of the generator.

Similarly, a generator obtains a *perfect cost-shifting hedge* if the rate of change of the hedge payoff with respect to the cost-shifting factors is equal to the (negative of the) rate of change of the cost-function with respect to the cost-shifting factors:

$$\frac{\partial H}{\partial \varepsilon}(P, \varepsilon) = -\frac{\partial C}{\partial \varepsilon}(Q(P, \varepsilon), \varepsilon)$$

Finally, given a profit-maximising supply curve $Q(P)$, consider the hedge contract defined as follows:

$$H(P) = \int^{P} Q(p) \mathrm{d}p$$

This has the property that the associated volume is just equal to $Q(P)$ and so yields a perfect hedge. We can also write this hedge contract in an equivalent way that suggests how we can create a perfect hedge using cap contracts:

$$H(P) = \int^{Q(P)} \left(P - \frac{\partial C}{\partial Q}(q, \varepsilon) \right) \mathrm{d}q$$

As we will see in Section 13.3, in the special case where there are no cost-shifting factors, a perfect hedge can be constructed using cap contracts alone.

13.3   The Construction of a Perfect Hedge: Specific Cases

13.3.1   Hedging by a Generator with no Cost Uncertainty

Let us now simplify to the case where the cost function of the generator is certain (there are no cost-shifting factors).

Consider first the case of a generator with a constant variable cost equal to c and a capacity of K. We know that the profit-maximising choice of rate of production of such a generator is to produce nothing when the spot price is below the variable cost, and to produce at capacity when the spot price is above the variable cost:

$$Q(P) = \begin{cases} 0, & P < c \\ 0 \leq Q \leq K, & P = c \\ K, & P > c \end{cases}$$

The profit of the firm therefore can be written as follows:

$$\pi(P) = (P - c)KI(P \geq c) - H(P)$$

Consider the following hedge contract, which is just a cap contract with a strike price equal to the variable cost of the generator and a volume equal to the capacity of the generator:

$$H(P) = (P - c)KI(P \geq c) = \mathrm{Cap}(P, c, K)$$

This is illustrated in Figure 13.1. Clearly, with this hedge contract, this generator with constant variable cost can completely eliminate the profit risk that it faces.

Result: A generator with constant variable cost and fixed capacity can obtain a perfect hedge of the risk that it faces by selling a cap contract with a strike price equal to the constant variable cost and a volume equal to the fixed capacity.

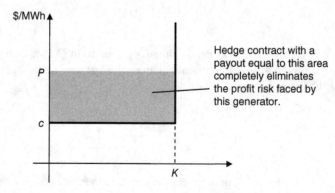

Figure 13.1   Hedge contract to eliminate profit risk for a simple generator with constant variable cost

More generally, let us suppose the generator's cost function took the form of a step function, with a variable cost of c_1 up to capacity K_1 and then a variable cost of c_2 up to capacity K_2 and so on. This generator could obtain a perfect hedge by purchasing a portfolio of hedge contracts. Specifically, this generator should purchase a volume of K_1 of a cap contract with a strike price of c_1, a volume of $K_2 - K_1$ of a cap contract with a strike price of c_2 and so on:

$$H(P) = \sum_{i=1}^{n} \text{Cap}(P, c_i, K_i - K_{i-1})$$

Here, we assume that $K_0 = 0$.

Let us now suppose that we have a range of cap contracts available with strike prices $S_0, S_1, S_2, \ldots$ These strike prices are assumed to be arbitrarily close together. Let us suppose we have a generator with an arbitrary cost function $C(Q)$. We know from Chapter 1 that the profit-maximising rate of production of such a generator is given by the inverse of the marginal cost curve:

$$Q(P) = C'^{-1}(P)$$

Let us suppose that $H(P)$ is the perfect hedge for this generator (i.e. the hedge contract that eliminates the variability in its profit). This generator can obtain an upper and lower bound to its perfect hedge by purchasing a portfolio of hedge contracts:

$$\sum_{i=1}^{n} \text{Cap}(P, S_i, Q(S_i) - Q(S_{i-1})) \leq H(P) \leq \sum_{i=1}^{n} \text{Cap}(P, S_{i-1}, Q(S_i) - Q(S_{i-1}))$$

Here, we are assuming that $Q(S_0) = 0$. This is illustrated in Figure 13.2. Moreover, as the strike prices become arbitrarily close together, the portfolio of hedge contracts can approximate arbitrarily closely the perfect hedge.

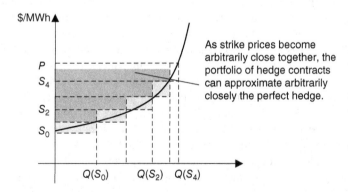

Figure 13.2 Given a set of strike prices, we can approximate the perfect hedge for a generator with an upward-sloping marginal cost curve

> *Result*: Given a set of cap contracts with strike prices arbitrarily close together, it is possible to form a hedge portfolio that approximates arbitrarily closely to the perfect hedge for a price-taking generator with an arbitrary supply curve $Q(P)$.

We saw earlier that a generator with constant variable cost and fixed capacity could achieve a perfect hedge with a single cap contract with a strike price equal to the variable cost and a volume equal to the capacity of the generator:

$$H(P) = \text{Cap}(P, c, K)$$

The volume associated with a single cap contract $\text{Cap}(P, S, X)$ is $V(P) = XI(P \geq S)$. This is equal to the profit-maximising output of the generator (except in the case where $P = S$, in which case the profit-maximising output of the generator is indeterminate).

Now consider the portfolio

$$\sum_{i=1}^{n} \text{Cap}(P, S_{i-1}, Q(S_i) - Q(S_{i-1}))$$

This has a volume $V(P) = Q(S_i)$, where S_i is the largest strike price just under P. As the strike prices become closer and closer together, this approaches the volume $Q(P)$ and therefore approximates the perfect hedge.

13.3.2 Hedging Cost-Shifting Risks

The previous analysis focuses on the case where a generator has a fixed and certain cost curve. We saw how such generators can, in principle, construct a portfolio that perfectly insulates them from the price risk they face. However, generators also face the risk of changes in the cost function, represented in the random cost-shifting factors ε.

For example, a generator might face the risk of changes in the price of its input fuel. Alternatively, a generator might experience an outage reducing its productive capability. What sort of instruments might a generator purchase to hedge against these risks?

It is worth emphasising that cost-shifting factors are often within the control of the generating firm. The amount that it pays for its input fuel and the reliability of its plant are factors over which, in principle, the generating firm has some control. If the generating firm could purchase a perfect hedge against these factors, it would have no incentive to take efficient actions – that is, no incentive to minimise its purchasing costs or to maximise its reliability. For this reason, such hedge products, where they exist at all, are usually carefully written to depend as much as possible on factors outside the control of the firm. Where they continue to depend on factors within the control of the firm there are typically partial exclusions that limit the extent of the hedge cover. These are equivalent to 'deductibles' and 'copayments' in insurance contracts.

Nevertheless, let us set aside these issues for a moment. Let us instead assume that factors, such as input prices or reliability, are outside the control of the generating firm. We will seek a perfect cost-shifting hedge that perfectly insulates the generator from these risks. This is a useful starting point for this analysis.

Let us focus first on the question of how a generator might go about hedging risks to its productive capacity. As before, let us focus on a generator with constant variable cost and fixed capacity. The capacity of the generator is normally K but may on occasion fall to $K - \varepsilon$. We can treat the capacity of the generator as a random variable. This generator can obtain a perfect hedge by constructing a hedge portfolio with the following payoff:

$$H(P, \varepsilon) = (P - c)(K - \varepsilon)I(P \geq c)$$

We can write this as the sum of two cap-like contracts. The first is a standard cap contract. The second contract pays out like a cap but on the drop in capacity of the generator.

$$H(P, \varepsilon) = \text{Cap}(P, c, K) - \text{Cap}(P, c, \varepsilon)$$

In other words, to hedge outage risk, this generator needs access to a hedge production that pays out like a standard cap contract but on a volume equal to the drop in capacity of the generator. This is a form of 'outage insurance'.

Now consider how a generator might go about hedging input fuel price risk. As before, we will focus on a generator with constant variable cost and fixed capacity. The constant variable cost of the generator is normally c, but on occasion the variable cost of the generator rises to $c + \varepsilon$ where $\varepsilon > 0$. As before, we can treat the variable cost of the generator as a random variable. This generator can obtain a perfect hedge by constructing a hedge portfolio with the following payoff:

$$H(P, \varepsilon) = (P - c - \varepsilon)KI(P \geq c + \varepsilon)$$

We can observe that this is equivalent to a cap-like contract that depends not on the spot price but on the spot price less ε:

$$H(P, \varepsilon) = \text{Cap}(P - \varepsilon, c, K)$$

This is not at all a standard cap contract. Rather, it is a cap contract that depends on the *difference* between the electricity spot price and the variable cost of the generator. If the variable cost of the generator is directly related to the input fuel price, then this is a cap contract that pays off on the difference between the electricity spot price and the input fuel spot price. The difference between the electricity spot price and, say, the natural gas spot price is known as the *spark spread*.

This analysis suggests that a generator that purchases natural gas on the spot market and that sells electricity on the spot market could obtain a perfect hedge using a financial instrument, which is like a cap but which depends on the difference between the electricity and the natural gas price.

But now let us suppose that such contracts do not exist. If this generator must hedge with a conventional cap contract, what sort of instrument should it purchase? We can write the required hedge contract as the sum of a conventional cap product and a less conventional product.

$$H(P, \varepsilon) = (P - c - \varepsilon)KI(P \geq c + \varepsilon)$$
$$= (P - c)KI(P \geq c) - (\min(P - c, \varepsilon)KI(P \geq c)$$
$$= \text{Cap}(P, c, K) - \min(P - c, \varepsilon)KI(P \geq c)$$

The additional hedge product required in this case is similar to a cap contract, but it pays out the lesser of the wholesale price P and the generator's cost $c + \varepsilon$, less the variable cost c, when the wholesale price is above this variable cost.

13.4 Hedging by Customers

The analysis in the previous section applies, by symmetry, to hedging by customers.

Here, we sketch out the parallels and identify the main results. As we have throughout this text, we will assume that the customer can be represented in the short-run by a utility function. The rate at which the customer receives utility, $U(Q, \varepsilon)$, depends on the rate of consumption, Q, and a set of random utility-shifting factors, ε.

As before, we will assume that this customer is a price-taker and faces an uncertain wholesale spot price for its output equal to P. The customer is assumed to purchase a portfolio of hedge contracts $H(P, \varepsilon)$, which depend on the spot price and the utility-shifting factors. The total hedged net utility of the customer is as follows:

$$\varphi(P, Q, \varepsilon) = U(Q, \varepsilon) - PQ - H(P, \varepsilon)$$

Using the fact that the customer chooses the rate of consumption to maximise the net utility given the price and the utility-shifting factors, $Q(P, \varepsilon)$, the overall rate at which the customer receives net utility is as follows:

$$\varphi(P, \varepsilon) = U(Q(P, \varepsilon), \varepsilon) - PQ(P, \varepsilon) - H(P, \varepsilon)$$

The conditions for perfect hedging are as follows: The customer can obtain a perfect price hedge if the rate of change of the hedge payoff with respect to the price is equal to the negative of the rate of consumption of the customer:

$$\frac{\partial H}{\partial P}(P, \varepsilon) = -Q(P, \varepsilon)$$

As before, we can say that the volume associated with the hedge contract must be equal to the (negative of the) profit-maximising rate of consumption.

Similarly, the customer can obtain a perfect utility-shifting hedge if the rate of change of the hedge payoff with respect to the cost-shifting factors is equal to the rate of change of the utility-function with respect to the utility-shifting factors:

$$\frac{\partial H}{\partial \varepsilon}(P, \varepsilon) = -\frac{\partial U}{\partial \varepsilon}(Q(P, \varepsilon), \varepsilon)$$

13.4.1 Hedging by a Customer with a Constant Utility Function

Let us now focus on the case where the utility function of the customer is fixed over time, resulting in the downward-sloping demand function $Q(P)$. From the previous analysis we know that, in principle, we can construct a perfect hedge from the demand function as follows:

$$H(P) = \int_P^{\overline{P}} Q(p)\,dp$$

Here, $\overline{P}$ is a price so high that the demand is zero $Q(\overline{P}) = 0$. Equivalently, as before, we can write the perfect hedge for a customer in a way that is close to a cap contract:

$$H(P) = \int^{Q(P)} \left(\frac{\partial U}{\partial Q}(q, \varepsilon) - P\right) dq$$

Let us suppose the customer has a demand function that takes a step function form, as set out next and as illustrated in Figure 13.3:

$$Q(P) = \begin{cases} Q_1, & V_2 \leq P < V_1 \\ Q_2, & V_3 \leq P < V_2 \\ Q_3, & P < V_3 \end{cases}$$

This customer can obtain a perfect hedge by purchasing a portfolio of hedge contracts with the following payoff:

$$H(P) = (V_1 - P)Q_1 I(P < V_1) + (V_2 - P)(Q_2 - Q_1)I(P < V_2) \\ + (V_3 - P)(Q_3 - Q_2)I(P < V_3)$$

This hedge payoff (up to a constant) can also be constructed from a portfolio of swaps and caps. Specifically, the required payoff can be obtained by selling a swap with volume Q_3 and buying a portfolio of three caps with different strike prices:

$$H(P) = -\text{Swap}(P, Q_3) + \text{Cap}(P, V_1, Q_1) + \text{Cap}(P, V_2, Q_2 - Q_1) + \text{Cap}(P, V_3, Q_3 - Q_2)$$

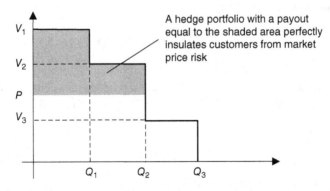

A hedge portfolio with a payout equal to the shaded area perfectly insulates customers from market price risk

Figure 13.3　Illustration of a perfect hedge contract for a customer with a simple demand function

We can check that this is the correct portfolio by differentiating with respect to the price. The resulting associated volume is as set out here, as required.

$$Q(P) = \begin{cases} Q_1, & V_2 \leq P < V_1 \\ Q_2, & V_3 \leq P < V_2 \\ Q_3, & P < V_3 \end{cases}$$

This result can clearly be generalised. Given enough cap contracts with a range of tightly spaced strike prices, it is possible to arbitrarily closely approximate the demand curve of any customer.

> *Result*: Given a set of cap contracts with strike prices arbitrarily close together, it is possible to form a hedge portfolio that approximates arbitrarily closely to the perfect hedge for a price-taking customer with an arbitrary demand curve $Q(P)$.

13.4.2 Hedging Utility-Shifting Risks

The previous analysis has focused on the case where the demand curve of the customer is fixed. However, in practice, the demand curve of a customer could vary with a variety of factors that affect his/her consumption of electricity, such as (in the case where the customer is itself a firm) the demand for the final good or service produced using the electricity or (in the case where the electricity is used for heating or cooling) the ambient temperature. These other factors could be expected to change the shape of the demand curve.

Let us consider hedging the simplest possible utility function in which the customer's demand curve is vertical at the level L up to some maximum valuation V, at which point the customer's demand drops to zero. Here, L represents the utility-shifting factor.

$$U(Q, L) = VQI(Q \leq L)$$

The net utility is therefore as follows:

$$\varphi(P, Q, L) = U(Q, L) - PQ = VQI(Q \leq L) - PQ$$

When we substitute in the utility-maximising rate of production, we find that the net utility as a function of the market price P and the utility-shifting factor L, is as follows:

$$\varphi(P, L) = (V - P)LI(P \leq V)$$

This is illustrated in Figure 13.4.

Since the volume of the contract varies with the utility-shifting factors, it is not possible to construct a perfect hedge for this generator using swaps and caps alone. In the case where the

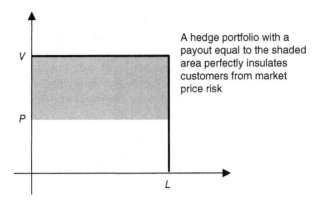

Figure 13.4   Illustration of a simple hedge contract to hedge a simple demand function

value V is sufficiently high that the customer always chooses to consume, the perfect hedge contract takes the following form:

$$H(P, L) = (V - P)L$$

This contract has a payoff similar to a standard swap contract, except that the volume varies with the utility-shifting factors. This form of contract is known as a *load-following hedge*. Let us define a 'standard' load-following hedge as having the following payoff:

$$\text{LFH}(P, V, L) = (V - P)LI(P \le V)$$

(Recall that, in this expression, the load L is uncertain at the time of writing of the contract, so this is not simply a variant of a standard cap contract).

13.5   The Role of the Trader

In previous sections, we have derived the conditions for generators and consumers to obtain a perfect hedge separately and independently. Now let us explore the conditions for equilibrium in the hedge market overall.

Let us introduce the concept of the *trader*. A hedge market trader purchases hedge contracts from generators and customers.

Let us suppose that we have a set of generators and consumers in a market. Let us suppose that each seeks a perfect hedge. The traders comply by purchasing hedge contracts $H_i^S(P, \varepsilon_i^S)$ from generators and $H_i^B(P, \varepsilon_i^B)$ from customers (loads).

Let us consider first the position of the set of all traders in the market. The total payout on all the hedge contracts is as follows:

$$\sum_i H_i^S(P, \varepsilon_i^S) + \sum_i H_i^B(P, \varepsilon_i^B)$$

Let us consider how this total payout varies with the wholesale spot price. Differentiating with respect to the price, we find the volume associated with each hedge contract. For a perfect price hedge, as we have seen, the volume associated with each contract matches the actual output of each generator and the actual consumption of each customer. At the level of the market as a whole, the total production is equal to the total consumption, so the total payout on the set of all the hedge contracts is independent of the market price:

$$\sum_i \frac{\partial H_i^S}{\partial P}\left(P, \varepsilon_i^S\right) + \sum_i \frac{\partial H_i^B}{\partial P}\left(P, \varepsilon_i^B\right) = \sum_i Q_i^S\left(P, \varepsilon_i^S\right) - \sum_i Q_i^B\left(P, \varepsilon_i^B\right) = 0$$

> *Result*: In an electricity market with no network congestion, if every generator and load acquires a perfect price hedge, the total payout on the sum of these hedges is independent of the spot price.

We can make a few more relevant observations. The first is that the energy balance equation uniquely determines the spot price. This price is a function of all the sources of uncertainty in the market – all of the cost-shifting factors for generators and all of the utility-shifting factors for customers – so we can write the spot price as $P(\varepsilon^S, \varepsilon^B)$, where ε^S is the vector of cost-shifting factors of the suppliers, and ε^B is the vector of utility-shifting factors of the buyers. This price must satisfy the following:

$$\sum_i Q_i^S\left(P\left(\varepsilon^S, \varepsilon^B\right), \varepsilon_i^S\right) = \sum_i Q_i^B\left(P\left(\varepsilon^S, \varepsilon^B\right), \varepsilon_i^B\right)$$

Note that there must be at least one source of uncertainty in the market to have any uncertainty in the spot price and therefore any need for hedging in the first place.

> *Result*: In a system with no network congestion, there must be at least one generator with an uncertain cost function, or one customer with an uncertain utility function for there to be uncertainty in the market price, justifying the need for price hedges.

It is important to recognise that even if every generator and every customer has a perfect hedge, there remains some risk in the industry that must be borne by the traders. This risk is the variability that arises due to the cost-shifting factors and utility-shifting factors, which translates into variability in the total hedge payout as follows:

$$\sum_i H_i^S\left(P\left(\varepsilon^S, \varepsilon^B\right), \varepsilon_i^S\right) + \sum_i H_i^B\left(P\left(\varepsilon^S, \varepsilon^B\right), \varepsilon_i^B\right)$$
$$= \sum_i U_i\left(Q_i^B\left(P\left(\varepsilon^S, \varepsilon^B\right), \varepsilon_i^B\right), \varepsilon_i^B\right) - \sum_i C_i\left(Q_i^S\left(P\left(\varepsilon^S, \varepsilon^B\right), \varepsilon_i^S\right), \varepsilon_i^S\right)$$

This remaining variability is known as *residual risk* and must be borne somewhere in the industry (by the traders, generators or customers).

> *Result*: Even if every generator and every customer receives a perfect hedge, as long as at least one generator has an uncertain cost function, or one customer has an uncertain utility function, there remains some volatility in the total economic welfare produced by the industry, which must be borne by some industry participant.

13.5.1 Risks Facing Individual Traders

Now let us focus on the position of an individual trader. An individual trader will purchase hedge contracts from generators and loads. In general, the volume associated with these hedge contracts need not sum to zero. However, *we will make the assumption that each trader seeks to avoid spot price risk and therefore chooses a hedge portfolio in which the associated volume sums to zero.*

In effect, this assumption implies that when the trader purchases contracts from generators, it seeks to balance those contracts with matching-volume contracts with customers. This practice eliminates the trader's exposure to spot-price risk (it does not eliminate all risk – since the trader may still face exposure to the cost-shifting factors, or the utility-shifting factors).

To see how this might operate in practice, consider the following simple example. In this example there is a single source of uncertainty. The consumption of customer C1 is either $L = 100$ MW in state A or $L = 200$ MW in state B. This difference in consumption is assumed to give rise to two different prices P_A and P_B where $P_A < P_B$. Customer C1 values each unit of consumption at the fixed amount V_1. By assumption, $P_A < P_B < V_1$.

Customer C2 has a certain utility function and consumes 80 MW in state A and 0 MW in state B. Customer C2 values each unit of consumption at the fixed amount V_2. By assumption, $P_A < V_2 < P_B$.

Generator G1 has a variable cost of c_1 and a capacity of $K_1 = 180$ MW. However, we will assume that the variable cost is sufficiently low so that this generator is producing at maximum capacity all the time $c_1 < P_A < P_B$. Generator G2 has a variable cost of c_2 and a capacity of $K_2 = 20$. This variable cost is such that this generator only produces in state B. $P_A < c_2 < P_B$.

The total output of the two generators is assumed to equal the total consumption of the two customers (which is 180 MW in state A and 200 MW in state B).

The raw or unhedged profit of each generator and customer is as set out in Table 13.1.

It is worth noting (as observed in the table) that the total unhedged profit and utility in the industry is independent of the price:

$$\frac{\partial \pi^{G1}}{\partial P} + \frac{\partial \pi^{G2}}{\partial P} + \frac{\partial \varphi^{C1}}{\partial P} + \frac{\partial \varphi^{C2}}{\partial P} = 180 + 20I(P(L) \geq c_2) - L - 80I(P(L) < c_2) = 0$$

Table 13.1   The profit of each generator and utility of each customer prior to hedging

Participant	Rate of Production or Consumption	Unhedged Profit or Utility
G1	180	$\pi^{G1}(P) = (P - c_1)180$
G2	$20I(P \geq c_2)$	$\pi^{G2}(P) = 20(P - c_2)I(P \geq c_2)$
C1	L	$\varphi^{C1}(P, L) = (V_1 - P)L$
C2	$80I(P < c_2)$	$\varphi^{C2}(P) = 80(V_2 - P)I(P < c_2)$

Table 13.2 Hedge contracts offered by trader alpha to the market participants

Participant	Trader Alpha Hedge Contract
G1	$H_\alpha^{G1}(P) = (P - c_1)(100 + 80I(P \geq c_2))$
G2	$H_\alpha^{G2}(P) = \mathrm{Cap}(P, c_2, 20) = (P - c_2)20I(P \geq c_2)$
C1	$H_\alpha^{C1}(P, L) = \mathrm{LFH}(P, V_1, L) = (V_1 - P)L$
C2	

Now let us suppose that we have two traders. Trader alpha seeks to provide a perfect hedge to customer C1 and trader beta seeks to provide a perfect hedge to trader C2.

Trader alpha offers the contracts to the market participants set out in Table 13.2.

Clearly, customer C1 is offered a perfect hedge. Note that with this set of contracts, for a given level of L, the payout of trader alpha is independent of the price:

$$\frac{\partial H_\alpha^{G1}}{\partial P} + \frac{\partial H_\alpha^{G2}}{\partial P} + \frac{\partial H_\alpha^{C1}}{\partial P} = 100 + 80I(P(L) \geq c_2) - L + 20I(P(L) \geq c_2) = 0$$

Trader beta offers the contracts to the market participants set out in Table 13.3.

Again, we see that customer C2 is offered a perfect hedge. In addition, the payout of trader beta is independent of the price:

$$\frac{\partial H_\beta^{G1}}{\partial P} + \frac{\partial H_\beta^{C2}}{\partial P} = 80I(P(L) < c_2) - 80I(P(L) < c_2) = 0$$

Finally, we need to show that generators G1 and G2 have a perfect hedge. This is easily checked:

$$H_\alpha^{G1}(P) + H_\beta^{G1}(P) = (P - c_1)80I(P < c_2) + (P - c_1)(100 + 80I(P \geq c_2))$$
$$= 180(P - c_1) = \pi^{G1}(P)$$

$$H_\alpha^{G2}(P) = (P - c_2)20I(P \geq c_2) = \pi^{G2}(P)$$

This example shows how traders may, by offering contracts with no price risk, collectively offer generators and loads the perfect hedges they require, while taking on no price risk on themselves. Of course, the traders collectively take on some risk – the residual risk defined earlier.

Table 13.3 Hedge contracts offered by trader beta to the market participants

Participant	Trader Beta Hedge Contract
G1	$H_\beta^{G1}(P) = (P - c_1)80I(P < c_2)$
G2	
C1	
C2	$H_\beta^{C2}(P) = (V_2 - P)80I(P < c_2)$

13.6 Intertemporal Hedging and Generation Investment

We have shown how, in principle, a generator or load can construct a portfolio of hedge contracts that perfectly insulates its risk. Where the cost function or utility function is constant over time, we have seen that the perfect hedge can, in principle, be constructed from a portfolio of standard cap contracts.

Let us now consider the position of an entity that is considering making an investment in a generation or load asset. This entity will also simultaneously purchase sufficient hedge contracts to ensure that the generation or load asset is fully hedged. The question we would like to consider is the following: Is it the case that this entity has efficient incentives to make investments?

Let us focus on the case of an investment in generating capacity. As in Chapter 9, let us focus on the simple case of generators with constant variable cost and fixed capacity. Furthermore, there are assumed to be constant returns to scale in generation, so that there is a constant cost of adding capacity. Finally, we will assume that generation capacity can be added in arbitrarily small lumps.

As in Chapter 9, let us assume that there are a finite number of different generation types. Generation type i is assumed to have a variable cost of c_i and a fixed cost of f_i.

From Section 8.3 we know that we have the optimal capacity of each generation type if for each generation type the expected gap between the price and the variable cost, at times when the price is above the variable cost, is equal to the unit cost of capacity:

$$E[(P - c_i)I(P \geq c_i)] = f_i$$

Now consider the position of an entrepreneur who is considering investing in a small amount of capacity of generation of type i. This entrepreneur is assumed to sell a hedge portfolio that makes the financial payments $H(P)$ and yields a fixed revenue P_H per unit of capacity. The hedged profit of the generation entrepreneur per unit of generation capacity is therefore as follows:

$$\pi(P) = (P - c_i)I(P \geq c_i) - H(P) + P_H - f_i$$

This entrepreneur is assumed to seek a perfect hedge. From the analysis in this chapter, we know that this entrepreneur can obtain a perfect hedge by selling a cap contract with a strike price of c_i and a volume equal to the volume of the new generation capacity. Let us suppose that cap contract has the price P_C per unit of capacity. The profit of the generation entrepreneur per unit of generation capacity is therefore as follows:

$$\pi(P) = (P - c_i)I(P \geq c_i) - \mathrm{Cap}(P, c_i) + P_C - f_i = P_C - f_i$$

In other words, a fully hedged generation entrepreneur will choose to invest if and only if the price of the corresponding cap contract exceeds the fixed cost per unit of capacity and vice versa.

But what can we say about the price of the cap contract? Let us assume that the market for cap contracts is competitive and there are a sufficient number of large, risk-neutral traders in that market so that the price for the cap contract is equal to the expected payoff. Then we have the following result: The price of the cap contract is equal to the expected gap between the spot price and the strike price at times when the spot price is above the strike price:

$$P_C = E[\text{Cap}(P, S)] = E[(P - S)I(P \geq S)]$$

Putting these results together, we find that the generation entrepreneur will invest if and only if the expected gap between the spot price and the variable cost at times when the spot price is above the variable cost exceeds the unit cost of adding capacity. In equilibrium, therefore, these two values must be the same:

$$P_C = E[(P - c_i)I(P \geq c_i)] = f_i$$

We have therefore proven that, at least in this simple case, the sale of a hedge contract does not distort the incentives for investment – the price of the hedge contract precisely reflects the relevant expected future price (in this case the area under the price–duration curve and above the variable cost of the generator).

Result: If the market for hedge contracts is competitive and the price for each hedge contract reflects the expected future payoff on that contract, then an entity facing a decision to invest in generation or load assets faces the correct economic signal for investment.

This is an important result. The volatile spot price provides each market participant with an important signal regarding the short-run use of a generation or consumption asset. However, it also exposes the generator or load to substantial risk. Hedge contracts can both offset that risk and provide the correct incentives for longer-term investment decisions in generation or consumption assets.

13.7 Summary

In previous chapters, we have seen how efficient operational and usage decisions in an electricity industry can be achieved through time-varying spot prices. However, this variability in spot prices exposes market participants to considerable risk. Market participants would often like to mitigate the risks they face through financial contracts known as hedge contracts. Although these contracts may be specially tailored bilateral arrangements between market participants, many hedge contracts take a standardised form that can be traded in over-the-counter markets or on a formal exchange. The standard forms of hedge contracts include swap contracts (also known as contracts-for-differences), cap contracts, floors, collars and a variety of other more exotic instruments, including hedge contracts that depend on weather or the outage of a generation plant.

For a price-taking generator, the variability in the profit received each period is due to variability in the wholesale price (which leads to variability in production) and variability in the cost function itself. This variability in profit can, in principle, be hedged through a hedge contract with an associated volume that is equal to the profit-maximising rate of production of the generator.

In the case of a generator with fixed variable cost, the variability in profit can be perfectly hedged through a cap contract with a strike price equal to that variable cost and a volume that matches the capacity of the generator. For a more general generator with an upward sloping marginal cost curve, the variability in the profit can be hedged arbitrarily closely by constructing a portfolio of cap contracts with varying strike prices.

Hedging factors that shift the cost function, such as changes in input costs or changes in capacity, require other, less common hedge products. Hedging variability in input costs can be achieved through a hedge contract that is similar to a cap contract that pays out based on the difference between the electricity spot price and the spot price of an input such as natural gas.

The same principles apply to hedging by electricity consumers. In principle a customer with a fixed downward sloping demand curve can hedge the spot price risk he/she faces by forming a portfolio of swap and cap contracts.

If the market for hedge contracts is competitive, the price per unit of the hedge contract reflects the expected future prices. In particular, the price of a cap contract reflects the area under the price–duration curve and above the strike price. As a result, despite insulating generating firms from spot price variability, an investor thinking about adding generation capacity that is perfectly hedged faces the correct incentives to do so.

Questions

13.1 A price-taking generator has a cost function given by $C(Q) = 10 + Q/1000$. Assume that the price is uniformly distributed between $10 and $100/MWh. If the generator has access to a range of cap contracts with strike prices separated by $5 (i.e. strike prices $0, $5, $10, $15, etc.), how closely can the generator eliminate the price risk it faces using a portfolio of cap contracts – in other words, what is the remaining volatility in the generator's profit function? If the gap in the strike prices reduced to $1, how much better can the generator eliminate its price risk?

13.2 What is the appropriate term of the cap contract that is used to signal the need for new investment in the network (Section 13.6)?

14

Managing Interlocational Price Risk

In Chapter 13, we implicitly assumed that network constraints could be ignored. This allowed us to focus on the problem of intertemporal hedging of wholesale spot prices. In this chapter, we reintroduce the potential for network constraints and explore what instruments we might require for hedging geographic variation of wholesale spot prices.

14.1 The Role of the Merchandising Surplus in Facilitating Interlocational Hedging

In Chapter 13, we introduced the concept of the hedge trader. A hedge trader purchases hedge contracts from generators and loads. We assumed that the individual generators and loads seek to achieve a perfect hedge. The traders seek to create portfolios that are independent of the spot price. Now let us generalise these ideas to the case of a system with network congestion.

In the presence of network congestion, there is a third set of factors that can affect the overall market outcomes – those factors that affect the state of the network. We might label the factors that affect network capability as ε^N. As long as there is uncertainty in the network capability, there can arise uncertainty in price and dispatch, even if every generator and load has a certain cost function and/or utility function.

As before, let us start by looking at the payout on the set of all hedge contracts in the entire market. Let us assume, without loss of generality, that generator i is located at pricing node i and faces nodal price P_i. The total payout on the hedge contracts held by this generator is assumed to be $H_i^S(P_i, \varepsilon_i^S)$. Similarly, load i also faces nodal price P_i and holds hedge contracts with a payout $H_i^B(P_i, \varepsilon_i^B)$. The total payout on all the hedge contracts is therefore as follows:

$$\sum_i H_i^S\left(P_i, \varepsilon_i^S\right) + \sum_i H_i^B\left(P_i, \varepsilon_i^B\right)$$

In Chapter 13, we saw that in the absence of network congestion this total payout is independent of the market price. Now let us ask the question: What additional flow of funds do traders collectively require in order to be able to collectively offer generators and loads a perfect hedge while taking no price risk on themselves? Let us call this additional flow of funds $T(P)$ where P represents the vector of nodal prices.

The Economics of Electricity Markets, First Edition. Darryl R. Biggar and Mohammad Reza Hesamzadeh.
© 2014 John Wiley & Sons, Ltd. Published 2014 by John Wiley & Sons, Ltd.

The total payout faced by traders collectively is the total payout on the hedge contracts plus this additional flow of funds:

$$\sum_i H_i^S\left(P_i, \varepsilon_i^S\right) + \sum_i H_i^B\left(P_i, \varepsilon_i^B\right) + T(P)$$

What form must this additional flow of funds take to ensure that traders face no price risk? Taking the partial derivative with respect to P_i, we find that the flow of funds must satisfy the following:

$$-\frac{\partial T}{\partial P_i} = Q_i^S\left(P_i, \varepsilon_i^S\right) - Q_i^B\left(P_i, \varepsilon_i^B\right) = Z_i\left(P_i, \varepsilon_i^S, \varepsilon_i^B\right)$$

Therefore, we can conclude that the flow of funds must take the following form:

$$T(P) = -\sum_i P_i Z_i\left(P_i, \varepsilon_i^S, \varepsilon_i^B\right) = MS\left(P, \varepsilon^S, \varepsilon^B\right)$$

In summary, in order for traders collectively to be able to offer a perfect hedge to all generators and loads while not taking on any price risk on themselves, they must have access to a flow of funds equal to the nodal price times the net injection at each node. However, this is just the *merchandising surplus* (introduced in Section 7.8).

Result: If, in a system with network congestion, hedge traders are collectively to be able to offer a perfect hedge to all generators and loads without taking on price risk on themselves, they must have access to a flow of funds equal to the merchandising surplus.

Intuitively, this result makes sense. Although generators and loads are, in a sense, natural counterparties for each other for transactions at the same pricing node, the merchandising surplus is the natural counterparty for transactions that occur across or between pricing nodes.

Note that when the traders have access to the merchandising surplus, if they provide a perfect hedge to all generators and loads, then the total payout to the traders collectively is equal to the total economic welfare to the industry:

$$\sum_i H_i^S\left(P_i, \varepsilon_i^S\right) + \sum_i H_i^B\left(P_i, \varepsilon_i^B\right) + MS\left(P, \varepsilon^S, \varepsilon^B\right)$$

$$= \sum_i U_i\left(Q_i^B\left(P_i, \varepsilon_i^B\right), \varepsilon_i^B\right) - \sum_i C_i\left(Q_i^S\left(P_i, \varepsilon_i^S\right), \varepsilon_i^S\right)$$

Result: If traders have access to the merchandising surplus and if traders provide a perfect hedge to generators and loads, then the remaining payoff to the traders is just equal to the total economic welfare in the industry.

14.1.1   Packaging the Merchandising Surplus in a Way that Facilitates Hedging

We have just demonstrated that traders collectively require access to a flow of funds equal to the merchandising surplus if they are collectively to provide a perfect hedge without taking any price risk on themselves.

In Chapter 13, we assumed that traders individually seek to provide a portfolio of hedges to generators and loads in which they do not take any price risk on themselves. Each trader will presumably require some flow of funds to offset the interlocational price risks on the hedge contracts that he/she provides.

This raises the question: In what form should we provide the merchandising surplus to the traders? In theory, in the absence of transaction costs, the parties could mutually negotiate to divide the merchandising surplus in a manner that allows each trader to offset the risks on the hedge contracts that he/she provides. However, in reality, this is simply impractical. Instead, we must find some mechanism that makes available to traders the merchandising surplus in a manner that they can easily manipulate to form the hedge contracts that they desire.

Specifically, we require some form of interlocational hedging instrument(s) that have the following property:

a. The payout on the hedging instrument(s) is collectively equal to the merchandising surplus; and

b. The instrument(s) can be combined to form the hedge contracts that traders desire.

Result: In designing interlocational hedging instruments, market designers should seek to make available an instrument that has the property that: (a) the payout on the hedging instrument(s) in equilibrium is collectively equal to the merchandising surplus, and (b) the instrument(s) can be combined to form the hedge contracts that traders desire.

14.2   Interlocational Transmission Rights: CapFTRs

Let us simplify the analysis by considering an electricity market with only one source of uncertainty – that uncertainty is the load at a specific node, which we will label the load node (there can be load at other nodes, but that load has a fixed demand curve). Let us suppose that the load node is node N.

With this assumption, the net injection at every node other than the load node is a simple function of the wholesale price at that node:

$$Q_i^S(P_i) - Q_i^B(P_i) = Z_i(P_i)$$

Since the supply curve is upward sloping and the demand curve is downward sloping, the net injection function $Z_i(P_i)$ is upward sloping. Let us make the assumption at this point that the net injection function $Z_i(P_i)$ is a true function (that is, it takes a single value for each price). This may not be the case where one or more generators have a constant variable cost. However, in this case we can assume that there is some slope (no matter how small) to the supply curve of each generator, which ensures that there is always some positive slope to the net injection function $Z_i(P_i)$.

Now, suppose that we have a financial instrument that pays out the price difference between the ith node and the load node (node N), multiplied by a fixed volume when the price at the originating node exceeds some threshold S. This instrument is a form of *financial transmission right* or FTR.[1] Since it resembles a standard cap contract, let us call this instrument a CapFTR:

$$\text{CapFTR}(P_i, P_N, S, V) = (P_N - P_i)VI(P_i \geq S)$$

We have seen earlier (Section 14.3) how any function of price can be approximated arbitrarily close using a portfolio of instruments of this kind. Specifically, given a set of instruments of this kind with strike prices arbitrarily close together, the traders can construct a contract with a volume that arbitrarily closely approximates the net injection function $Z(P)$:

$$\sum_{i=1}^{n} (Z(S_i) - Z(S_{i-1}))I(P \geq S_i) \leq Z(P) \leq \sum_{i=1}^{n} (Z(S_i) - Z(S_{i-1}))I(P \geq S_{i-1})$$

Therefore, let us make the assumption that the traders have formed a portfolio of CapFTRs with a volume that matches the injection function $Z_i(P_i)$ at each node. The total payout on the CapFTRs at node i is therefore

$$T_i(P_i, P_N) = (P_N - P_i)Z_i(P_i)$$

The total payout on all of these hedge contracts at every node is equal to the merchandising surplus:

$$\sum_i T_i(P_i, P_N) = \sum_i (P_N - P_i)Z_i(P_i) = -\sum_i P_i Z_i(P_i) = \text{MS}$$

Result: In the case where there is a single source of uncertainty at a node known as the load node, if the system operator makes available a set of financial transmission rights from each node to the load node with a volume that depends on the price at the originating node, traders can construct a portfolio of these transmission rights that mirrors the net injection at each node other than the load node, and collectively the payout on this portfolio of transmission rights matches the merchandising surplus.

Moreover, a trader can use CapFTRs to hedge not only the total net injection at each node, but the individual production or consumption of any generator and load at that node, and can therefore form part of a hedging transaction involving any subset of generators and loads. As

---

[1] Note that although the terminology is financial transmission right, there is no reason why we need to restrict attention to the transmission network. Perhaps a better label would be financial network right?

long as the total net injection of the transaction is zero, the trader can use CapFTRs to ensure that overall it faces no price risk.

CapFTRs can also provide a hedge against network outages, even when there is no uncertainty in the cost or utility functions of the generators and loads.

14.3   Interlocational Transmission Rights: Fixed-Volume FTRs

Many liberalised electricity markets make available to market participants a financial instrument known as a fixed-volume financial transmission right (or often just financial transmission right, FTR, for short).[2] A standard FTR from pricing node i to pricing node j pays out the price difference multiplied by a fixed volume:

$$\text{FTR}\left(P_i, P_j, V\right) = \left(P_j - P_i\right)V$$

Although in principle we could define a separate FTR for every pair of nodes in the network, it is sufficient to define a separate FTR from every node to a designated central or *hub node*. Let us suppose that the hub node is node N. An FTR from any pair of nodes i and j can then be constructed as an FTR to and from the hub node:

$$\text{FTR}\left(P_i, P_j, V\right) = \left(P_j - P_i\right)V = (P_N - P_i)V - \left(P_N - P_j\right)V$$
$$= \text{FTR}(P_i, P_N, V) - \text{FTR}\left(P_j, P_N, V\right)$$

In other words, in a network with n nodes there are at most $n - 1$ independent FTRs.

As we have seen from Section 14.2, the key questions for us to answer are the following:

a. Do these FTRs in equilibrium collectively have a payout equal to the merchandising surplus?
b. Can these FTRs be combined to form the hedge contracts that traders require?

14.3.1   Revenue Adequacy

Let us look first at the question of whether or not the FTRs in equilibrium collectively have a payout equal to the merchandising surplus.

Let us suppose that in equilibrium we have a set of FTRs, with the volume on link l denoted V_l. This set of FTRs implies a set of injections at each node as follows:

$$\forall i, \ \hat{Z}_i = \sum_{l:\text{org}(l)=i} V_l - \sum_{l:\text{term}(l)=i} V_l$$

By definition, these net injections sum to zero: $\sum_i \hat{Z}_i = 0$.

---

[2] These are sometimes known as FTR obligations to distinguish them from FTR options.

The total payout on these FTRs is then equal to the value of the implied injections:

$$\sum_l \text{FTR}\left(P_{\text{org}(l)}, P_{\text{term}(l)}, V_l\right) = \sum_l \left(P_{\text{term}(l)} - P_{\text{org}(l)}\right) V_l$$

$$= \sum_i P_i \left(\sum_{l:\text{term}(l)=i} V_l - \sum_{l:\text{org}(l)=i} V_l \right) = -\sum_i P_i \hat{Z}_i$$

We will say that the FTRs are collectively *simultaneously feasible* if they imply a set of net injections that satisfy the network constraints:

$$\forall l, \sum_l H_{li} \hat{Z}_i \leq K_l$$

Let us assume that the implied set of injections $\hat{Z}_i$ is a feasible alternative dispatch. Using the result in Section 8.3, we know that the merchandising surplus (using the optimal dispatch prices) in this feasible alternative dispatch is less than or equal to merchandising surplus at the optimal dispatch:

$$-\sum_i P_i \hat{Z}_i \leq \text{MS}$$

The condition that the payout on the collective set of FTRs is less than or equal to the merchandising surplus is known as *revenue adequacy*. We have proven that if the FTRs correspond to a set of injections that are feasible on the network at the time of dispatch, the collective payout on these FTRs will be less than or equal to the merchandising surplus. In other words, simultaneous feasibility of FTRs implies revenue adequacy.

Result: Given a set of FTRs, if those FTRs correspond to a set of net injections that are feasible on the network at the time of dispatch, the total payout on the FTRs will be less than or equal to the merchandising surplus.

This result is encouraging, but it is not sufficient to achieve the objective we desire. This analysis asserted that in order to allow traders to hedge the contracts they wish to provide, they must have access to a flow of funds equal to the merchandising surplus. We have shown above that the flow of funds arising from a set of FTRs is *less than* or *equal to* the merchandising surplus.

If there were no uncertainty in the market, this problem could be resolved – we could simply set the FTRs in a manner corresponding to the actual dispatch. Then the value of the FTRs would be exactly equal to the merchandising surplus. But, if there were no uncertainty in the market, there would be no need for hedging instruments at all – no swaps, caps or FTRs.

In the case where the only uncertainty in the market is due to network outages, it is clear that FTRs cannot provide the flow of funds that traders require – even if the FTRs are feasible and yield a payout equal to the merchandising surplus on the original network, they will typically not do so following an outage.

But even setting aside network outages, can we choose the FTRs in such a way as to yield a payout equal to the merchandising surplus? It turns out that the answer is no. It is easy to construct simple examples in which there is uncertainty in the market and it is not possible to construct a set of fixed-volume FTRs that have a payout that is always equal to the merchandising surplus.

Intuitively, this arises for the following reason. In order for a set of FTRs to have a payout that is equal to the merchandising surplus, it must correspond to a set of flows that is equal to the maximum flow on any link that is operating at its maximum in the optimal dispatch. We have seen that there are only as many independent FTRs as there are nodes in the network. However, in a typical network, there can be many more transmission links than there are nodes in the network. The FTRs must be fixed in advance, whereas the constraints that are binding will vary with network conditions that arise *ex post*. No matter how the FTRs are set in advance, a different set of constraints may be binding *ex post*. As a consequence, no matter how the FTRs are set in advance, we cannot guarantee that they will yield a payout equal to the merchandising surplus *ex post*.

14.3.2 Are Fixed-Volume FTRs a Useful Hedging Instrument?

We have just seen that collectively, it is not always possible to choose a set of FTRs that has a payout equal to the merchandising surplus. However, there is a much simpler objection to the use of fixed-volume FTRs: They do not allow traders to provide hedges to generators and loads while taking on no price risk on themselves.

As we have seen, most generators have an upward-sloping supply curve; most customers have a downward-sloping demand curve. If there is any price responsiveness in the rate of production of generators or the rate of consumption of customers, and if there is any price uncertainty in the market, then there will be some price-dependent volume variation in the output of generators or the consumption of customers. It is not possible to hedge a price-dependent volume variation with a hedging instrument with a price-independent volume.

Result: A fixed volume FTR can be used to hedge the price risks incurred by a trader in hedging a transaction between a generator and a load where the volume of the transaction is independent of the price. However, nearly all generators and loads have a rate of production or consumption that depends on the local nodal price. In order to provide a perfect hedge to generators and loads, traders must provide a hedge contract with a volume that matches the actual rate of production or consumption of the generator or load. Fixed-volume FTRs do not allow traders to offset the price risks they face when providing perfect hedges to generators and loads. If fixed-volume FTRs are the only interlocational hedging instrument available, either traders will not provide perfect hedges or traders will face price risks or both.

14.4 Interlocational Hedging and Transmission Investment

In Chapter 13, we saw that conventional hedges (swaps and caps) can provide a useful signal of the need for generation investment. Is there some link between transmission rights and signals for network investment?

14.4.1 Infinitesimal Investment in Network Capacity

Let us focus first on the case of a very small expansion in the capacity of an existing network element. From the theory in Chapter 18, we will see that it is efficient to make a small expansion in the capacity of an existing network element if and only if the area under the constraint–marginal-value curve (which is also the average constraint marginal value) exceeds the cost of expansion.

Let us suppose that we have a link l with capacity K_l with a corresponding constraint–marginal-value μ_l. Let us suppose that the cost of expanding the link by a small amount is f_l. The theory says that it is efficient to expand link l by a small amount if and only if

$$E(\mu_l) \geq f_l$$

Let us assume that the traders have access to the merchandising surplus and that they provide a perfect hedge to generators and loads. With these assumptions, using the result in Section 14.1, the collective payoff to the traders is just equal to the total economic welfare in the industry. Let us denote the total welfare in the industry as $W(K, \varepsilon^S, \varepsilon^B)$. Here, K is the vector of link capacities, ε^S is the vector of cost-shifting factors and ε^B is the vector of demand-shifting factors.

Now let us suppose that traders pay for a small increase in the capacity of link l, $K_l \rightarrow K_l + \Delta K_l$. The expected net change in the payoff to traders is:

$$\Delta K_l \frac{\partial}{\partial K_l} E\left[W(K, \varepsilon^S, \varepsilon^B)\right] - \Delta K_l f_l = \Delta K_l E\left[\frac{\partial W}{\partial K_l}(K, \varepsilon^S, \varepsilon^B)\right] - \Delta K_l f_l = \Delta K_l \left(E[\mu_l] - f_l\right)$$

Clearly, the traders will choose to invest in additional capacity if and only if $E(\mu_l) \geq f_l$ as required.

Result: If (a) the traders provide a perfect hedge to generators and loads and have access to the merchandising surplus, and (b) the traders collectively choose to augment the transmission network by a small amount (incurring the cost of the augmentation), then they will choose to do so if and only if it is economically efficient to do so.

14.4.2 Lumpy Investment in Network Capacity

Let us now consider a larger change in network capacity. This change in network capacity may affect the prices and/or the dispatch of one or more market participants. The change in network capacity may also affect the payout on transmission rights (such as the CapFTRs discussed earlier).

Let us suppose that a network entrepreneur is able to do the following:

a. Purchase new hedge contracts from all the affected parties whose volume is affected by the change in network capacity; and
b. Purchase new financial transmission rights for the change in volume brought about by the change in network capacity.

Then the network entrepreneur faces the correct incentive to augment the capacity of a network link.

To see this, let us suppose as before that the traders have purchased perfect hedges from generators and loads and have access to the merchandising surplus. The total payout to the traders is then equal to the economic welfare from the market. This can be expressed as follows (here we have suppressed the cost-shifting factors and utility-shifting factors to make the presentation as tidy as possible):

$$W(K) = \sum_i H_i^S(P_i) + \sum_i H_i^B(P_i) + \text{MS}(P) = \sum_i U_i(Q_i^B(P_i)) - \sum_i C_i(Q_i^S(P_i))$$

Now suppose that a network entrepreneur is considering expanding the capacity of a network link $K \rightarrow K'$. This results in a change in prices $P \rightarrow P'$ and a change in rates of production and consumption $Q^B \rightarrow Q^{B'}$ and $Q^S \rightarrow Q^{S'}$.

The total increase in economic welfare can be written as follows:

$$W(K') - W(K) = \sum_i \int_{Q^B}^{Q^{B'}} \frac{\partial U_i}{\partial Q_i^B}(q)\mathrm{d}q - \sum_i \int_{Q^S}^{Q^{S'}} \frac{\partial C_i}{\partial Q_i^S}(q)\mathrm{d}q$$

As before, we can express this in a way that more closely resembles a cap contract:

$$W(K') - W(K)$$

$$= \sum_i \int_{Q^B}^{Q^{B'}} \left(\frac{\partial U_i}{\partial Q_i^B}(q) - P_i' \right)\mathrm{d}q - \sum_i \int_{Q^S}^{Q^{S'}} \left(P_i' - \frac{\partial C_i}{\partial Q_i^S}(q) \right)\mathrm{d}q$$

$$- \sum_i P_i'(Z_i' - Z_i)$$

Therefore, we can conclude that the network entrepreneur faces the correct incentive to make the network upgrade provided that

a. The network entrepreneur can purchase additional hedge contracts from generators corresponding to the change in volume brought about by the network augmentation, with the payout

$$\int_{Q^S}^{Q^{S'}} \left(P_i' - \frac{\partial C_i}{\partial Q_i^S}(q) \right)\mathrm{d}q$$

b. The network entrepreneur can purchase additional hedge contracts from loads corresponding to the change in volume brought about by the network augmentation, with the payout

$$\int_{Q^B}^{Q^{B'}} \left(\frac{\partial U_i}{\partial Q_i^B}(q) - P_i' \right)\mathrm{d}q$$

c. The network entrepreneur can purchase transmission rights with a payout equal to the change in the net injection valued at the new prices:

$$-\sum_i P'_i \Delta Z'_i$$

14.5 Summary

In Chapter 13, we explored how to hedge the risks faced by generators and loads in a system with no network congestion. In this chapter, we extended these ideas to the case of network congestion.

If traders provide perfect hedges to all generators and loads, we saw in the previous chapter that in the absence of network congestion, the traders collectively face no price risk. In the presence of network congestion, the traders can still achieve a position where they face no price risk. However, to do that, the traders must have access to a flow-of-funds equal to the merchandising surplus. When traders have access to the merchandising surplus, the total payout on the portfolio of hedges held collectively by the traders is equal to the total short-run economic welfare created by the market.

In this chapter, we propose that the market design should make available the merchandising surplus to traders through a set of interlocational hedging instruments, which (a) collectively have a payoff equal to the merchandising surplus, and (b) can be used by traders to hedge the risks of the individual transactions in which they engage.

In the case where the only source of uncertainty is at one node (referred to as the load node), the net injection at every other node is simply a function of the local nodal price. We showed that in this case we need a simple financial instrument with a payout equal to the price difference between that node and the load node and a volume equal to the net injection at that node.

This financial instrument can be constructed from a set of financial rights known as CapFTRs. A CapFTR has a payout equal to the price difference between two nodes multiplied by a fixed volume, but only when the price difference in the originating node exceeds a threshold value. By forming a portfolio of these CapFTRs with a range of varying strike prices, a trader can arbitrarily closely approximate the net injection as a function of price at any node.

Moreover, a trader can use CapFTRs to eliminate the price risk it faces in a transaction involving a subset of generators and loads in which the overall net injection of the generators and loads is zero. Finally, when the CapFTRs have been chosen to match the net injection at each node, the total payout on the CapFTRs is equal to the merchandising surplus.

Historically, liberalised electricity markets have made available an instrument for hedging interlocational price risk with a fixed volume, known as a fixed-volume FTR or simply an FTR. For any given set of FTRs on a network, there is a corresponding set of net injections. It is straightforward to show that if that set of net injections is physically feasible on the network, then the total payout on the FTRs collectively is less than or equal to the merchandising surplus. This is expressed in the phrase 'simultaneous feasibility implies revenue adequacy'.

However, it is also straightforward to construct examples where it is not possible to construct a set of FTRs for which the payout (in all realisations of the uncertainty) is *equal* to the merchandising surplus. Therefore, FTRs do not achieve the first of the two objectives for an interlocational hedging instrument mentioned earlier.

In addition, FTRs, as a fixed-volume instrument, are not a useful tool for hedging the price-dependent volume of most generators and loads (many generators have an upward-sloping supply curve and most loads have a downward-sloping demand curve). Therefore, FTRs do not achieve the second of the two objectives for an interlocational hedging instrument.

It is straightforward to demonstrate that for an infinitesimal increase in the capacity of a network element, the expected change in the merchandising surplus is equal to the expected constraint–marginal-value for that network element. This shows that traders collectively face the correct incentives to expand network capacity at the margin.

Finally, if a network entrepreneur wishes to make an incremental (noninfinitesimal) increase in the capacity of a network element, it is possible to show that the entrepreneur may (a) provide new hedge contracts to the affected generators and loads, and (b) obtain new CapFTRs to hedge the change in net injection at the affected nodes; the resulting expected change in the payout to the trader is just equal to the economic welfare improvement from the increase in the capacity of the network element. In other words, in principle, it is possible to decentralise the network investment decision using CapFTRs and hedge contracts.

Questions

14.1 What is the role of the merchandising surplus in facilitating hedging of interlocational price risks? What does the theory say about how or why the merchandising surplus should be made available to traders?

14.2 What are the drawbacks of fixed-volume transmission rights when it comes to hedging interlocational price risk?

14.3 Can traders offer hedges to generators and loads that completely eliminate the risk faced by generators and loads without taking on risk on themselves?

14.4 How should a trader construct a portfolio of CapFTRs in order to hedge a transaction involving a generator and a load?

14.5 What is the role of the merchandising surplus in facilitating hedging of interlocational price risk?

Further Reading

There is a useful textbook by Rosellón and Kristiansen (2013) that goes into much more detail on financial transmission rights.

Part VII

Market Power

Control room of the Waikaremoana hydro station, New Zealand, 1928 (Source: Neil Rennie)

15

Market Power in Electricity Markets

Electricity markets have many features that make them prone to the exercise of market power at certain times. This chapter explores the definition of market power, the measurement of market power and the factors that affect market power, in the context of a power system with no network constraints and therefore a single spot price for electricity. In Chapter 16, we extend these ideas to the case when network constraints are binding.

15.1 An Introduction to Market Power in Electricity Markets

What exactly is market power? And why is it harmful? These are important preliminary questions that should be clearly addressed at the outset.

15.1.1 Definition of Market Power

What is market power? Across the economics literature there are a range of definitions of market power. For example, Productivity Commission (2002) states that a firm has market power 'if it can profitably sustain prices above the efficient cost of supply for a significant period of time'. Baumol and Blinder (2008) define market power as the power to 'prevent entry of competitors and to raise prices substantially above competitive levels'. Church and Ware (2000), in a widely used economics textbook, say that 'A firm has market power if it finds it profitable to raise prices above marginal cost'.

However, there is a broad strand of the economics literature that simply defines a firm as having market power if it is not a price taker. There is a broad consensus that a price-taking firm – that is, a firm that has no influence on the market price – has no market power. Many, perhaps most, economists take this as their starting point and define a firm as having market power if it has some influence over the market price.[1]

---

[1] This is also the approach taken in Wikipedia: 'A firm with market power has the ability to individually affect either the total quantity or the prevailing price in the market. Price makers face a downward-sloping demand curve'.

The Economics of Electricity Markets, First Edition. Darryl R. Biggar and Mohammad Reza Hesamzadeh.
© 2014 John Wiley & Sons, Ltd. Published 2014 by John Wiley & Sons, Ltd.

In this textbook, a generator will be said to have market power if it can, by varying its rate of production, affect the wholesale market price that it is paid. Similarly, a load has market power if it can, by varying its rate of consumption, affect the wholesale price it must pay.

In a conventional large power system, most small or medium-sized electricity consumers (including all residential loads and most commercial loads) and the smallest generating units (up to a capacity of, say, a few MW) have no impact on the wholesale market price and therefore have no market power, even at off-peak times. On the other hand, in some electricity markets, the largest generators have a degree of market power, especially at peak times.

It is useful to make a distinction between *possessing* market power and the *exercise* of that market power. For the reasons discussed further later, even if a generator or load has some ability to influence the wholesale market price, it will not necessarily choose to do so. In most instances the change in the wholesale market price that would result from a feasible variation in the output of the generator or load is just too small to warrant any change in the actual behaviour of the firm. A generator may have some market power – that is, the ability to have some influence on the wholesale market price – but only under particular circumstances would it find it profitable to deliberately alter its offers to the market in order to influence the wholesale market price. We can say that a generator or load is *exercising market power* when it alters its bids or offers to the market in a manner that is deliberately designed to alter the wholesale market price.

In principle, market power can be exercised by any electricity generator or consumer. Market power can also, in principle, be exercised by network service providers that participate in the wholesale market, such as an HVDC link.

Furthermore, market power can, in principle, be exercised in any of the markets associated with the wholesale market for electrical energy. As we have seen, wholesale markets for energy are often associated with a variety of related markets, such as markets for balancing services (such as the markets for frequency regulation and contingency services, discussed in Part V), forward markets for electricity (such as markets for hedge products, discussed in Part VI), or markets in locational risk management products or network capacity rights (Chapter 14). Some of these issues are discussed later. For the most part, however, we will focus on the exercise of market power in the wholesale energy market.

15.1.2 Market Power in Electricity Markets

Wholesale electricity markets are prone to the exercise of market power. In fact they are more prone to market power than most other markets. Frank Wolak (Wolak, 2005), a well-known economist and expert in electricity markets, writes

> It is difficult to conceive of an industry more susceptible to the exercise of unilateral market power than electricity. It possesses virtually all of the product characteristics that enhance the ability of suppliers to exercise unilateral market power[2]

There are several reasons why electricity markets are prone to the exercise of market power:

- First, at present (although this may change in the future) the wholesale demand for electricity is typically very insensitive to the wholesale price in the short-run. In the language of

---

[2] Wolak (2005), page 4. Similarly Twomey et al. (2005): 'There are sound theoretical reasons (and supporting evidence) for suspecting that electricity markets may be unusually susceptible at times to the exercise of market power, compared to other markets' (page 49).

economics, the short-run price elasticity of demand for wholesale electricity is very low.[3] This very low price elasticity of demand arises from a number of sources. The primary reason is that many electricity consumers (especially small consumers) are simply not exposed to the wholesale price – that is, they are on retail contracts for which the price they pay is independent (in the short-run) of the wholesale spot price – obviously, these consumers have no incentive at all to respond to the wholesale spot price.

- Even for those consumers who are or could be exposed to the wholesale spot price, the opportunities for intertemporal substitution for electricity demand are often limited. In most uses, such as electricity for lighting or computing, consumers require electricity to be available at a specific time. It is often hard to substitute for electricity consumption at a later time. The scope for intertemporal substitution of electricity consumption is primarily limited to circumstances where electricity is used for heating or cooling and where there is some thermal inertia. In addition, although there is some limited scope for pumping water uphill that can later be used to generate hydro electricity, on the whole electricity cannot easily be stored. These factors, in combination, imply that wholesale demand for electricity is very inelastic in the short-run.

- At the same time, the stock of generation assets is fixed in the short-run and generators have a finite productive capacity. Generators typically have limited scope to 'squeeze more output' when the wholesale price is high. As a result, as demand approaches the maximum capacity of the system to supply, the supply curve becomes very steep. The combination of very inelastic demand and very inelastic supply contributes to an extreme sensitivity of price to small fluctuations in the supply–demand balance at peak times. At such times even very small generators may have a significant influence on the wholesale market price.

- Binding network constraints limit, from time to time, the size of the area over which generators can compete with one-another, giving rise to the scope for localised market power. In extreme cases a small number or even a single generator, in what is known as a 'load pocket', can effectively charge any price up to the price ceiling.

- Generators interact with each other repeatedly in the dispatch process. Using the terminology of game theory introduced in Section 1.8, the dispatch process is a *repeated game* between generators. The repeated nature of this game gives opportunities for generators to learn from each other, to develop reputations, to signal their intentions, and to establish implicit or tacit co-operative or collusive arrangements.

These factors are present to a greater extent in the wholesale electricity market than in most other markets in the economy. As a result, even those wholesale electricity markets that appear, on the surface, reasonably competitive by conventional competition measures may exhibit large amounts of market power at specific times.

Result: Wholesale electricity markets tend to be prone to the exercise of market power, particularly at times of tight supply/demand balance. The primary reason is the difficulty of storing electricity, the steepness of the supply curve at peak times and the lack of responsiveness of most customers to the wholesale price.

---

[3] Recall that elasticity is defined as the derivative of the logarithm of the price with respect to the logarithm of the quantity.

15.2   How Do Generators Exercise Market Power? Theory

In Section 5.2, we observed that under specific conditions price-taking participants in a market submit bid and offer curves that match their marginal cost and marginal value curves. Specifically, we saw that a market participant that (a) has no impact on the wholesale spot price (i.e. is a price-taker), (b) is always dispatched to a price–quantity pair that falls on its offer curve, (c) is not bound by ramp-rate limits or minimum generation constraints, and (d) is not part of a wider collusive arrangement, will offer its output to the market at a price that closely reflects its short-run marginal cost (SRMC) curve (or its demand curve for a load) – at least in that range of prices that it expects will arise with positive probability. Put simply, a price-taking profit-maximising generator (in the absence of any other market distortions) will offer its output in a way that mimics its marginal cost curve.

Therefore, putting aside intertemporal constraints such as ramp-rate limits or energy limits, a generator can be said to exercise market power when it systematically submits an offer curve that departs from its true, underlying, short-run marginal cost curve in order to influence the wholesale spot price it is paid and is therefore dispatched to a price–quantity combination that does not fall on its short-run marginal cost curve.

> *Result*: In a nodally priced market, a generator is exercising market power if, in the absence of intertemporal constraints, it submits an offer curve that results in it being dispatched to a price–quantity combination that does not fall on its short-run marginal cost curve.

But how exactly does a generator exercise market power? How do generators that have the ability to influence the wholesale price choose to offer their output to the market?

Let us focus on the unilateral action of a single generator (this analysis also applies to a group of generators who are able to establish a tight collusive arrangement). We will deal briefly with the case of an energy-limited generator later in this section.

A generator with market power faces a downward-sloping *residual demand curve* (introduced in Section 1.7). That is, for each level of output that may be chosen by the generator there is some corresponding wholesale spot price; the higher the level of output of the generator, the lower the resulting wholesale spot price. We will explore shortly how a generator learns about the shape of the residual demand curve it faces.

15.2.1   The Price–Volume Trade-Off

A generator that can influence the wholesale price faces what is commonly known as a *price–volume trade-off* – that is, it can choose to be dispatched for a larger amount, receiving a lower wholesale spot price (gaining volume at the expense of the market price), or it can be dispatched for a lesser amount, receiving a higher wholesale spot price (losing volume in exchange for a higher price). A generator that faces a downward-sloping residual demand curve does not necessarily have an incentive to alter its offer away from its short-run marginal cost curve. It may be that the reduction in volume necessary to achieve a material increase in the price is simply too large to justify altering the offer in any way. This is illustrated in Figure 15.1.

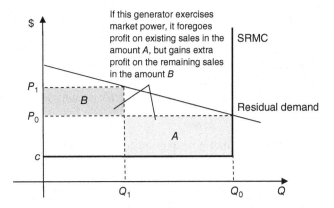

Figure 15.1 Price–volume trade-off in the incentive to exercise market power

An exercise of market power always involves a trade-off between the margin earned on the output of the generator and the level of output of the generator. Let us suppose we have a generator with a simple stylised SRMC curve as shown in Figure 15.1. If this generator is dispatched to produce at the rate Q and paid the price P, it receives profit at the rate

$$\pi(P, Q) = (P - c)Q$$

If this generator offers its output to the market in a manner reflecting its SRMC curve, it will be dispatched to produce at the rate Q_0, will receive the price P_0, and will receive profit at the rate $\pi(P_0, Q_0) = (P_0 - c)Q_0$. This generator may, however, choose to offer in such a way that it will be dispatched to produce at a lower rate of production Q_1, will receive the price P_1, and profit at the rate $\pi(P_1, Q_1) = (P_1 - c)Q_1$. It is straightforward to check that this latter profit is larger than the former profit if and only if the change in price times the new quantity is larger than the change in quantity multiplied by the price–marginal cost margin at the original price:

$$\pi(P_1, Q_1) > \pi(P_0, Q_0) \Leftrightarrow (P_1 - P_0)Q_1 > (P_0 - c)(Q_0 - Q_1)$$

In other words, if the generator produces at a lower rate of production, it foregoes the profit it earns on the extra sales (reflected in the area A in Figure 15.1), but gains extra profit on the remaining sales (area B). Whether or not this generator has an incentive to distort its offer curve away from its SRMC depends on the relative size of these two areas. If area B is larger than area A, this generator has an incentive to exercise market power – that is, to alter its offer to the market in such a way as to be dispatched to the price and quantity (Q_1, P_1).

The relative size of these two areas depends on factors, such as the following:

a. The slope of the residual demand curve (or technically, the elasticity of the residual demand curve) – the steeper the slope, the greater the incentive to exercise market power. This depends in turn on the slope of the market demand curve and the slope of the offer curve of other generators in the market – that is, the number and capacity of the generators that are able to expand their output in response to a price increase – which depends in turn on the nature and extent of any transmission constraints. These are all factors that are discussed in detail later.

b. The size of the generator and, in particular, the unhedged capacity of the generator (the larger the size of the unhedged sales of the generator, the greater the incentive to exercise market power for a given slope of the residual demand curve – we will discuss later the impact of hedging on the incentive to exercise market power); and

c. The level of wholesale price relative to the variable cost of the generator. The lower the variable cost of the generator, the greater the profit on the existing sales that is foregone by a given reduction in output.

The key factors that affect the incentive to exercise market power are discussed further in Section 15.2.2.

15.2.2 The Profit-Maximising Choice of Rate of Production for a Generator with Market Power

Another, slightly more sophisticated, way to view the actions of a generator with market power is as follows. As we saw in Chapter 1, a price-taking generator has an incentive to be dispatched up to a point where the marginal cost of the generator intersects the residual demand curve. In contrast, a generator with market power has an incentive to be dispatched up to the point where the marginal cost of the generator intersects the marginal revenue curve derived from that residual demand curve. This result was demonstrated in Section 1.7.

Mathematically, if a generator has a profit function given by

$$\pi(Q) = P(Q)Q - C(Q)$$

then the profit maximising rate of production is where the marginal revenue is equal to the marginal cost:

$$P(Q) + QP'(Q) = C'(Q)$$

This is illustrated in Figure 15.2. This diagram illustrates the profit-maximising rate of production for a fixed marginal cost curve, but varying residual demand.

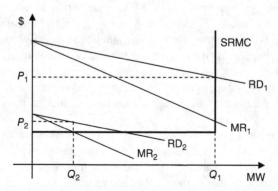

Figure 15.2 A generator with market power chooses to be dispatched to a quantity where marginal revenue intercepts SRMC

When the residual demand is given by RD_2 in Figure 15.2, the profit-maximising level of output for the generator is where the marginal revenue curve, MR_2, intersects the SRMC curve. The profit-maximising combination of price and output is (Q_2, P_2). Since this price–quantity combination is not on the marginal cost curve, this generator is exercising market power. On the other hand, when the residual demand curve is higher (but has the same slope), at RD_1, the profit-maximising price and output combination is (Q_1, P_1). Since this price–quantity combination is on the marginal cost curve, this generator is not exercising market power.

15.2.3 The Profit-Maximising Offer Curve

The preceding analysis focussed on the question of the profit-maximising choice of rate of production for a generator, given its marginal cost curve and the residual demand curve it faces. However, in practice, participants in a wholesale electricity market do not directly choose their rate of production – instead they submit an offer curve to the market operator and receive back instructions as to how much to produce.

How, exactly, does a generator manipulate its offer curve to exercise market power?

Let us start with the case of no uncertainty about the marginal cost curve or residual demand curve. In principle, if there was no uncertainty, each generator could compute the profit-maximising level of output and the resulting wholesale price – that is, each generator could compute the price–output combination that maximises its overall profit. But how can the generator achieve this price–output combination using an offer curve?

In the absence of uncertainty, once the optimal price–output combination is computed, there are literally an infinite number of ways of designing the offer curve that achieves that point. Let us suppose the pair of price and output (Q_1, P_1) in Figure 15.3 is the profit-maximising combination. The generator can achieve this combination of price and output by submitting any offer curve that intersects this point. The generator can offer its output using offer curve A, or offer curve B or any of the remaining infinite number of curves that intersect the point (Q_1, P_1). Put another way, if the generator knew the residual demand curve with certainty, it could achieve any price–output combination that it wanted by either limiting the amount it offered to the market (i.e. economic withholding, discussed further later), or by raising the price at which

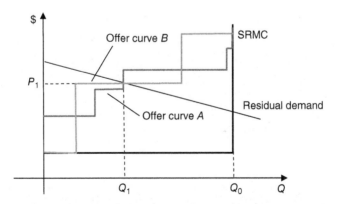

Figure 15.3 If the residual demand is known with certainty, a large range of offer curves will achieve the profit-maximising price–output combination

it offers its output to the market (i.e. 'pricing up', also discussed further later). Both approaches are entirely equivalent.

In practice, the residual demand curve depends on the offer curves of every other generator in the market, the state of the network and the level of demand. As a result the precise residual demand curve a generator will face in the future is at least partially uncertain. Therefore, in practice, a generator must submit an offer curve that attempts to maximise its profit over a range of possible residual demand curves. What does this imply for the offer curve of a generator that is seeking to exercise market power?

In certain circumstances, computing the optimal offer curve when faced with a range of possible residual demand curves is fairly straightforward. For example, in the case where the residual demand curve has a constant slope, but only varies in its intercept with the vertical axis, the resulting locus of profit-maximising price–quantity combinations (one for each residual demand curve) is upward sloping – so the generator can simply submit an offer curve that passes through each one of these optimal combinations. No matter which residual demand curve arises in practice, the generator will be dispatched to a profit-maximising level of output. This situation is illustrated in Figure 15.4. Note that this optimal offer curve lies above the generator's true SRMC curve (although for very high levels of demand it matches the generator's SRMC curve).

Unfortunately, however, things become more complicated when the residual demand curve varies not only in the intercept with the vertical axis, but also in slope. In this case, the locus of points that is profit-maximising for the generator may be backward-sloping in parts, as illustrated in Figure 15.5. In Australia the market rules do not permit a generator to submit a backward-sloping offer curve. The best the generator can do in this circumstance is to submit an offer curve that is vertical at the point of the bend and, if necessary, to submit a rebid if the price goes even higher.

The precise details as to how a generator with market power will construct its offer curve is complex (and beyond the scope of this textbook). Intuitively, however, a generator with market power will generally exercise that market power by submitting an offer curve that differs from its marginal cost curve, resulting in a price–quantity dispatch combination that is not on its marginal cost curve (and is usually above).

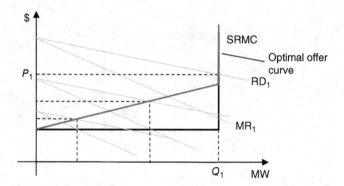

Figure 15.4 Where there is uncertainty in the vertical position of the residual demand curve, the locus of profit-maximising points falls on an upward-sloping line, which is then the generator's optimal offer curve

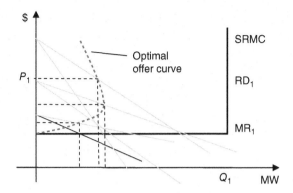

Figure 15.5   Where there is uncertainty in the residual demand curve, it may happen that the locus of profit-maximising points is backward sloping

---

Result: A profit-maximising generator with market power would like to produce at a price–quantity combination where its marginal revenue is equal to its marginal cost. Where there is uncertainty in the residual demand curve, the intersection of the marginal revenue and marginal cost curves traces out a path of price–quantity combinations. As long as that path of price–quantity combinations is upward sloping, the generator with market power can submit this path as an offer curve to the dispatch engine and will be dispatched to the profit-maximising quantity for different realisations of the residual demand curve.

15.3   How do Generators Exercise Market Power? Practice

15.3.1   Economic and Physical Withholding

In the simplest terms, generators usually exercise market power by changing their offer curves to either reduce the quantity offered to the market at a given price, or, equivalently, by increasing the price at which the generator is prepared to produce at a given rate of output. Either way, the generator exercising market power is dispatched to a price–quantity combination such that, if the residual demand curve were horizontal at that market price, the generator would willingly choose to produce more. There is a sense in which the exercise of market power involves the voluntary withholding of production capability.

It is common to distinguish two different forms of withholding: *economic withholding* and *physical withholding*.

A generator engages in *economic withholding* when it submits an offer curve that leads it to be dispatched for a price–quantity combination that is above its short-run marginal cost curve. This could be achieved, as we have seen, either by reducing the capacity offered at a given price, or increasing the price at which it offers a given quantity.

A generator engages in *physical withholding* when it technically makes some proportion of its plant physically unavailable (perhaps by shutting it down).

Under most circumstances economic withholding and physical withholding have very similar effects on the market – they both result in a higher wholesale spot price and lower dispatch by the generator engaging in withholding. In fact in markets without a market price cap, economic withholding and physical withholding are essentially indistinguishable.

However, small differences may arise between economic and physical withholding in the presence of a market price cap. The difference is that, at times when the price rises to the market price cap, a generator engaging in economic withholding may be called on to supply some of the capacity it has priced at or near the price cap. In contrast, a generator engaging in physical withholding will not normally (except perhaps in extreme circumstances when the system operator can overrule the offer curve of a firm) be called on to supply any more output than it is announced it is capable of producing. In other words, physical withholding, unlike economic withholding, may give rise to a circumstance in which there is insufficient generation available to meet the load without involuntary load shedding.

In the presence of a market price cap, a generator exercising market power will never (in the absence of other interventions) have an incentive to physically withhold its output. Once the price has risen to the market price cap, the price–volume trade-off no longer applies (there is no scope for further increases in price). As long as the market price cap is above the marginal cost of production, once the price reaches the market price cap there is no longer any incentive to withhold output. At that point, the profit-maximising decision of the generator is to produce as much as required (as long as the price remains at the price ceiling). This can be achieved by offering that additional capacity to the market at the wholesale price cap.

In other words, a generator can always do just as well – and will sometimes do better – by engaging in economic withholding rather than physical withholding. The exercise of market power, although undesirable for the reasons discussed below, does not normally give rise to system reliability issues. This observation might change if certain forms of withholding were made illegal. In that case, generators may find that they can more easily justify engaging in physical withholding (which implies that plant is simply unavailable perhaps for technical reasons) than economic withholding (which implies that plant is still available – but just at a very high price).

As an illustration of possible economic withholding of capacity, Figure 15.6 shows the offer curves of a particular generator in the Australian market on a particular day in the summer of

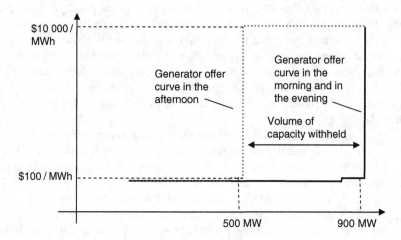

Figure 15.6 Economic withholding by a generator in the NEM, January 2010

2010. Earlier in the day and later the same day, this power station was offering around 900 MW to the market. During the afternoon peak, however, it priced around half of that capacity at the price ceiling (which was $10 000/MWh at that time). The relevant local wholesale spot price, which was between $50–$100/MWh during the morning, reached the price ceiling of $10 000/ MWh and remained above $9000/MWh until around 5:30 p.m. Around 7:00 p.m., this generator again offered around 900 MW to the market at a price of less than $300/MWh.

15.3.2 Pricing Up and the Marginal Generator

It is not strictly necessary for a generator to raise the offer price for some of its capacity all the way up to the price cap to be engaging in economic withholding – in fact, repricing a proportion of its capacity to any price above the out-turn equilibrium wholesale spot price will have the same effect.

There is a particular circumstance where a generator may want to raise the offer price for all or part of its output, rather than reduce the quantity that is offered. This arises, in particular, when the generator is said to be the *marginal generator*.

As we saw in Chapter 4, in a wholesale electricity market, the wholesale spot price at any specific point in time is almost always a function of the bids and offers of a very small number of certain market participants. Since, in practice, generators are the predominant participants in the wholesale market, often the wholesale spot price will depend on just one generator's marginal offer. The generator(s) whose offer directly affects the wholesale spot price at a given point in time is/are said to be the 'marginal' generator(s).

A generator does not need to be marginal to have an incentive to exercise market power. A generator engaging in economic withholding, for example, will not normally be the marginal generator. However, a generator that is a marginal generator does have a clear incentive to distort its offer – at least up to the level of the next-highest offer in the market. This is known as *pricing up* and is illustrated in Figure 15.7.

Let us suppose that the offer curves of all the other generators in the market are known. Let us suppose that when the generator in question offers its output to the market at a price equal to its variable cost, it finds that its own offer is at the point where supply and demand intersect. This results in a wholesale price equal to the marginal offer of the generator in question. However,

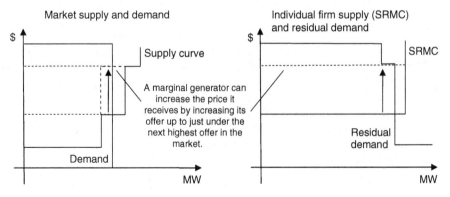

Figure 15.7   A marginal generator has an incentive to increase its marginal offer to the level of the next highest offer in the market

because this generator is the marginal generator, it can raise its offer at least to the point where this offer is just under the next-highest offer in the market. This increases the wholesale spot price and, depending on the elasticity of demand, can be a profitable strategy. If demand is inelastic as illustrated in Figure 15.7, it is always profitable for a marginal generator to increase the price of its marginal offer to the level of the next-highest offer in the market.[4]

The second diagram in Figure 15.7 illustrates the same effect using a residual demand curve. Given the offers of the other generator in the market, a marginal generator has an incentive to increase the price of its marginal offer to a point just under the next-highest offer in the market. Obviously, if the next-highest offer in the market is very close to the existing offer of the marginal generator, the scope for pricing-up is limited.

Result: A generator with market power will usually exercise that market power to raise the wholesale price. It can do this by reducing the capacity offered at low prices (economic withholding) or capacity offered at all (physical withholding). In addition, a generator that happens to be the 'marginal generator' can increase the price it is paid by increasing the price at which it offers its (marginal) units of capacity, up to the next step in the market supply curve.

15.4   The Incentive to Exercise Market Power: The Importance of the Residual Demand Curve

Even amongst those generators that occasionally face a downward-sloping residual demand curve, the incentive to exercise market power varies considerably from one time period to another depending on factors, such as the following:

a. The slope of the residual demand curve. Other things being equal, the larger the price increase for a given reduction in production, the larger the incentive to exercise market power. Since the supply curve of other generators typically becomes very steep as output approaches its physical maximum, the residual demand curve facing generators typically increases in slope significantly as demand approaches the maximum capacity of the system to supply. This is discussed further later.
b. The hedge position of the generator. The incentive to raise the spot price only applies to the output of the generator for which the generator is paid the spot price. As discussed in Section 15.5, the larger the share of output that a generator has 'presold' at a fixed price in the forward market, the smaller the incentive to exercise market power.
c. The presence of network congestion. The slope of the residual demand curve facing a generator depends, amongst other things, on the ability of other generators to expand their output in response to a price increase. Network constraints may limit the ability of other generators to respond to a local price increase giving rise to significant local market power. This is discussed in Chapter 16.

---

[4] Of course it may be possible to increase profit even more by increasing the marginal offer even further (so that it is no longer the marginal offer), but this would then be called economic withholding rather than pricing up.

15.4.1   The Shape of the Residual Demand Curve

Let us focus first on the importance of the shape of the residual demand curve. The residual demand curve shows, for each rate of production of the generator, the corresponding market price. If we ignore network constraints, the residual demand curve facing a generator is equal to the market demand curve less the supply curve of all the other generators in the market.

Strictly speaking, the supply curve of a generator is only well-defined when the generator is a price-taker. Where there is more than one generator with market power in a market, the market is best modelled as a situation of oligopoly. However, even in an oligopoly it is possible to form a view as to the supply curve of other generators in the market. For the moment let us assume that we have some notion of the supply curves of all the other generators in a market. Given the market demand curve (the aggregate demand for electricity), we can construct the residual demand curve as the market demand less the supply of other generators.

15.4.2   The Importance of Peak Versus Off-Peak for the Exercise of Market Power

In practice, the supply curve of other generators in an electricity market typically takes the form of a 'reverse hockey stick' – the supply curve is very flat at low levels of demand (economists say that supply is very *elastic*); in contrast, at times of high demand the supply curve becomes very steep (very *inelastic*).

As a direct consequence, a reduction in the output of a single generator (the generator in question) will typically only have a very limited impact on the market price at off-peak times. On the other hand, at peak times, the residual demand curve facing a generator can be very steep. Therefore, the potential for market power is greatest at times when demand is high relative to the total generation capacity.[5]

We can express the same idea another way. The impact on the market price of a reduction in output by the generator in question depends on the willingness and ability of other generators in the market to increase their output to offset that withdrawal. At times when other generators in the market have plenty of spare capacity they can usually increase their output without requiring much increase in the local market price. But at times when other generators in the market are operating at or very near their capacity, they cannot increase their output any further no matter how high the market price is. In this case a single generator can have substantial market power.

As a general rule, therefore, the scope for the exercise of market power tends to be greater at times when the level of demand is high relative to the total capacity of the industry. In fact, in the case where demand is perfectly inelastic, even very small generators may exercise very significant market power at peak times (i.e. times when there is little spare capacity in the system). In particular, if a generating firm controls $X\%$ of the total generating capacity in the industry, on a day when demand exceeds $100 - X\%$ of the total available generating capacity, such a generator potentially has the ability to raise the market price to an arbitrarily high level (or to the market price cap).

---

[5] In addition, in some markets the supply curve may take the form of a step function. The steps in this function may be large in regions where all the capacity of one generation type (e.g. coal-fired plant) is used up and an increase in price is needed to bring a new generation type (e.g. gas-fired plant) online. The opportunity to exercise market power may also be large in these steep or inelastic regions of the supply curve.

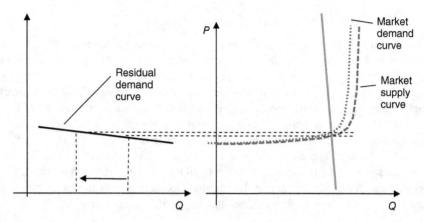

Figure 15.8 At times of low demand, a given reduction in output is likely to have a small impact on the market price

As a general rule, when we ignore network constraints, opportunities for market power tend be greatest on high-demand days. However, the available generator capacity in the market also varies from day to day due to planned and unplanned generator outages. Opportunities for market power may also arise on days when demand is not at its peak but at times when generator capacity is reduced due to generator outages (Figures 15.8 and 15.9).

Result: In general, there tends to be significantly more scope for the exercise of market power at times when the level of spare capacity in the system is small. If demand is inelastic, at times when the level of spare capacity in the system is small, a generator with even a small share of the total capacity may have very substantial market power.

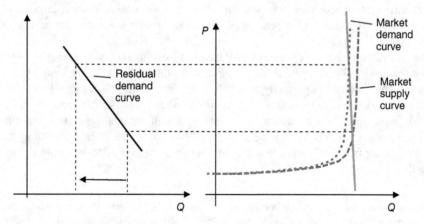

Figure 15.9 At times of high demand, a given reduction in output is likely to have a large impact on the market price

15.4.3 Other Influences on the Shape of the Residual Demand Curve

Other factors can also have an important impact on the shape of the residual demand curve and therefore on the incentive to exercise market power.

One of the more important factors is the slope of the market demand curve. The residual demand curve (as we have noted many times) is the market demand curve less the supply curve of other generators. Therefore, the more the price responsiveness in the market demand curve (technically, the more elastic the market demand curve), the lower the slope of the residual demand curve and therefore the less the opportunities to exercise market power.

Increasing the ability and incentive of customers to respond to electricity market conditions is therefore, in the long run, one of the most important ways of reducing the susceptibility of the electricity market to market power.

Another potentially important factor affecting the shape of the residual demand curve is the presence of any price caps. Once the market price cap is reached there is no further incentive to exercise market power (no further price–volume trade-off is possible).

The final important factor affecting the shape of the residual demand curve is the presence of network congestion, which is discussed further in Chapter 16.

15.5 The Incentive to Exercise Market Power: The Impact of the Hedge Position of a Generator

As mentioned Section 15.4, one of the most important influences on a generator's incentive to exercise market power is its hedge position – that is, the volume of output the generator has presold in the forward or hedge markets. With certain exceptions that are discussed further later, the sale of a hedge contract effectively locks in the price a generator will receive for the corresponding volume of its output. It no longer benefits from an increase in the wholesale spot price over that presold volume of output. Any benefit from exercising market power is limited to the unhedged portion of its output.

Let us start with the simplest case in which a generator has a fixed volume price hedge with a volume $\hat{Q}$. The hedged profit of the generator is now as follows (ignoring a constant related to the fixed price of the hedge):

$$\pi(P, Q) = (P - c)Q - P\hat{Q}$$

As before, if the generator offers its output to the market at marginal cost, it will be dispatched for a price–output combination (P_0, Q_0), and if it chooses to produce at a reduced rate of production, it may achieve the price–output combination (P_1, Q_1). The latter combination will yield a higher hedged profit if and only if

$$(P_1 - P_0)(Q_1 - \hat{Q}) > (P_0 - c)(Q_0 - Q_1)$$

This is illustrated in Figure 15.10, which is a variant on Figure 15.1. Given that the generator has, in effect, presold a volume $\hat{Q}$ of its output on the hedge market, an increase in the wholesale price only benefits the generator over the unhedged output. As a result the total gain from increasing the price on the remaining sales (area B) is greatly reduced in size relative to the case of the unhedged generator, reducing the likelihood the generator will have an incentive to

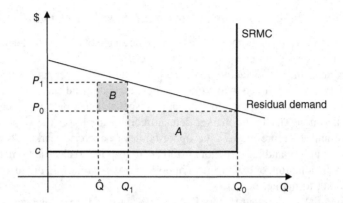

Figure 15.10 Effect of hedge level on the incentive to exercise market power

exercise market power. The higher the hedge level of a generator, the lower the incentive to exercise market power and the smaller the extent to which market power is exercised when the generator chooses to do so. In the limit, if the generator is fully hedged (i.e. faces a hedge level equal to its output $\hat{Q} = Q_0$), the generator has an incentive to offer all of its output at its marginal cost no matter what residual demand curve it faces.

More generally, let us suppose that a generator has sold hedge contracts with a payout $H(P)$. The hedged profit of the generator is then

$$\pi(Q) = P(Q)Q - C(Q) - H(P)$$

The profit-maximising rate of production for this generator satisfies the following condition:

$$P(Q) + \left(Q - \hat{Q}(P)\right)P'(Q) = C'(Q)$$

Here, $\hat{Q}(P) = H'(P)$ is the volume associated with the hedge contract.

From this condition, we can see that if, at the profit-maximising rate of production Q, and the corresponding market price $P(Q)$, the generator happens to be hedged precisely to that level $\hat{Q}(P(Q)) = Q$, then the generator behaves as though it has no market power (it produces at a price–quantity combination on its marginal cost curve). Similarly, if the generator happens to be over-hedged at its profit-maximising rate of production $\hat{Q}(P(Q)) > Q$, then it will choose to produce at a price–quantity combination that is below its marginal cost curve. The exercise of market power can result in a firm depressing the price below its marginal cost curve. Finally, if the generator is under-hedged at its profit-maximising rate of production $\hat{Q}(P(Q)) < Q$, then it will produce at a price–quantity combination above its marginal cost curve (the classic exercise of market power).

It is worth noting that a generator who only expects to be able to exercise market power at higher-price times could choose a hedge portfolio in which it is fully hedged or almost fully hedged at low price outcomes, but is essentially unhedged at high price outcomes. This generator could then enjoy the benefits of hedging (i.e. lower profit volatility) for lower prices and the benefits of the exercise of market power at higher prices.

> *Result*: The hedge portfolio of a generator has a very strong influence on its incentive to exercise market power. If the volume associated with the hedge portfolio at the profit-maximising level of output is well below the profit maximising level of output, the generator has a strong incentive to raise the market price above the generator's marginal cost curve. If the volume associated with the hedge contract at the profit-maximising level of output is above the profit-maximising level of output, the generator has an incentive to depress the market price below the generator's marginal cost curve. The generator can choose the hedge contract to have different levels of hedge cover at different prices.

15.5.1 Short-Term Versus Long-Term Hedge Products and the Exercise of Market Power

It is widely accepted that the presence of long-term fixed-price hedge contracts has a substantial impact on the incentive of a generator to exercise market power. However, there are some caveats to this central result. Many (probably most) hedge contracts are of relatively short duration – between 1 and 3 years. When a hedge contract expires, it is normally replaced with a new contract. The price of that contract at the time of renewal will depend on future price expectations, which are likely to be influenced, at least to a certain extent, by the wholesale spot price outcomes in the past.

Put simply, a generator with a large volume of short-term hedge contracts may take the view that an exercise of market power will increase future wholesale price expectations and therefore will raise the price it receives for future hedge contract sales. In this context the generator will discount its volume of short-term hedge contracts when calculating whether or not to exercise market power. In the limit, if an exercise of market power flows directly through into increased future hedge prices, this generator can ignore its short-term hedge position when deciding whether or not to exercise market power – it is as though it is entirely unhedged.

In other words, although hedge contracts have a strong impact on the incentive to exercise market power, calculating the impact of a given level of hedge contracts on the incentive of a generator is not simply a matter of calculating the hedge position of the generator at that point in time. As already noted, where some of those contracts are likely to be renewed in the future, exercising market power today may increase the price received for future sales of those contracts, increasing the incentive to exercise market power.

15.5.2 Hedge Contracts and Market Power

We have seen that the level of hedge contracting is a strong determinant of the incentive to exercise market power. But how do generators choose what level of hedge contracting to choose?

We can model this as a two-stage game. In the first stage the generator chooses its hedge level. In the second stage the generator plays in the spot market.

Biggar and Hesamzadeh (2011) point out that, under certain conditions, where a generator has market power the optimal choice of the hedge level in the first stage is an all-or-nothing decision. Either the generator will choose to be very highly hedged (even to the point of wanting to depress the market price) or the generator will choose to not be hedged at all (or even to be negatively hedged) in order to raise the market price.

In fact, when there is a generator with market power, it is unsurprising that the hedge market may be very illiquid or may fail to exist at all. After all, whatever position the hedge traders take in the forward market, the generator with market power can move the market *ex post* in such a way that the hedge traders lose out.

In other words, enhancing competition has a mutually reinforcing effect – by reducing the impact of any one generator on the price, the hedge market is likely to be more liquid and therefore generators and traders are more willing to trade in hedge contracts, which further reduces the incentive to exercise market power.

One final point can be made about the link between market power and hedge contracts: network rights of various kinds – such as inter-regional settlement residues or financial transmission rights – also have a substantial impact on a generator's incentive to exercise market power. These are discussed further in Chapter 16.

15.6   The Exercise of Market Power by Loads and Vertical Integration

The previous sections have focused on the exercise of market power by generators. However, naturally, market power can in principle be exercised by electricity consumers (loads). This is not normally a problem – since the vast majority of electricity consumers are so small as to have no practical impact on the wholesale spot price for electricity. However, some loads may on occasion be large enough to influence the wholesale spot price.

Let us consider the position of a load with utility function $U(Q)$, which faces a market price that depends on its own rate of consumption $P(Q)$. Here, increasing the rate of consumption is assumed to increase the price. Let us suppose the load has sold a hedge portfolio with a payout $H(P)$. The hedged net utility of the load is as follows:

$$\varphi(Q) = U(Q) - P(Q)Q - H(P)$$

The net-utility-maximising rate of consumption for this load satisfies the following condition:

$$U'(Q) = P(Q) + (Q - \hat{Q}(P))P'(Q)$$

Here, $\hat{Q}(P) = -H'(P)$ is the volume associated with the hedge contract. As before, we can say that the customer chooses to produce at a rate where the marginal revenue curve (after adjusting for the hedge position) is equal to the demand curve. This results in a price–quantity combination that lies below the demand curve.

This is illustrated in Figure 15.11 for the case where the load is unhedged. As can be seen, the load chooses to produce at a price–quantity combination that lies below the demand curve.

The higher the hedge level of the load, the lower the incentive to reduce its consumption to lower the market price (that is, the lower the incentive to exercise market power).

Result: Loads typically exercise market power by reducing their consumption and lowering the market price, resulting in a price–quantity combination that lies below the load's demand curve. The hedge level of the load strongly influences this effect – the higher the hedge level, the lower the incentive for the load to exercise market power. For a high enough hedge level, the load may pursue a price–quantity combination that lies above the demand curve.

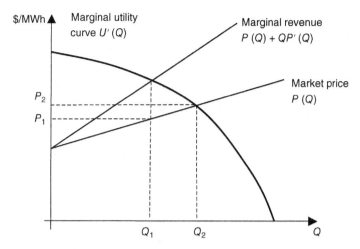

Figure 15.11   The utility-maximising choice of the rate of consumption for a customer with market power

15.6.1 Vertical Integration

Vertical integration between generators and loads is, in practice, an important substitute for vertical contractual arrangements. It is therefore worth exploring the impact of vertical integration on the incentive to exercise market power.

Let us suppose we have an unhedged generator that merges with an unhedged load. Let us assume the combined entity does not purchase any additional hedges.

Let us suppose the combined entity has a degree of market power. This market power is reflected in a downward-sloping residual demand curve. But now the residual demand curve is a function of the net injection of the combined entity into the market. When the combined entity generates at the rate Q^S and consumes at the rate Q^B, the wholesale market price (which depends only on the net injection) is $P(Q^S - Q^B)$ and the overall profit is as follows:

$$\pi(Q^S, Q^B) = P(Q^S - Q^B)(Q^S - Q^B) - C(Q^S) + U(Q^B)$$

Maximising this with respect to the rate of production and the rate of consumption separately, we find that the profit-maximising rate of production satisfies

$$P(Z) + (Q^S - Q^B)P'(Z) = C'(Q^S)$$

Here, $Z = Q^S - Q^B$ is the net injection. By inspection we can observe that the impact of the vertical integration is to moderate the incentive to exercise market power relative to the unhedged generator – in exactly the same manner as if the generator had purchased a hedge contract with a volume equal to the rate of consumption of the load $\hat{Q} = Q^B$.

Similarly, the profit-maximising rate of consumption satisfies

$$U'(Q^S) = P(Z) + (Q^S - Q^B)P'(Z)$$

Again we see that the impact of the vertical integration is to moderate the exercise of market power relative to the unhedged load – in exactly the same manner as if the load had purchased a hedge contract with a volume equal to the rate of production of the generator.

Result: Vertical integration of an unhedged generator and an unhedged load moderates the incentive to exercise market power, in exactly the same manner as if the generator had purchased a hedge contract with a volume equal to the consumption of the load and vice versa.

Does this mean that vertical integration will always reduce the incentive to exercise market power? Not necessarily. This analysis suggests that vertical integration will moderate the incentive to exercise market power relative to the case of an unhedged generator and load. Generators and loads will almost always choose to enter into hedging arrangements. To complete the analysis, we would need to say something about the impact of the merger on the hedge position of the combined entity.

15.7   Is the Exercise of Market Power Necessary to Stimulate Generation Investment?

In the electricity industry it is common to hear the claim that a degree of market power is necessary to ensure that all generators can cover their fixed costs. Some commentators argue that if all generators simply offered their output to the market in a manner that reflected their marginal cost curve, some generators would be unable to cover their fixed costs. The exercise of market power is necessary, it is argued, to ensure that the industry as a whole is able to cover its costs. Is the recovery of fixed costs incompatible with a highly competitive market? Put another way, is the exercise of market power necessary to cover fixed costs?

This claim is incorrect. We saw in Chapter 9 that under certain conditions, if all generators are price-takers and even if all generators submit an offer curve that matches their marginal cost curve, the resulting market price outcomes will ensure that there is an efficient mix of generation assets and all of those assets will be able to earn a non-negative economic profit. It is not true that market power is necessary to ensure adequate generation capacity.

Borenstein and Bushnell (2000, page 50) make this point clearly:

> There is simply no support in theory or practice for the claim that firms – even firms in capital-intensive industries – must exercise market power in order to cover their costs . . . Economic theory does not support an argument that price must exceed the competitive level for firms to break even. In fact, under reasonable conditions, the absence of market power leads to normal returns on investment with exactly the socially optimal quantity of electricity generation capacity.

It is important to emphasise that the exercise of market power does not imply that generators will necessarily be earning excess returns or monopoly rents. In a situation where there is general overcapacity in the market, generators may exercise market power when there are opportunities to do so but still the overall annual average price (or, more strictly, the price–duration curve) may only be barely sufficient to allow all the existing generators to remain in the market. In this case, however, economic efficiency would say that some plant should be retired from the market as it is surplus to requirements. In this case the exercise of market power has the effect of inefficiently delaying that process of retirement of surplus capacity.

It is useful to note that once market power is established in an industry, thereby inducing inefficient entry and overcapacity in the generation sector, the removal of that market power would lead generators to earn inadequate returns until such time as that exercise capacity is eliminated. Thus, while it is not true that market power is necessary to allow all generators to earn adequate returns, starting from an efficient mix of generation, it is also true that starting from an inefficient oversupply of generation, continued market power may be necessary to preserve existing generator returns (however, doing so is socially inefficient).

15.8   The Consequences of the Exercise of Market Power

We have discussed how generators and loads might exercise market power and the factors affecting the incentive to exercise market power. But what, exactly, is the harm from market power?

This question is of fundamental importance. Earlier, we noted that the term market power can have different meanings to different writers. However, the nature and extent of the harm from market power tends to be glossed over to an even greater extent.

In our view, the problem of market power (like all other public policy problems) should be judged on its impact on economic welfare. In Section 1.1, we drew a distinction between short-run and long-run economic efficiency considerations. The short-run economic efficiency impacts relate to the efficiency in the use of an existing set of assets and the allocation of consumption and production across generators and consumers. In the long run, economic efficiency relates to the efficiency of the investment decisions in new production and consumption assets.

15.8.1   Short-Run Efficiency Impacts of Market Power

As we noted in Section 1.7, in the short-run the exercise of market power reduces allocative and productive efficiency. Specifically, customers have a marginal utility for extra consumption at the margin that exceeds the marginal cost of production for at least some generators, suggesting that economic welfare could be improved. In addition, the marginal cost of production of some generators is higher than the marginal cost of production of other generators, suggesting that the same amount of output could be produced at lowest cost.

Ideally, given the stock of generation assets in the market at any given point in time, those assets should be used as efficiently as possible. This implies, amongst other things, that generators should be called on in accordance with their merit order – cheaper generators should be called on to produce, and should have their capacity exhausted, before calling on more expensive generators. One of the effects of market power is that it may give rise to out-of-merit-order dispatch. That is, lower-cost generation may have its output reduced, while other, more expensive generation is called on to make up for the deficit.

For example, let us suppose a baseload generator with a variable cost of \$10/MWh capable of producing, say, 2000 MW, reduces its output from, say, 1800 MW to, say, 1200 MW. Let us suppose that, as a result, a peaking generator with a cost of \$210/MWh is required to turn on to make the shortfall of, say, 600 MW. This results in a social waste of $200 \times 600 = \$120\,000$ for every hour that this market power persists (not counting any start-up costs).

Furthermore, and perhaps even more significantly, the exercise of market power may result in inefficient use of demand-side response. The exercise of market power may result in particularly high prices that may cause loads that have the incentive and ability to do so to reduce their consumption. For example, let us suppose a particular load is willing to reduce its consumption by 300 MW when the wholesale spot price reaches $5000/MWh. If a baseload generator (with a variable cost of, say, $10/MWh) reduces its output and successfully raises the wholesale spot price above $5000/MWh, inducing this load to shutdown, there is a loss in economic value equal to $(5000 - 10) \times 300$ or approximately $1 500 000 per hour that the exercise of market power persists.

15.8.2 Longer-Run Efficiency Impacts of Market Power

Higher spot prices may also have an important impact on generator investment or retirement decisions. In particular, the exercise of market power may inefficiently delay the closure of inefficient plants, or may bring forward investment in new generation capacity before it would otherwise be needed. In this case the harm from market power is the increase in the present value of the cost of the path of future generation investment due to the bringing forward of new investment or the deferment of the retirement of existing capacity.

Perhaps most importantly, the exercise of generator market power can deter valuable investment by customers. An increase in the average level of the wholesale spot price reduces the value of investments by customers. In addition, the threat of further increases in the wholesale spot price reduces the willingness of customers to make investments that increase their reliance on electricity as an energy source.

For example, let us suppose that an aluminium smelter is considering constructing a new facility. It may be concerned that once it has built its facility prices will rise due to the exercise of market power. The aluminium smelter may choose to forego its investment or switch to another energy source. Either way, long-run economic welfare suffers. In our view, the primary reason for control of market power is to provide an assurance to the customer that the value of investments they make in reliance on a supply of electricity will not be expropriated through higher prices *ex post*.

Importantly, these economic consequences of market power do not depend on whether or not the market participants are earning above-normal profits or economic rents.

15.8.3 A Worked Example

Some insight into the consequences of market power can be gained by considering a simple concrete example. Let us consider a simple network in which there is 1000 MW of generation with a variable cost of $10/MWh and 200 MW of capacity with a variable cost of $40/MWh. The total industry capacity is therefore 1200 MW. Demand is inelastic up to the price of $1000/MWh and varies between 700 and 1300 MW. Above the price of $1000/MWh the demand drops to zero.

As a benchmark, let us first consider the price and quantity outcomes when all generators are price-takers. The market outcomes in this case are illustrated in Figure 15.12. As we can see, the market price is either $10, $40 or $1000/MWh, and, in each case, the price does not rise to the next level until all of the capacity at the existing level has been exhausted.

Now let us suppose that amongst the generation with a variable cost of $10/MWh there is a single generator with a capacity of 300 MW. This leaves 700 MW of other generators at $10/MWh and 200 MW at $40/MWh. Let us suppose that all the other generators are very small (so small that they are effectively price takers). Let us also assume that the largest generator has not entered into any forward contracts.

What will be the behaviour of the dominant generator in this case? When demand is only 700 MW, the dominant generator cannot affect the price. Even if it produces nothing, the

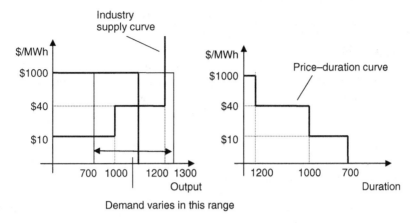

Figure 15.12 Derivation of the price–duration curve assuming a competitive industry

remaining $10/MWh generators can make up the deficit. However, if the demand increases to 701 MW, the remaining $10/MWh generators can no longer make up the deficit. If the output of the dominant generator reduces to just under 1 MW, the price increases to $40/MWh. This increases the dominant generator's profit to $40 − $10 = $30/MWh (when producing 1 MW), which is greater than zero (its profit when the price is $10/MWh), so the generator will reduce its output to 1 MW.

Similarly, if the demand increases to 702 MW, the dominant generator can increase its output to 2 MW and the price will still be $40/MWh. This continues to the point where demand is 901 MW and the dominant generator is producing 200 MW. At this point, if the dominant generator reduces its output to just under 1 MW, the remaining generators cannot make up the difference and the price increases to $1000/MWh.

The resulting price–duration curve is illustrated in Figure 15.13.

In the short run, the impact of the exercise of market power in this example is to cause some customers to choose not to consume, even though the marginal cost of generation (for at least some generators) is lower than the marginal value for these customers, and some generators to be dispatched out of merit order. As a consequence, the total cost of generation is higher than it would be in the absence of market power – in the earlier example, whenever

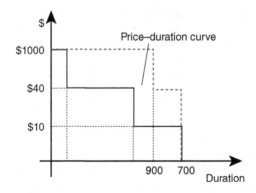

Figure 15.13 Illustration of the impact of market power on the price-duration curve

demand increases above 900 MW, 200 MW of $40/MWh generation is turned on, even though there is spare capacity of $10/MWH generation – essentially raising the total cost of generation by $6000.

The last two effects are collectively referred to as allocative inefficiency. The exercise of market power almost always leads to allocative inefficiency in the short-run.

But what about the long run? As we have seen, in the long run, generators and loads have the potential to change the stock of production and consumption assets. What effect does the exercise of market power have on these decisions?

Let us look first at the incentive for investment in generation assets. Compared to the case where every generator is a price-taker, with the exercise of market power, the area under the price–duration curve is much higher than before. As a consequence new generators will be induced to enter this market. The entry of additional capacity of $10/MWh and $40/MWh generation will lower the price–duration curve. Since by assumption, there was an optimal mix of generation at the outset, this additional capacity in the industry is inefficient and wasteful.

Now let us turn to the impact of market power on customer-side investment. First, the higher price–duration curve (relative to the case where all generators were price-takers) itself reduces the value of investments by customers. When making new investments, or replacing existing assets, customers switch to other energy sources, reducing the demand for electricity over time. Furthermore, customers are less willing to make investments in reliance on a continuing supply of electricity at a stable price. In fact, any investments that increase the customer-value for electricity run the risk of leading to a situation where the dominant firm seeks to increase the price, extracting some of the value of those investments.

Result: The exercise of market power in a generation market has the following harmful effects on the overall economic welfare:

a. In the short-run, customers choose to reduce their consumption even though their marginal value is above the marginal cost of some generators. Similarly, some generators choose to increase their production even though their marginal cost is above the marginal cost of some generators. Overall, there is a reduction in allocative efficiency – that is, the total output could be produced at a lower cost and the total economic welfare could be increased by increasing output.

b. In the longer run, some generators may choose to invest in expanding capacity in the market even though it is inefficient to do so. Similarly, customers may be reluctant to invest in assets that increase their reliance on electricity as an energy source. These effects undermine dynamic efficiency.

15.9 Summary

The term market power is often used without definition or, where it is defined, is defined in a variety of ways. We prefer the definition that a generator or load has market power if it is not a price-taker – that is, if it can, by varying its production or consumption, have some impact on the local market price. We can draw a distinction between possessing market power and exercising market power. A generator possesses market power if it faces a downward-sloping residual demand curve. Similarly, a load possesses market power if it faces an upward-sloping residual demand curve. Even if a participant possesses market power, it does not necessarily have an incentive to exercise market power.

Electricity markets tend to be prone to the exercise of market power, due to inelastic demand for electricity, inelastic supply of electricity at peak times, difficulties of storing electricity, and the presence of network congestion that can significantly limit the number of parties that can compete with each other.

Market participants that exercise market power face a 'price–volume trade-off'. In the case of a generator exercising market power, the generator must cut its rate of production (losing some volume) in order to raise the market price (increasing the profitability of the remaining production). Generators exercise market power by offering their output in such a way as to induce a price–quantity combination (that is, a rate of production and a corresponding market price) that does not lie on its marginal cost curve.

In practice, generators exercising market power reduce their output by either increasing the price at which they offer a portion of their capacity ('economic withholding') or physically shutting down some plant ('physical withholding'). In some cases a generator will also have an incentive to raise the price at which it offers its output – this occurs particularly when a generator is the 'marginal generator'.

The incentive to exercise market power depends very strongly on the shape of the residual demand curve. At peak times the residual demand curve tends to be much steeper than at off-peak times. Opportunities to exercise market power tend to be much more significant at peak times. Even quite small generators can have significant market power at peak times.

Another key factor in determining the incentive to exercise market power is the hedge position of the market participant. Increasing the hedge level of a market participant significantly reduces the incentive to exercise market power. In fact if a generator is 'over-hedged' (that is, hedged to a level that exceeds its profit-maximising level of output), it has an incentive to depress the price to achieve a price–quantity combination below its marginal cost curve.

Loads exercise market power by reducing consumption in order to lower the market price. As with generators, this incentive can be particularly strong at peak times. The exercise of market power by loads is also affected by the hedge level. Vertical integration (that is, the combination of a generator and a load) reduces the incentive to exercise market power in exactly the same way as taking a hedge position.

It is sometimes suggested that the exercise of market power is essential to allow generators sufficient incentives for investment. This is not true. A competitive market yields an efficient mix of generation investment without the exercise of market power. However, if there is a surplus of generation capacity in the market, the exercise of market power will help to ensure that generators earn a normal return. However, this leads to both inefficient short-run outcomes and inefficient delay in the retirement of capacity from the market.

It is important to be clear as to the economic harm from market power. Market power has both short-run and long-run impacts on economic welfare. In the short-run, market power reduces allocative efficiency and productive efficiency (increasing the cost of producing a given amount of electric power). In the long run, market power distorts investment decisions by generators (over-incentivising entry or discouraging timely exit). Most importantly, market power can also deter sunk investment by customers in assets that rely on a reliable supply of electricity at a reasonable price. These consequences arise whether or not the firm's exercising market power are earning above-normal profits or 'monopoly rents'.

Questions

15.1 Prove the result in the text, if the profit function of a generator is given by $\pi(P, Q) = (P - c)Q$, show that

$$\pi(P_1, Q_1) > \pi(P_0, Q_0) \Leftrightarrow (P_1 - P_0)Q_1 > (P_0 - c)(Q_0 - Q_1)$$

15.2 The short-run responsiveness of customers to the wholesale price is typically very low. Does this mean that there is no economic harm from the exercise of market power in the wholesale electricity market?

15.3 A large industrial electricity user is choosing where to build a large factory. It has narrowed the search down to two towns. The towns are identical in every respect except that one has a number of competing electricity generators and the other town has a single generator. The towns are connected by an electricity network. What are the factors the electricity user should take into account when making its location decision? What does this tell us about the possible harm from the exercise of generator market power?

15.4 How do we know when a generator has exercised market power in the past? What factors would we need to look for? What information do we need? Is it enough to know that the generator has produced at a price–quantity combination above its historic marginal cost curve? Why or why not?

15.5 Under what circumstances would a generator exercise market power by lowering the price below its SRMC curve?

15.6 In the Australian market, a price cap kicks in after approximately 10 h of pricing at the market price ceiling (which is above \$13000/MWh). This price cap significantly constrains prices (to a maximum of \$300/MWh) for at least several days. Faced with the threat of this price cap, how do you think generators would respond when faced with multiple opportunities to raise the price to the price ceiling?

15.7 Can you list five policy actions that mitigate market power?

Further Reading

There are many general textbooks on the economics of market power in general (i.e. not specific to the electricity industry). Church and Ware (2000) is one useful reference. There is a relatively large literature on the economics of market power in electricity markets. Twomey et al. (2005) is worth reading. See also Wolak (2005) and Borenstein and Bushnell (2000). Bushnell, Mansur and Saravia (2008) provide interesting evidence of the importance of vertical arrangements in mitigating market power.

16

Market Power and Network Congestion

Chapter 15 explored the issues of market power in the absence of network congestion. Importantly, network congestion interacts strongly with market power. As we will see, network congestion may either enhance or reduce the potential to exercise market power. At the same time, the exercise of market power may either alleviate or exacerbate network congestion.

16.1 The Exercise of Market Power by a Single Generator in a Radial Network

Let us focus first on simple radial networks. In addition, we will focus on the case where a single generator is acting alone (rather than acting jointly with another generator at another location).

We have seen that when a generator attempts to withhold capacity in order to increase the local market price, the impact on the local market price depends on the willingness and ability of other generators in the market to increase their output in response to the withdrawal, together with the willingness and ability of loads to reduce their demand.

In Section 15.4, we saw that when other generators in the market are operating at their physical limit, they no longer have the ability to offset the withdrawal of capacity – this tends to increase the market power of the generator in question.

In exactly the same way, transmission constraints can serve to limit the ability of generators at other locations to increase their output in response to a generator exercising market power at a particular location. This tends to increase the scope for market power.

We can express this another way: Using the language of competition policy, we can say that transmission constraints reduce the geographic scope of the relevant market – reducing the number of generators in direct competition with the generator exercising market power.

The Economics of Electricity Markets, First Edition. Darryl R. Biggar and Mohammad Reza Hesamzadeh.
© 2014 John Wiley & Sons, Ltd. Published 2014 by John Wiley & Sons, Ltd.

16.1.1 The Exercise of Market Power by a Single Generator in a Radial Network: The Theory

When a generator withdraws capacity from the market, the net injection function at that location shifts to the left – in other words, at each level of the local spot price the net injection $Z_i(P_i)$ is smaller. This automatically implies an increase in the flow on the network towards this node (or a reduction in the flow away from this node).

If the flow is initially at the limit in the exporting direction, a reduction in the output of a generator has a tendency to alleviate the constraint (i.e. reduce the constraint marginal value) or to cause the constraint to become unbinding.

If the flow is initially close to the flow limit in the importing direction, the reduction in output may cause the import flow to reach the import limit on a network element. At this point, the generators in the other regions can no longer increase their output in response to any further withdrawal by the generator with market power. The only generators who can now serve to mitigate the effects of the market power are generators (or loads) internal to that region. If there are few such generators (or if they are already operating at their physical limit), the generator withdrawing capacity may have very substantial market power.

Mathematically, we can express this as follows. Let us suppose that there is just one potentially binding constraint in a radial network. This one binding constraint potentially divides the network into two pricing regions (see Section 7.7). Let us suppose that we have a generator in pricing region 1 that is seeking to exercise market power. Let us suppose that the output of this generator is G. Let us suppose that the net injection of all the other market participants in this pricing region is $Z_1(P_1)$. Let us suppose that the net injection of all the market participants in the other pricing region is $Z_2(P_2)$. Finally, let us suppose that the network link joining the two pricing regions has a fixed capacity of, say, K.

There are two states to consider: Where the network limit is binding (and the prices in the two regions are different) and where the network limit is not binding (in which case the prices in the two regions are the same).

We are interested in the slope of the residual demand curve. In Section 15.2, we saw that, other things being equal, the steeper the slope of the residual demand curve, the greater the incentive to exercise market power. The slope of the residual demand curve in this case is dP_1/dG.

In the case where the network constraint is not binding, the market price is common across both regions and is given by the intersection of supply and demand across both regions:

$$Z_1(P_1) + Z_2(P_1) + G = 0$$

This implies that the slope of the residual demand curve depends on the responsiveness of the net injection to price across all locations:

$$\frac{dP_1}{dG} = \frac{-1}{Z_1' + Z_2'}$$

In the case where the network constraint is binding, the market price in each region is determined by the net injection in that region alone. In this case

$$Z_1(P_1) + G + K = 0$$

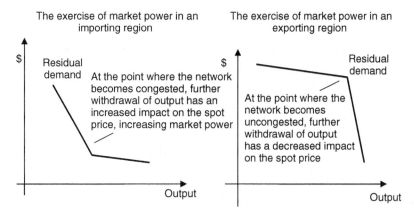

Figure 16.1   The exercise of market power in importing and exporting regions in a radial network

Therefore, the slope of the residual demand curve now depends only on the responsiveness of the remaining market participants in the region with the generator exercising market power:

$$\frac{\mathrm{d}P_1}{\mathrm{d}G} = \frac{-1}{Z_1'}$$

Comparing these last two equations, we can see that as long as there is some price-responsiveness of the generators in the other region (i.e. as long as $Z_2' > 0$), moving from the state in which the network is unconstrained to a situation in which the network is constrained will result in an increase in the (absolute value of the) slope of the residual demand curve, increasing the incentive to exercise market power.

If there are very few other price-responsive market participants in the same region as the generator exercising market power (i.e. if Z_1' is small), moving from the state in which the network is unconstrained to a state in which the network is constrained can result in a very significant increase in market power. This is known as a *load pocket* and is discussed further later.

There are potentially two cases to consider. The first is where the exercise of market power takes place in an importing region. Let us consider first the case where the network flows in the importing region are near their limits. In this case the withdrawal of output by a generator will increase flows into the region. Initially, before the network limit is binding, the prices will increase across the whole network. However, further withdrawal of output may cause the network limit to bind. As we noted earlier, once the network limit is binding, the residual demand curve becomes steeper – perhaps substantially steeper. At this point the incentive to exercise market power can become significantly greater. This is illustrated in Figure 16.1.

The second case to consider is where the exercise of market power takes place in an exporting region. Let us suppose that flows are initially binding in the exporting direction. In this case the withdrawal of output will tend to decrease the export flows. Initially, as long as the network limit is binding, as we noted earlier, the residual demand curve will tend to be steeper. However, further withdrawal of output may relieve the binding network limit. At this point, as we noted earlier, the residual demand curve will become less steep – perhaps substantially less steep. At this point the incentive to exercise market power becomes lower. Again, this is illustrated in Figure 16.1.

As a general rule, generators with market power in exporting regions will seek to exercise their market power to reduce output just to the point where the constraints are alleviated, equalising prices across the regions.

Result: The exercise of market power tends to be closely associated with the presence of binding network limits. On a radial network, when flows are near their import limits, the exercise of market power by a generator in the importing region will tend to cause the constraints to bind. Furthermore, the generator in the importing region will likely have a greater incentive to exercise market power once the network limit is binding. When flows are binding in the export direction, the exercise of market power by a generator in the exporting region will tend to alleviate the network constraints and the incentive to exercise market power is likely to be lower once the constraint is removed.

To summarise, market power can both exacerbate and alleviate network constraints. Market power tends to exacerbate network constraints on import flows into regions with significant market power. At the same time, market power tends to be used to alleviate network constraints on export flows from regions with significant market power.

One immediate consequence is that even if there are no observed binding network constraints in a market, it does not imply that the network constraints are not affecting the behaviour of market participants – generators with market power could be using that market power to ensure that export constraints do not, in fact, bind.

Result: Caution has to be taken when assessing the nature and extent of observed network constraints as (for example) a signal of the need for new investment in the network. In the presence of market power the absence of potential network constraints does not necessarily imply that network constraints are not having an adverse effect on the market. At the same time, the presence of severe network constraints does not necessarily imply that network augmentation is justified.

Let us consider a simple network with two nodes and one link. The first node has a large amount of generation with a variable cost of \$20/MWh. The second node has 1000 MW of generation with a variable cost of \$10/MWh, and a large amount of generation with a variable cost of \$1000/MWh. There is a variable, price-insensitive load at the second node.[1] The network link has a flow limit of 1000 MW. This is illustrated in Figure 16.2.

---

[1] This situation is equivalent to asserting that there is demand with a constant value of \$1000/MWh for electricity up to some rate of consumption.

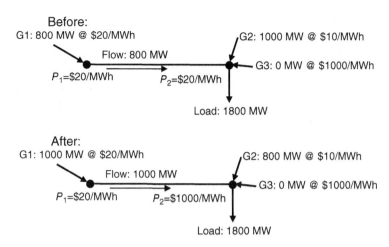

Figure 16.2 Import transmission constraints can result in significant market power

Let us suppose that the load is initially 1800 MW. The optimal dispatch involves the generator at node 2 being dispatched to 1000 MW and the generator at node 1 being dispatched to 800 MW. The network limit is not binding and the common price is \$20/MWh.

Now consider what happens when the generator at node 2 withdraws capacity from the market. As this generator reduces its output from 1000 MW down to 800 MW, the output of the generator at node 1 increases to offset – there is no impact on the market price. However, further withdrawals of output by the generator at node 2 cause the network limit to be binding. At this point the price at node 2 increases to \$1000/MWh.

If the generator at node 2 produces at full production (1000 MW), it receives the price of \$20/MWh on all of its output, earning a profit of $(20 - 10) \times 1000 = \$10\,000/h$. On the other hand, if the generator at node 2 produces at 800 MW, the network constraint is binding, the local price is \$1000/MWh and it earns a profit of $(1000 - 10) \times 800 = \$792\,000/h$. This generator clearly has a strong incentive to exercise market power.

16.2 The Exercise of Market Power by a Single Generator in a Meshed Network

Now let us examine the impact of the exercise of market power in a meshed network. Let us focus on a simple three-node, three-link network. Without loss of generality there will be assumed to be a potentially binding constraint between nodes 1 and 2, in the direction of node 2. The flow limit on this link is assumed to be K. As throughout this text, we will assume the electrical characteristics of the three links are identical. Therefore, when the constraint does bind, the price at node 3 is equal to the average of the prices at nodes 1 and 2 (see Section 7.4). There are three cases to consider: where the generator with market power is at node 1, 2 or 3.

Let us suppose we have a generator that is considering exercising market power. As before, we are interested in the slope of the residual demand curve – that is, how the local price varies with changes in the output of this generator.

Let us suppose the output of this generator is G. Let us suppose that the net injection of the other market participants at each of the three nodes is $Z_1(P_1)$, $Z_2(P_2)$ and $Z_3(P_3)$. By the energy balance equation, the sum of these net injections plus the output of the generator in question must sum to zero:

$$Z_1(P_1) + Z_2(P_2) + Z_3(P_3) + G = 0$$

In the case where the network constraint is not binding there is a common price P on the network. From the energy balance equation, the responsiveness of this price to the output of the generator in question is as follows:

$$\frac{dP}{dG} = \frac{-1}{Z_1' + Z_2' + Z_3'}$$

When the network constraint is binding, we know that $P_3 = \frac{1}{2}(P_1 + P_2)$. Therefore, the responsiveness of the prices at nodes 1 and 2 to the output of the generator in question must satisfy the following:

$$Z_1' \frac{dP_1}{dG} + Z_2' \frac{dP_2}{dG} + \frac{1}{2} Z_3' \left(\frac{dP_1}{dG} + \frac{dP_2}{dG} \right) + 1 = 0$$

From this expression, we can see that if the responsiveness of the prices to the output of the generator in question is the same at nodes 1 and 2 (i.e. if $dP_1/dG = dP_2/dG$), then the responsiveness of these prices to the output of the generator in question is exactly the same in the case where the network constraint is not binding.

Now let us take the case where the generator in question is located at node 2. When the constraint is binding, the following condition must hold (taking node 2 as the swing bus):

$$\frac{2}{3} Z_1(P_1) + \frac{1}{3} Z_3(P_3) = K$$

In addition, we have that $P_3 = \frac{1}{2}(P_1 + P_2)$. Therefore, we can write the following:

$$\frac{2}{3} Z_1' \frac{dP_1}{dG} + \frac{1}{6} Z_3' \left(\frac{dP_1}{dG} + \frac{dP_2}{dG} \right) = 0$$

We now have two equations in two unknowns (dP_1/dG and dP_2/dG). It is relatively straightforward to show that the responsiveness of the price at node 2 to the output of the generator dP_2/dG has a larger slope (in absolute value) than the case where the link was unconstrained ($dP/dG = -1/(Z_1' + Z_2' + Z_3')$) provided that the determinant on the matrix of coefficients in this set of simultaneous equations is positive. Demonstrating this is left as one of the exercises at the end of the chapter.

In summary, although the mathematics is more complicated, we find the same result as earlier: The slope of the residual demand curve facing a generator becomes steeper when a network constraint is binding. Moreover, this is true whether the generator is located at node 1, 2 or 3.

Result: In a three-node network, the slope of the residual demand curve facing a generator is steeper when a network constraint is binding relative to the case where the network constraint is not binding.

16.3   The Exercise of Market Power by a Portfolio of Generators

The previous analysis focused on the exercise of market power by a single generator with market power in a simple radial network. However, generating firms typically own plants in more than one location. Is it the case that owning a portfolio of plants might exacerbate opportunities for market power?

It turns out that owning a portfolio of generating plants in different locations may exacerbate opportunities for market power. Specifically, by owning plants at different locations, a generating firm may have an incentive to increase output at one location, in order to induce a constraint to bind, and to reduce the amount by which output must be reduced at another location.

Let us consider the case of a simple two-node, one-line network. Let us suppose that a generating firm owns generating assets in both locations. Let us suppose that the output of the generating assets at each location is given by G_1 and G_2, respectively. We know from the earlier analysis that the following condition must hold:

$$Z_1(P_1) + Z_2(P_2) + G_1 + G_2 = 0$$

Here, $Z_1(P_1)$ is the net injection of all the other generating and consuming assets at node 1 and $Z_2(P_2)$ is the same for node 2. There are two cases to consider – where the constraint is not binding and where the constraint is binding. When the constraint is not binding, there is a common price across the nodes and it is only the total production of the generating company $G_1 + G_2$ that matters and that affects the wholesale price. In this case, the common price satisfies

$$Z_1(P) + Z_2(P) + G_1 + G_2 = 0$$

In the event that the constraint is binding, the price at node 1 and node 2 satisfies

$$Z_1(P_1) + G_1 = K \text{ and } Z_2(P_2) + G_2 = -K$$

Here, K is the flow limit on the network link.

In this context the generating firm may have a choice. It can seek to exercise market power on the combined market (with no network constraint) in which case only its total production matters. Alternatively, the firm can shift production from node 2 to node 1. This may cause the network constraint to bind, lowering the price at node 1. At this point the firm may be able to exercise market power by withdrawing output at node 2, raising the price at node 2.

This is illustrated in Figure 16.3. For any given value of the total production of the firm $G_1 + G_2$, if the constraint is not binding, there is a common price at both nodes. However, by shifting its production to node 1, the generating firm may be able to cause the constraint to bind, lowering the price at node 1 and raising the price at node 2. This may be more profitable than exercising market power when the constraint is not binding.

Result: When a generating firm owns generating plants at more than one location, it may have more opportunities to exercise market power. By shifting production around within the portfolio, the generating firm may be able to change the pattern of constraints on the network creating opportunities to exercise market power.

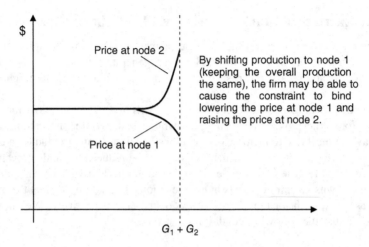

Figure 16.3 Exercise of market power by a portfolio of generating assets

16.4   The Effect of Transmission Rights on Market Power

In Chapter 15, we saw that the hedge position has a strong impact on the incentive to exercise market power. The same is potentially true for transmission rights.

In Chapter 14, we introduced the concept of transmission rights. A transmission right between two specific locations on the network pays out an amount equal to the price difference between the two locations on the network multiplied by a volume (which could, itself, depend on the price at one or other node).

In Chapter 14, we envisaged that generators and loads would purchase the hedges they require referenced to their own local nodal price. These hedges were provided by a counter-party referred to as a trader. In this context, transmission rights are a tool for traders to enable them to provide the hedge products that generators and loads require.

Here, however, we consider what might happen if a generator with market power was able to directly acquire one of these transmission rights. What effect does a transmission right have on incentives to exercise market power? It turns out that purchase of a transmission right in the importing direction increases local generator market power, whereas the purchase of a transmission right in the exporting direction decreases local market power.

From Section 14.2, we know that a generic transmission right from node i to node j has the following payoff:

$$T(P_i, P_j) = (P_j - P_i)V$$

Here, V is the volume associated with the transmission right.

The profit of a generator at node i holding this transmission right is as follows:

$$\pi(Q) = P(Q)Q - C(Q) + (P_j - P(Q))V$$

Therefore, the profit-maximising rate of production of a generator at node i holding this transmission right is

$$P(Q) + (Q - V)P'(Q) = C'(Q)$$

We can conclude that the holding of an exporting transmission right reduces the incentive to exercise market power in exactly the same manner as holding a hedge contract. In the same way, holding an importing transmission right increases the incentive to exercise market power.

16.5 Summary

Network congestion has a significant impact on the incentive to exercise market power. In a radial network, the slope of the residual demand curve facing a generator is steeper when network constraints are binding, due to the fact that remote generators are no longer willing or able to increase output in response to a local withdrawal of generation. A generator in an importing region with little or no other price responsive generation or load can have very substantial market power. This is sometimes known as a load pocket. For this reason generators may seek to withdraw capacity in order to induce import constraints to bind. Conversely, generators in an exporting region may exercise market power in order to raise the local price and to alleviate binding export limits.

In a meshed network the analysis is more complicated, but the same result arises: The slope of the residual demand curve is steeper when network constraints are binding. In a meshed network, network constraints can limit, but not entirely prevent, remote generators responding by increasing their injection in response to a local price increase.

In a meshed network, where the generator has plants at different pricing locations, a situation can arise where the generator exercises market power by increasing output at some locations and reducing output at others.

Financial transmission rights (just like other hedge products) can have a significant impact on the incentive to exercise market power. The purchase of an importing financial transmission right reduces the incentive to exercise market power, while the purchase of an exporting transmission right increases the incentive to exercise market power.

Questions

16.1 True or False: A generator in a load pocket has no market power if its capacity is less than the spare (unused) capacity on the importing transmission link?

16.2 Can the diagram in Section 16.3 (Figure 16.3) be generalised to the case of a meshed network and a generator with production assets at three or more nodes?

16.3 Demonstrate the result in Section 16.2: In a meshed network, the residual demand curve facing a generator with market power is steeper when a network constraint is binding than when there are no network constraints. Does your result depend on the location of the generator with market power?

Further Reading

There are a few articles on the topic of market power and network constraints, including Joskow and Tirole (2000), Borenstein, Bushnell and Wolak (2000) and Gilbert, Neuhoff and Newbery (2004).

17

Detecting, Modelling and Mitigating Market Power

The previous chapters have explored the theory and practice of market power. In this chapter, we explore how market power can be detected or forecasted and what can be done to mitigate market power.

17.1 Approaches to Assessing Market Power

When it comes to market power, there are two common issues that confront policymakers:

- How to detect whether or not market power has been exercised in the past.
- How to forecast the impact of market developments (such as a merger of generators or an expansion of the network) on the level of market power that may occur in the future.

There are three categories of approaches for detecting market power episodes in the past and forecasting potential market power in the future. The first set of approaches looks at actual price and dispatch outcomes in the past and compares those outcomes with assessed demand and supply curves of market participants. These approaches are simple, relatively transparent, straightforward to implement, have relatively limited data requirements and require relatively few assumptions. The primary disadvantage of this set of approaches is that they can only be used for assessing where market power might have arisen in the past. They are of no use in forecasting the likely impact of market developments on market power in the future.

The second set of approaches uses a range of strong simplifications of varying degrees of sophistication to approximate market outcomes. These approaches differ in the information they require and the degree of computation involved. They all have in common that they do not seek to directly estimate the market price and dispatch outcomes. These approaches can usually be applied to market outcomes in the past (thereby inferring in the extent of market power exercised in the past) and to potential market outcomes in the future (thereby forecasting the potential for market power in the future). These approaches are more straightforward and transparent than full-scale market modelling (discussed later), but typically rely on substantial

The Economics of Electricity Markets, First Edition. Darryl R. Biggar and Mohammad Reza Hesamzadeh.
© 2014 John Wiley & Sons, Ltd. Published 2014 by John Wiley & Sons, Ltd.

simplifications and many *ad hoc* assumptions. Some of these approaches are controversial. Nevertheless, these approaches are useful as simple screening tools.

The third set of approaches uses full-scale market modelling techniques to forecast future price and dispatch outcomes. These approaches tend to require relatively sophisticated modelling, often involving substantial computation. These models almost always require substantial assumptions about, say, generator cost characteristics and hedging positions. As a result, these models often appear nontransparent and 'black box' to outsiders. The primary advantage of these approaches is that they can, in principle, answer a range of questions about how market outcomes might change in response to a change in market conditions in the future.

Result: Policymakers use a range of approaches to detect and forecast the exercise market power, ranging from relatively simple observations of outcomes in the past, through simple indicators that seek to very roughly capture market outcomes, to full-scale market modelling. Each of these approaches has its pros and cons.

The following sections examine each of these approaches in turn.

17.2 Detecting the Exercise of Market Power Through the Examination of Market Outcomes in the Past

How can we tell whether or not market power has been exercised at a specific point in time in the past?

As discussed earlier, under the broadest definition, market power has been exercised by a generator when the generator submits an offer curve that has the effect that the generator is dispatched for a price–quantity combination that is not on its marginal cost curve for other than purely technical reasons (such as start-up costs and ramp rate constraints). In principle then, the detection of market power *ex post* is primarily a matter of comparing a generator's offer curve (or, more strictly, its price–dispatch quantity combination) to its marginal cost curve.

Of course, an immediate problem with this approach is that a generator's marginal cost curve is not typically publicly available information and may not even be available to market monitors. This question is discussed further later.

Nevertheless, let us assume for the moment that we have some proxy or estimate of a generator's marginal cost curve at a given point in time. In addition, let us suppose that we can observe the price and dispatch outcome for that generator in the market at that point in time.

Given the actual or observed price–quantity outcome in the market, there is a question as to how to quantify the departure from the benchmark or proxy cost curve. There are two conventional approaches:

a. Approaches that measure the reduction in output relative to the output on the marginal cost curve at the same price (the distance A in Figure 17.1); and
b. Approaches that measure the difference in price between the out-turn price and the benchmark marginal cost at the same rate of production (the distance B in Figure 17.1).

For both of these approaches, there are issues regarding how the measure should be aggregated over time, and how the proxy or benchmark marginal cost curve should be determined.

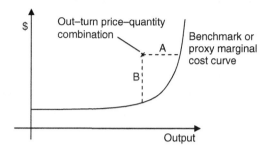

Figure 17.1 Market power can be measured as the deviation in quantity or price from the SRMC curve

17.2.1 Quantity-Withdrawal Studies

Many authors have proposed measuring market power as the gap between the actual dispatch and a measure of the amount the generator should have been willing to produce at the same point in time. These are known as quantity-withdrawal studies.[1]

As with all of these approaches, a key question is how to estimate the proxy or benchmark cost curve. Although basic information, such as the capacity of a generator, is usually available, a generator may be unwilling to produce at its capacity unless the price increases to a very high level. In some cases, we may be able to obtain enough information (on say the input fuel costs and the 'heat rates' of a generator) to be able to estimate its marginal cost curve. However, there still arises the potential for the generator's marginal cost curve to change – perhaps due to outages, or changes in input costs that could not have been forecasted in advance.

Another approach, which is closer to a 'revealed preference' approach, relies on the observation that at times when the market is reasonably competitive, generators have an incentive to offer their output to the market in a manner that broadly reflects their short-run marginal cost (SRMC) curve. Under this approach a particular time is chosen when the market is assumed to be reasonably competitive, and the generator offer curves at that time are assumed to broadly reflect their SRMC. The price–quantity combinations of each generator at some other times are then compared to this benchmark to detect the exercise of market power.

Of course, the marginal cost curve of a generator could still change over time. However, the closer in time the benchmark curve is to the potential episode of market power. the less likely it is that an outage is driving the outcome. For example, if a time can be found *on the same day* when the market appears to be relatively competitive, the generator's offer curve at that time could be taken as a benchmark or proxy cost curve for its offers later the same day.

To illustrate this approach, Figure 17.2 shows the behaviour of a particular generator in the Australian National Electricity Market (NEM) on one day in the summer of 2010. At around 11:00 a.m., this power station was offering around 915 MW to the market at a price less than $300/MWh. This offer curve might be taken as a benchmark or indicative offer on this day. The dots represent the price–quantity combinations for this generating plant later the same day. As can be seen for many half-hour intervals on this day, this plant was dispatched to a quantity, and

---

[1] This approach was introduced by Joskow and Kahn (2002) and has been advocated by Brennan (2003, 2005) amongst others. Twomey *et al.* (2005, page 35) conclude that they 'see the potential for this tool to become a standard technique of market power analysis'.

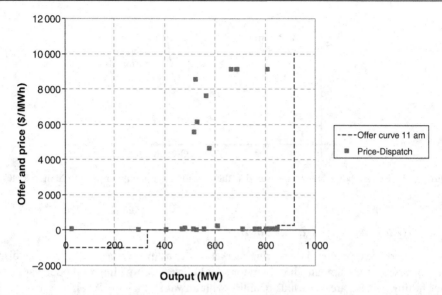

Figure 17.2 Benchmark offer curve and price–dispatch outcomes for a generator in the NEM in summer 2010

paid a price well above this benchmark level. This could be indicative of the exercise of market power.

The quantity-withdrawal approach is adequate for detecting the exercise of market power where a reasonably credible estimate of the generator's marginal cost curve is available. As noted earlier, this might be the case where the generator makes competitive offers on the same day. However, even in this case, the cost curve of a generator might change (e.g. due to a unit outage). But, even more importantly, there are many generators that only offer to the market at peak times. These generators may not offer their output to the market earlier the same day.

In some US markets, generators are constrained to submitting an offer that reflects their marginal cost curve. In this case the quantity-withdrawal strategies discussed earlier are not feasible. However, such generators may have the flexibility to declare outages of plant, to achieve the same effect. Some studies have attempted to measure outage rates to determine if physical plant failure is more common at times when market power is likely to be present.[2]

A second key issue that arises with quantity-withdrawal studies relates to how to aggregate over multiple time periods to present a picture of the overall scope for the market power of a generator.

Opportunities to exercise market power typically only arise when market conditions are favourable – such as when network constraints are binding, when demand is high, or when there are outages of generation or network plant. Is it possible to develop a measure that reflects the aggregate or average level of market power of a generator over time?

A simple average of the quantity withdrawal may be highly misleading. If a generator behaves competitively 99% of the time, the average quantity withdrawn may be very small indeed (certainly smaller than the error in the estimation of the marginal cost curve). Yet in an

[2] Joskow and Kahn (2002) find evidence of physical withholding, but this has been questioned by Hogan, Harvey and Schatzi (2004).

energy-only market, such as the Australian NEM, exercising market power 1% of the time is sufficient to have a very large impact on the annual average price. (The exercising of market power 1% of the time could easily more than triple the annual average wholesale price).

At the same time, the simple volume withdrawn may not indicate the extent of market power. After all, on some days (particularly days of very high demand), the generating firm may be able to achieve a very high price with a small quantity withdrawn – in this case, the smaller quantity withdrawal reflects a higher degree of market power, rather than smaller. To our knowledge, there is no straightforward way to aggregate quantity-withdrawal episodes to find an overall aggregate or average level of market power other than simply counting the number of episodes.

17.2.2 Price–Cost Margin Studies

Many studies propose to detect the existence of market power by measuring the bid–cost margin (i.e. the distance B in Figure 17.2).

For theoretical reasons, this margin is usually measured as the Lerner index – that is, the price–marginal cost margin, defined as the distance between the spot price and the marginal cost (the distance B) divided by the price:

$$L = \frac{P - \text{SRMC}}{P}$$

From Section 1.7, we know that an unhedged profit-maximising generator should choose a level of output where the Lerner index is equal to the reciprocal of the elasticity of the residual demand curve. This theory suggests a link between the Lerner index and the elasticity of the residual demand curve faced by a generator. The smaller the Lerner index, the lower the level of market power.

One of the problems with this approach is that as the output of a generator approaches its available capacity, the SRMC curve will usually become very steeply sloped. As a consequence, the price–cost margin will be very sensitive to the precise location of the SRMC curve chosen – potentially making the indicator of market power unreliable. Twomey et al. (2005) conclude that

> Given all of these issues, even if a study uncovers a large price–cost margin, it is still difficult to say conclusively whether this is due to abuse of market power or estimation error. This was well illustrated in the highly contentious hearings to determine the refunds to utilities from suspected market power abuse by a number of generators during the California crisis, 2000–2001.[3]

Price–cost margin studies have tended to be controversial in practice.[4] We consider this approach to be unreliable and therefore not worth pursuing further.

As with the quantity-withdrawal studies, there remains the problem of how to aggregate measures of the price–marginal cost margin over many episodes. Again we find that if a generator behaves competitively 99% of the time, the simple average of the Lerner index could

---

[3] Twomey et al. (2005), page 23.
[4] See, for example, Mansur (2007).

be close to zero on average despite the fact that the generator is able to have a very material impact on the average wholesale price.

17.3 Simple Indicators of Market Power

The approaches discussed in Section 17.2 relied on actual market outcomes. They therefore cannot be used to forecast what might happen in the future when market conditions change. In this section, we explore a range of indicators of market power that vary in sophistication but that all seek, in various ways, to forecast market outcomes from a limited range of information without full-scale formal modelling of the market. These approaches include the following:

a. Market-share-based measures including concentration ratios and the Hirschmann–Herfin-dahl Index (HHI);
b. Simple models based on supply–demand scarcity, such as the pivotal supplier indicator (PSI), the residual supply index (RSI) and their variants;
c. Measures of the elasticity of residual demand.

17.3.1 Market-Share-Based Measures and the HHI

The simplest indicators of market power are based purely on information on the market share of the various generating firms. The market share of a firm at a point in time is the share of that firm's production in the total output. Suppose there are n generating firms in the electricity industry. The ith firm produces at the rate Q_i. The total rate of production is $Q = \sum_i Q_i$. The market share of the ith firm is

$$s_i = \frac{Q_i}{Q}$$

Let us suppose that these market shares are ordered from largest to smallest. The four-firm concentration ratio is defined as the sum of the market share of the largest four firms in the industry:

$$CR_4 = \sum_1^4 s_i$$

The HHI is defined as the sum of the squares of the market shares of the generating firms in the market:

$$HHI = \sum_i s_i^2$$

But what can market shares tell us about the potential exercise of market power? There are simple economic models in which market shares are *one* factor that influences a measure of market power (such as the Lerner index). For example, the Cournot oligopoly model and the dominant-firm-competitive fringe model (see Section 1.7) show that there may be a role for market share in affecting the level of market power. However, there are no models in which market share *alone* is a key determinant of market power.

The HHI has some theoretical foundation. From the analysis of a Cournot game, we know that for an individual firm the Lerner index for that firm is equal to the market share for that firm divided by the elasticity of market demand:

$$L_i = \frac{P - C_i'}{P} = \frac{s_i}{\varepsilon}$$

Multiplying both sides by the market share of the firm and summing over the firms, we find that in a Cournot oligopoly the average Lerner index is equal to the HHI divided by the elasticity of market demand:

$$\sum_i s_i L_i = \frac{\sum_i s_i^2}{\varepsilon} = \frac{\text{HHI}}{\varepsilon}$$

Market-share-based indicators of market power suffer from several serious defects:

1. They ignore the importance of the elasticity of the demand curve. Even if a firm has a very high market share, at times when the demand side of the market is highly responsive to price increases that firm may have little or no market power.
2. They ignore the fact that generators that are not currently producing (and therefore have zero market share) may start producing if the price increases by a small amount thereby significantly limiting market power. Even a generator with a very high share of production at a point in time will have little or no market power if, when it seeks to raise the price by a small amount, a large number of other generators increase their production. To address this problem, market shares are typically computed using the total generation capacity, rather than the total volume of production at a point in time. This is acceptable for measuring market power at peak times when all generators should be producing; however, it can overlook opportunities for market power at off-peak times.
3. They ignore the fact that network constraints or capacity constraints limit the ability of some generators to increase their output in response to a price increase. If all other generators in the market are already operating at their capacity, they cannot increase their output in response to price increases – in this case, even a very small generator (measured by market share) may have substantial market power. Similarly, a generator that has a very small market share, when taking the industry as a whole, may be located in a load pocket and may have substantial market power.
4. Finally, the market share approach ignores the impact of hedge levels. We have seen that generators that are very highly hedged have little or no incentive to exercise the market power that they otherwise might possess, even if they have a very high market share.

In our view, these defects are serious. We consider that in the context of the electricity market, market share indicators should be treated with extreme caution where they are used at all.

Some variations on the basic indicators just presented have been proposed to overcome some of these defects. For example, a variant of the HHI has been proposed to account for the fact that, at any one time, a proportion of the generation plant will be either operating at full capacity or located behind a network constraint unable to respond to an increase in price. Let us suppose that there are m generators that are unconstrained – that is, able to increase their output in response to a price increase. The capacity-constrained HHI is defined as follows:

$$\text{HHI}^{\text{CC}} = \sum_{i=1}^{m} s_i^2 + \left(1 - s^C\right) \frac{s^C}{m}$$

Here, s^C is the share of the capacity held by constrained generators. In principle this indicator could take into account generators who are limited in their ability to respond due to network constraints.

In principle, market share indicators should be computed relative to the relevant geographic market at any point in time. Due to changing patterns of network congestion, the relevant geographic market will change from one time interval to the next. In principle, there could be a different market share indicator at each time interval. This raises the further problem of how to aggregate this information over time to present an overall picture of the market power of a generator.

Market-share-based indicators tend to be widely used in practice. But in our view they can be highly misleading. Since the demand curve tends to be much more inelastic in the electricity industry, even moderate levels of, say, the HHI, can correspond to significantly higher levels of market power than in many other markets. Conversely, the tendency to ignore hedge levels means that measures of market power based on market share indicators alone tends to overstate the potential for market power. As mentioned earlier, market-share-based indicators should be used with caution where they are used at all.

17.3.2 The PSI and RSI Indicators

Can we do better? Are there simple alternative indicators that provide something of an indication of market outcomes without full-scale market modelling?

A number of indicators make use of the observation that the typical supply curve in an electricity industry has a 'hockey stick' shape (Section 15.4). That is, the industry supply curve tends to be very flat at off-peak times and very steep at peak times. As a rough approximation, therefore, we really only need to be concerned about market power at peak times. Let us therefore focus on those peak times.

The PSI takes this approach to its logical extreme. Focussing on a single generator, the generator is said to be pivotal in the market at a certain point if the total demand could not be satisfied without at least some output from that generator.

The PSI is a simple binary indicator – equal to 0 if market demand could be met by the output of other generators alone, and equal to 1 if market demand could not be met by the output of other generators alone.

Let us assume that we have an electricity market (or a submarket where we can identify the relevant network constraints) with perfectly inelastic demand Q. The supply curve of generators is assumed to have the hockey stick shape – that is, is assumed to be essentially flat up to the total capacity of the generators. Let us suppose that the total capacity of the generation in the market (including importing capacity on network connections into the market) is K^{T}.

Now let us focus on a specific generator. Let us suppose the capacity of that generator is K_i. The PSI is defined as equal to 1 if the demand exceeds the capacity of all the other generators (and network links) in the market and 0 otherwise:

$$\text{PSI}_i = I\left(Q - K^{\text{T}} + K_i\right)$$

In principle, a different PSI is calculated for each point in time (corresponding to different realisations of the demand). These different PSIs are then aggregated in some way, perhaps by taking the simple average. The simple average of the PSI defined earlier is just the point on the load–duration curve (Section 5.6).

$$E(\text{PSI}_i) = \Pr\left(Q \geq K^{\text{T}} - K_i\right)$$

The idea here is that when a generator is pivotal it has the ability to increase the wholesale price to very high levels. Therefore, the proportion of time that a generator is pivotal is something of an indicator of the degree of market power of an individual generator. We would hope that the PSI is somehow linked to market outcomes – if the market has a generator with a high PSI, the average market price would be higher than if all generators have a low PSI.

The problem is that the simple PSI indicator is both under-inclusive and over-inclusive. Experience shows that many generators can exercise substantial market power when demand is high even when they are not necessarily pivotal. To this extent the PSI indicator tends to underestimate market power. Conversely, the PSI indicator suggests that a generator is pivotal even when just 1 MW of its output is required to balance supply and demand. However, in practice generators will not sacrifice very large volumes to achieve an increase in the spot price (unless the spot price is allowed to increase to very high levels indeed). In practice, market power is exercised through the withdrawal of relatively modest volumes (say less than 50% of a generator's capacity). Even more importantly, most generators have some level of hedge cover and will not reduce output below their level of hedge cover even if doing so increases the wholesale price. For this reason the PSI tends to overestimate the extent of market power.

An alternative indicator, the RSI recognises that market power is not quite so binary as the PSI suggests. The RSI is similar to the PSI but is measured on a continuous scale rather than a binary scale. Intuitively, the RSI for a company is the ratio of other generating capacity available to meet demand to the total demand. An RSI less than 100% implies that demand cannot be met by the output of other generators alone, suggesting that the generator in question has a strong potential to exercise market power.

Formally, using the preceding definitions, the RSI for generating firm i in a relevant market can be defined as the ratio of all other generating capacity to the total demand:

$$\text{RSI}_i = \frac{K^{\text{T}} - K_i}{Q}$$

There is a simple relationship between the PSI and RSI at a given point in time. The PSI is the indicator function applied to 1 minus the RSI:

$$\text{PSI}_i = I\left(Q - K^{\text{T}} + K_i\right) = I\left(1 - \frac{K^{\text{T}} - K_i}{Q}\right) = I(1 - \text{RSI}_i)$$

There is a theoretical link between the RSI and the Lerner index. Specifically, if we assume the market demand has constant elasticity ε, then in a Cournot oligopoly the Lerner index for a firm must exceed 1 minus the RSI divided by the elasticity:

$$L_i = \frac{s_i}{\varepsilon} \geq \frac{1}{\varepsilon}\left(1 - \frac{K^{\mathrm{T}} - K_i}{Q}\right) = \frac{1 - \mathrm{RSI}_i}{\varepsilon}$$

As with the PSI, the idea is that the RSI would be calculated at different points in time and then aggregated over time in some way. Ideally, there would be a clear link between the level of the aggregate RSI indicator and indicators of the exercise of market power in a market, such as the average price level.

Both the PSI and RSI are relatively easy to compute from publicly available information. In addition, these measures are relatively simple and transparent. However, both have the drawbacks. Both ignore the impact of hedging on incentives to exercise market power, and both require (in principle) case-by-case assessment of the relevant geographic market. Finally, both implicitly assume a hockey-stick shape for the supply curve (and therefore the residual demand curve facing a generator). With increasing demand-side market participation, it seems plausible that this hockey-stick shape will erode, potentially undermining the usefulness of these indicators.

Despite these reservations, in our view both the PSI and RSI are useful indicators that, if used carefully, can shed some useful light on the potential for market power.

17.3.3 Variants of the PSI and RSI Indicators

One of the drawbacks of the RSI and PSI indicators mentioned earlier is that they cannot easily be applied in the presence of network constraints – instead, they must be applied on a case-by-case basis.

Some commentators have proposed generalising the notion of the PSI to the notion of the *must-run generation*. Must-run generation is the volume of output of a given generating plant that must be called on to balance supply and demand. We can see that the PSI is equal to one if and only if the must-run generation for a particular generating firm is greater than zero.

The advantage of the must-run generation approach is that it can easily be generalised to the case of network constraints. Intuitively, for any generating plant, the must-run generation at a point in time is the level of output for which this plant would be dispatched even if it offered its output to the market at a very high price.

Formally, we can define the must-run generation as the lowest level of output for one generating firm consistent with (a) the energy balance equation, (b) the production limits on all the other generators and (c) the flow limits on the network.

However, when we take into account network constraints, as we have seen earlier (Section 16.3), we find that generating firms with a portfolio of generating plants may be able to exercise market power by increasing output in some locations, and reducing output in other locations. Therefore, our definition of must-run generation should reflect a 'worst case scenario' in which a generating firm chooses to increase output at some locations and reduce output at others.

Specifically, let us suppose that a generating firm has a portfolio of generators P. Let us focus on a subset of those generators $S \subseteq P$. Let us define $-S$ to be the generators that are in the original portfolio but not in the subset (S and $-S$ are a partition of P so that $S \cup -S = P$). Given a rate of production of all the generators in $-S$ and given a rate of consumption at each load node (Q_i^{B}) we can, in principle, compute the minimum feasible rate of production from the generators in S consistent with meeting all of the physical limits of the network. We can define the must-run-generation for a subset of generators, given a rate of production of the other generators in the portfolio as follows:

$$\text{MRG}(S| - S) = \min_{Q_i^S, i \notin -S} \sum_{i \in S} Q_i^S$$

subject to the following conditions:

a. The energy balance constraint $\sum_i Z_i = \sum_i (Q_i^S - Q_i^B) = 0$;
b. The production limits on each generator $\forall i, 0 \leq Q_i^S \leq K_i$; and
c. The physical limits on the network, expressed as $\forall l, \sum_i H_{li} Z_i \leq K_l$.

We can define the must-run-generation for a subset as the largest must-run-generation achievable by varying the production of the other generators in the portfolio:

$$\text{MRG}(S) = \max_{Q_i^S, i \in -S} \text{MRG}(S| - S) = \max_{Q_i^S, i \in -S} \min_{Q_i^S, i \notin -S} \sum_{i \in S} Q_i^S$$

subject to (as earlier) the energy balance constraint, production limits and the physical network limits.

Having defined the must-run-generation for a subset of generators in this way, we can then convert the indicator into a binary indicator, such as the PSI. Since this version of the PSI includes network constraints, it has been called the *transmission-constrained PSI* or TCPSI. The TCPSI for a subset of generators is equal to 1 if the must-run-generation is positive and 0 otherwise.

$$\text{TCPSI}(S) = I(\text{MRG}(S))$$

An alternative approach is to define an indicator that expresses the share of the capacity of the generation in the subset that must run at any one time. We might call this the *transmission-constrained must-run ratio* or TCMRR:

$$\text{TCMRR}(S) = \frac{\text{MRG}(S)}{K_S}$$

Here, $K_S = \sum_{i \in S} K_i$ is the total capacity in the subset S.

Computing these indicators requires solving a constrained-optimisation problem, which is very like the optimal dispatch task set out in Part III. The difference here is that, essentially all cost and demand information is ignored. The only information that is retained about generation and load is the capacity of the generating units, on the one hand, and the (fixed) load at each node.

The TCPSI and TCMRR are somewhat more sophisticated than the PSI and RSI. In particular, these approaches take explicit account of the impact of network constraints, and the potential for interaction between different generators in a portfolio. However, some of the other problems noted earlier still remain, such as the fact that the hedge positions of generators are ignored. In addition, there remains the question of how to aggregate these measures over time.

Despite these limitations, our view is that where the computation facilities are available, the TCPSI and TCMRR offer some promise as offering a picture of the scope for market power by a group of generators, without fullscale modelling.

17.3.4 Measuring the Elasticity of Residual Demand

The approaches we have considered so far (market share indicators, and the PSI and RSI, and their variants) rely on relatively little information on the nature of supply and demand in the market. In fact, all that was required was information on the total capacity of generation and the total (inelastic) demand at a point in time. This is an acceptable approximation where demand is inelastic and generators all have more-or-less similar variable costs. However, this may not always be a reasonable assumption. As noted earlier, in the future it is likely that there will be increasing responsiveness of demand to the wholesale price and there may arise a greater range of generation technologies with a wider range of variable cost.

For this reason it may be worthwhile considering slightly more sophisticated approaches. Specifically, let us now assume that we have information about the supply curve of generators or the demand curve of customers. This information could be derived from actual market bids or offers at a point in time in the past, or it could be constructed from underlying information on generator cost curves.

If we have access to more complete supply and demand information, how should we use this information to make predictions about market power outcomes short of full-scale market modelling?

One approach is to conduct an estimate of the residual demand curve facing an individual generator. Recall that the residual demand curve facing a generator is equal to the market demand curve less the supply curve of the other generators in the market. In principle, if a generator faces a very steep (inelastic) residual demand curve, it may have a strong incentive to exercise market power.

Calculation of the shape of the residual demand curve facing a generator is known as *residual demand analysis*. This approach has the advantage of a strong theoretical foundation. As we have seen, the exercise of market power involves a price–volume trade-off. The shape of the residual demand curve determines the nature of that price–volume trade-off.

The simplest form of residual demand analysis involves the relatively straightforward construction of the residual demand curve from the supply and demand curves in a given market.

However, in the presence of network constraints, this approach would require determination of the relevant geographic market on a case-by-case basis. A more sophisticated approach takes explicit account of network constraints by essentially solving a form of the optimal dispatch problem.

Specifically, the residual demand curve could, in principle, be computed as follows: Let us suppose we are seeking to determine the residual demand curve for an individual generator i located at pricing node i. Let us suppose that the output of this generator is Q_i^S. The corresponding price on the residual demand curve is the nodal price P_i, which is found by solving the conventional optimal dispatch problem:

$$\max_{Q_i^S, Q_i^B} \sum_i B_i(Z_i)$$

subject to the normal constraints (the energy balance constraint, the production limits on each generator, and the physical limits on the network).

This approach determines the residual demand curve for an individual generator at a specific location. In the case of a generation portfolio with plants at different locations, we could approach the problem differently – perhaps by finding the combination of production that maximises profit for the generation portfolio given the supply and demand of the other generators and consumers in the market.

Having determined the residual demand curve, the question remains how to determine a measure of the extent of market power. In addition, since the residual demand curve is different at every point in time, the question remains as to how to aggregate this measure of market power over time.

There is also an even more fundamental problem. In many wholesale power markets, including the Australian NEM, generator offer curves and customer bid curves take the form of a step function. As a consequence, the residual demand curve also takes the form of a step function. We have seen that the scope for market power depends on the 'steepness' or elasticity of the residual demand curve. However, a step function has elasticity, which is either 0 (indicating no market power) or infinite (indicating extreme market power). To make some sense out of the elasticity of the residual demand curve at an individual point, it is necessary to approximate the residual demand curve with a smooth function.

This is illustrated in Figure 17.3. Unfortunately, experience shows that the manner in which that smoothing is carried out can have a substantial influence on the resulting measures of market power.

Moreover, there is an even more fundamental criticism. There is no easy way to convert knowledge of the residual demand curve into a measure of market power. Even if the residual demand curve is flat over much of its range, there remains the possibility that the residual demand curve is sufficiently steep in the remaining part of its range such that a large withdrawal of capacity could have a substantial impact on the market price. Conversely, even if the residual demand curve is very steep at a point, it may not be profitable for the generator in question to withdraw capacity.

Residual demand analysis has an important theoretical foundation, and has the potential to yield useful insights. However, in our experience, residual demand analysis gives rise to serious methodological issues in practice.

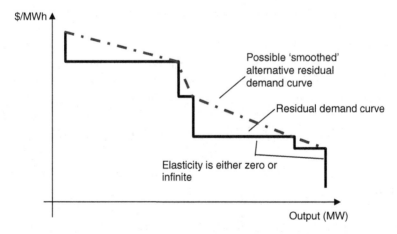

Figure 17.3   Smoothing a step function to obtain a more meaningful measure of elasticity

17.4 Modelling of Market Power

Residual demand analysis (discussed in Section 17.3) is relatively straightforward when there is a single dominant generating firm. If all the other generating firms can be assumed to behave competitively, then it is reasonable to assume that they offer their output to the market at their short-run marginal cost curve.

But almost inevitably, where there is one generator with market power, there is likely to be more. In this case the residual demand curve facing any individual generator depends on the strategic behaviour of the other firms with market power. In this context it is no longer sufficient to look at each firm in isolation; we must model the market as a strategic game.

As explained in Section 1.8, to describe a strategic game, we must specify a set of actions for each of the players, an objective function or payoff for each of the players, and a solution concept.

In the case of short-run modelling of market outcomes in the electricity market, the players are the generators and customers (or, at least, the larger generators and customers that may have market power). For the sake of simplicity, let us assume that only generators have market power. All customers are assumed to be price-takers. The action for each player is usually the rate of production of electricity – that is, the most common modelling approach in the electricity market is to assume that the market participants play a Cournot game. The objective function or payoff for electricity generators is typically the short-run profit. Finally, the standard solution concept is the Nash equilibrium.

As discussed in Section 1.8, a Nash equilibrium is a set of actions Q^{S*} (one for each player), which has the property that, for each player, the action maximises the payoff of that player given the actions of the other players.

Let us suppose we have a set S of strategic generators and $-S$ of nonstrategic generators (i.e. generators that are price-takers). Let us focus on a specific strategic generator i located at node i. Given the output of all the other strategic generators, $Q_j^S, j \in S, j \neq i$, for each output of the generator in question, Q_i^S, we can in principle solve the optimal dispatch problem to determine the corresponding local nodal price $P_i(Q_i^S|Q_{-i}^S)$ (in exactly the same way that we saw in Section 17.3 in the calculation of the residual demand curve). Here, the optimal dispatch task is (as before)

$$\max_{Q_i^S, Q_i^B, i \in -S} \sum_i B_i(Z_i)$$

subject to the normal constraints (the energy balance constraint, the production limits on each generator and the physical limits on the network).

Once the price is determined, we can compute the profit of the generator in the usual way:

$$\pi_i\left(Q_i^S|Q_{-i}^S\right) = P_i\left(Q_i^S|Q_{-i}^S\right)Q_i^S - C_i\left(Q_i^S\right)$$

A particular action Q_i^{S*} is a *best response* to the actions of the other generators if it satisfies the following:

$$Q_i^{S*} \in \text{argmax}_{Q_i^S} \pi_i\left(Q_i^S|Q_{-i}^S\right)$$

A combination of outputs Q^{S^*} is a Nash equilibrium if for each individual player the specified output for that player is a best response to the specified actions of the other players:

$$\forall i, Q_i^{S^*} \in \mathrm{argmax}_{Q_i^S} \pi_i \left(Q_i^S | Q_{-i}^{S^*} \right)$$

In principle, full-scale market modelling of this kind offers the most promise to capture and reflect the strategic interaction between generators and therefore to reflect likely market outcomes.

However, there are certain problems. One problem is that there may exist more than one Nash equilibrium. In this case there arises the question of how to interpret the results. What weight should be put on the different equilibria? Are they all equally likely to arise or are none of them likely to arise? How can we use information on the different equilibria to make a forecast of what is likely to happen in the market?

Another problem relates to the hedging choices of the generators. We have seen that the hedge position of a generator has a significant impact on its incentive to exercise market power. But how do generators choose their level of hedging? Presumably, this should be included in a more sophisticated model.

In addition, there are a couple of practical problems. In many cases full market simulation models are relatively complex and nontransparent. They therefore are of limited use in contested environments in which transparency is required.

Finally, there is the problem that these models tend to be highly computation intensive. As a consequence there are practical limits to the number of strategic generators that can be considered and the number of possible actions allowed to each strategic generator. This is discussed further later.

17.4.1 Modelling of Market Power in Practice

How can the model presented earlier be implemented in practice? A common way to proceed is to specify a set of strategic generators and a set of possible actions (in this case, rates of production) for each strategic generator. For example, a generator may be given a choice of, say, 10 or 20 different output levels.

The analysis then proceeds in two steps. In the first step, the profit of each generator is computed under each possible combination of output levels. In the second step, these results are then examined to find the Nash equilibria.

This approach has the advantage that it finds all the Nash equilibria, but it is extraordinarily computation intensive. If there are 10 strategic generators, each with 10 possible actions, the optimal dispatch task must be carried out 10^{10} times. Computing optimal dispatch in a full-scale network might take, say, 100 ms. Finding the Nash equilibria for just one market scenario would then take 31.7 years. Clearly this is completely infeasible. And yet, real markets might have well many more than 10 strategic generators.

The common response is to significantly limit the number of strategic generators, the number of actions available to each generator and the sophistication of the modelling of the network. With five strategic generators and five actions, all of the Nash equilibria can be found for one market scenario in 5.2 min.

This approach is often used in practice. However, there remain certain problems. One of the primary problems is the tendency for a proliferation of Nash equilibria. Restricting the possible levels of output of each generator to just a small number of choices tends to introduce new Nash equilibria. This arises because limiting the output of a generator to a small number of levels prevents a generator from undercutting another generator by increasing its output slightly. The question of how to handle multiple Nash equilibria remains unanswered.

17.4.2 Linearisation

An approach has been suggested to this modelling task that reduces the significance of the problems mentioned previously. This approach involves using linearisation to convert the task into a single-stage mixed-integer linear optimisation problem. At least for relatively small scale problems, this approach significantly reduces the computation time required. This approach is described further in Hesamzadeh and Biggar (2012) and Moiiseva, Hesamzadeh and Biggar (2014).

17.5 Policies to Reduce Market Power

There are a number of policies that could be pursued to reduce the level of market power in an electricity network. These include the following:

- Policies to reduce market concentration or prevent increases in concentration. In all modern economies competition laws prevent mergers between competitors that substantially lessen competition. In the context of the electricity market, merger control is especially important, especially in local geographic markets or load pockets. In addition to merger control active divestiture of generating capacity can improve market outcomes. It is believed that divestiture of electricity generating capacity in the United Kingdom was a factor in lowering wholesale prices at the time of the transition to the New Electricity Trading Arrangements (NETA) market design.
- Policies to increase the responsiveness of demand to the wholesale price. One of the primary reasons that wholesale electricity markets are prone to market power is the lack of responsiveness of electricity customers to wholesale prices. However, this is likely to change in the future as customers increasingly take up smart devices and smart appliances that are able to respond, directly or indirectly, to wholesale prices. Increasing the demand-side responsiveness increases the elasticity of the residual demand curve, potentially significantly limiting opportunities for market power.
- Policies to reduce congestion on the network. Network congestion is another primary determinant of opportunities to exercise market power. Networks should be operated and planned in a way that takes into account the potential for market power. This may mean increasing the capacity of a network link merely to increase competition between generators.
- Policies to increase average hedge levels. We have seen that the hedging positions of generators are one of the primary determinants of the incentive to exercise market power. By facilitating hedging, market designers may be able to reduce the exercise of market power. In some cases it may be appropriate to mandate or require minimum levels of hedge cover.
- Control on the bidding of generators. Many markets around the world, especially in the United States, place controls on the offers that generators can make in the wholesale electricity market. In our view, it is preferable to enhance competition through changes to market structure or reducing congestion before imposing behavioural rules on generators. However, in some cases behavioural rules may be required. These could be targeted to

generators that are expected to have opportunities to exercise market power. Where behavioural rules are imposed, they should not prevent the price rising to high levels at times of genuine tight supply–demand balance.

- Price caps. Price caps limit the incentive to exercise market power. However, they also cause other serious distortions. We do not advocate the use of price caps to control market power except when no other alternatives are available.

17.6 Summary

Policymakers dealing with market power often want to know when market power has been exercised in the past, or how market power is likely to change following a change in the market, such as a merger or the expansion of a network element.

The simplest approaches to detecting market power rely on observations of market price and dispatch outcomes. In principle, market power arises when the price–quantity combination for which a generator is dispatched does not fall on its marginal cost curve. Since the marginal cost curve is not normally observable, the marginal cost curve must be estimated. There are two potential ways to measure the extent of market power: (a) the degree of quantity withdrawal (i.e. the gap between the rate of production chosen by the generator and the corresponding rate of production at the same price on the marginal cost curve); and (b) the degree of price–marginal cost mark-up – the gap between price and marginal cost divided by the price. However, since the marginal cost curve is typically very steep as output approaches total capacity, the latter approaches are unreliable.

When we are interested in predicting when market power might change in the future, we must rely on some form of market model that (perhaps in some highly stylised way) draws a link between basic market information and market outcomes. The simplest of these models rely only on market share information. The most well-known of these market share approaches is the HHI. The HHI has some theoretical foundation, but is unreliable in the context of the electricity industry where demand is inelastic and supply becomes very inelastic at peak times.

Other indicators have been proposed that more closely reflect key features of the electricity market. If we assume that demand is inelastic and generators have more-or-less the same variable cost, then we need only be concerned with the exercise of market power at peak times. At peak times the output of nearly all generators is required to balance supply and demand. A generator can be said to be 'pivotal' if at least some of its output is required to balance supply and demand. The pivotal supplier index is equal to 1 if a generator is pivotal and 0 otherwise. The higher the average PSI of a generator, the greater its opportunities for market power. The residual supply index of a generator is a related measure that reports the ratio of all-other-capacity to the total demand. An RSI measure less than 1 indicates that a generator is pivotal.

The PSI and RSI measures suffer from the drawback that the relevant geographic market must be separately determined for each time interval in the period over which the measure is applied. This process can be automated by variants of these measures, which take into account network constraints. One approach carries out an optimal-dispatch-like constrained optimisation to find the level of must-run generation for a portfolio of plants. When the must-run generation is positive the TC-PSI measure is equal to 1 (and 0 otherwise). The TC-MRR expresses the proportion of must-run generation as a share of the capacity of the generation in the portfolio.

The PSI and RSI measures and their variants ignore the possibility for price responsiveness of demand, and ignore the potential for cost variation between generators. If supply and demand information is known, we can use that information to more directly compute the potential for market power by estimating the shape of the residual demand curve. If network constraints may be binding, the residual demand curve has to be calculated on a case-by-case basis. This process can be automated through a modelling process that very closely replicates the optimal dispatch task. In principle, these techniques allow the entire residual demand curve to be determined. Problems arise, however, in assessing how to aggregate different residual demand curves, or different measures of market power based on the residual demand curve, over time.

Analysis of the residual demand curve assumes that we have some knowledge of participant bid and offer curves. However, where there two or more market participants with market power, these offer curves can only be determined jointly. Where there are two or more market participants with market power, the market must be modelled as a strategic game. The typical approach is to model the choices of rates of production of electricity generators (i.e. a Cournot game) assuming that each generator seeks to maximise its own profit.

In practice the task of finding the Nash equilibrium outcome is highly computationally intensive, especially when there are more than a few strategic players in the market, each with more than just a few choices of output. In addition, full-scale market modelling faces problems of nontransparency and a proliferation of Nash equilibria. The computation time problem can be partially mitigated by converting the problem into a single-stage optimisation by linearising the KKT conditions that characterise the optimal dispatch outcome in each case.

Electricity markets are prone to market power, but there are several policies that can be undertaken to reduce market power, such as policies for breaking up large generators, policies to promote demand-side responsiveness, policies to reduce network congestion, policies to increase hedge levels, controls on generator offer curves, and (in extreme cases) some forms of price caps.

Questions

17.1 What are the strengths and weaknesses of each of the following: market share indicators, PSI and RSI, residual demand analysis, and full market modelling?

17.2 If there is a problem with market power, what are the pros and cons of imposing a market price cap?

17.3 If a generator reduces the amount it offers to the market and the wholesale price increases significantly, does this mean the generator is exercising market power – why or why not?

Further Reading

For more information on the TC-PSI indicator see Hesamzadeh, Biggar and Hosseinzadeh (2011). There is a useful summary of market power indicators in Newbery et al. (2004).

Part VIII

Network Regulation and Investment

A Wellington Municipal Electricity Department pole gang erect a hardwood pole, New Zealand, 1935
(Source: Neil Rennie)

18

Efficient Investment in Network Assets

In Part IV, we explored the conditions for efficient investment in generation and load assets. Now, in this part, we extend those conditions to include efficient investment in network assets.

18.1 Efficient AC Network Investment

Network investment decisions are inherently long-run decisions. Therefore, we need a long-run investment model. Since generation capacity can also be adjusted in the long run, we need to consider network investment decisions in combination with generation capacity decisions.

As with the discussion of generation investment in Chapter 9, let us start from the assumption that network investment is neither sunk nor lumpy. In addition, we will assume (as throughout this text) that losses can be ignored.

To proceed, let us consider a slight generalisation of the problem set out in Chapter 9. In that problem, we have a set of potential generation and/or consumption locations indexed by i. In state s of the world, the generation of type t and location i has a cost function $C_{its}(Q_{its}^S, K_{it})$. We can then determine the benefit function $B_{is}(Z_{is}, K_{it})$.

Now let us suppose that the we have a set of possible network configurations indexed by θ. Network θ involves a set of network links L^θ and a matrix of power-transfer distribution factors H^θ. Let us suppose that, when the capacity of link $l \in L^\theta$ is given by K_l^θ, the cost of the network is $C(\theta, K_l^\theta)$. For the moment we will assume that the network is reliable, so we do not need to consider stochastic variation in network configuration or network capacity.

The overall problem of finding the optimal dispatch and the optimal mix and level of investment in both generation and network capacity is as follows:

$$W = \max_{Z_{is}, K_{it}, \theta, K_l} E\left[\sum_i B_i(Z_i, K_{it})\right] - C\left(\theta, K_l^\theta\right)$$

$$= \max_{Z_{is}, K_{it}, \theta, K_l} \sum_{i,s} p_s B_{is}(Z_{is}, K_{it}) - C\left(\theta, K_l^\theta\right)$$

The Economics of Electricity Markets, First Edition. Darryl R. Biggar and Mohammad Reza Hesamzadeh.
© 2014 John Wiley & Sons, Ltd. Published 2014 by John Wiley & Sons, Ltd.

subject to the following conditions:

$$\forall s, \sum_i Z_{is} = 0 \quad \leftrightarrow p_s \lambda_s$$

$$\forall l, s, \sum_i H^\theta_{li} Z_{is} \le K^\theta_l \quad \leftrightarrow p_s \mu_{ls}$$

For any given network configuration, the KKT conditions for K^θ_l yields the fundamental equation for determining the optimal level of network capacity on a network link: The optimum level of network capacity is the point where the expected value of the constraint-marginal-value is equal to the marginal cost of adding capacity (taking into account the optimal mix of generation at each location):

$$E\left[\mu_l\right] = \sum_s p_s \mu_{ls} = \frac{\partial C}{\partial K^\theta_l}$$

Result: For any given network configuration the optimal level of network capacity is where the expected value of the constraint-marginal-value on the network flow limit constraint is equal to the marginal cost of adding capacity (taking into account the optimal mix of generation at each location)

Earlier, we saw that (under certain specific assumptions) the optimal level of investment in a specific type of generation is where the area under the price–duration curve and above the variable cost of that generation type is equal to the fixed cost of adding capacity of that generation type. We can interpret the preceding equation in a similar way: The optimal level of investment in network is where the area under the constraint-marginal-value–duration curve is equal to the fixed cost of adding capacity.

It may be worth noting that as long as there is some cost to adding network capacity (i.e. as long as $(\partial C / \partial K^\theta_l) > 0$) then the preceding expression shows that in the optimally configured network, congestion always occurs with some positive probability even after allowing for adjustment to the optimal mix of generation at each location. Such network congestion may not happen in normal operation however – it may only occur when there are, say, outages of other generation or network plant.

Result: As long as there is some positive cost to adding network capacity, and as long as network capacity can be added in arbitrarily small increments, in the optimally configured network there is always some positive probability that any given network link will be congested. It is not optimal to build the network to the point where any given network link is never congested.

18.2   Financial Implications of Network Investment

Who gains and who loses from a nontrivial expansion in network capacity?

The total economic welfare of the power system can be expressed as the sum of the expected profit to generators, the expected net utility of customers, and the expected merchandising surplus less the cost of the network.

$$E\left[\sum_{i,t} \pi_{it}\right] + E\left[\sum_t \varphi_i\right] + E[\text{MS}] - C(K_l)$$

$$= \sum_{i,t} E\left[P_i Q_{it}^S\right] - \sum_{i,t} E\left[C_{it}\left(Q_{it}^S, K_{it}\right)\right] + \sum_i E\left[U_i\left(Q_i^B\right)\right] - \sum_i E\left[P_i Q_i^B\right]$$

$$- \sum_i E\left[P_i\left(\sum_t Q_{it}^S - Q_i^B\right)\right] - C(K_l)$$

$$= \sum_i E\left[U_i\left(Q_i^B\right)\right] - \sum_{i,t} E\left[C_{it}\left(Q_{it}^S, K_{it}\right)\right] - C(K_l) = W(K_{it}, K_l)$$

From this we can see that a change in network capacity has benefits spread over three parties: The customers, the generators and the merchandising surplus. The impact on generators depends on the shape of the long-run cost curve for generation. If the cost of adding generation capacity is upward sloping then, even after allowing for an adjustment in the mix of generation at each node, generators at some nodes (nodes with higher prices) will be better off after the change in network capacity (that is, will have a higher expected profit) and generators at other nodes will be worse off after the change in network capacity (that is, will have a lower expected profit). Similarly, customers at some nodes will be better off and others will be worse off. Finally, there will be a change in the merchandising surplus. The merchandising surplus may be larger or smaller as a result of the network augmentation.

In general, all three effects will be present. However, in the special case in which there is a flat long-run cost curve for generation (i.e. where generation capacity of each type can be added at a constant fixed cost), then generator profits are zero both before and after the change in network capacity (after allowing the mix of generation to adjust at each node). In this case the consequences of the change in network capacity are restricted to the impact on customers and the impact on the merchandising surplus. If, in addition, demand is inelastic, the impact on customers is due entirely to the change in the prices paid.

18.2.1 The Two-Node Graphical Representation

It is difficult to visualise these effects in general. However, in the special case of a two-node network, there is a relatively straightforward graphical representation of some of the financial implications of network augmentation.

In a two-node, one-link network there are potentially two different nodal prices. For each level of capacity on the network link there is a corresponding optimal mix of generation capacity at each node and a corresponding expected nodal price. Figure 18.1 shows how the expected nodal price at each location might vary with changes in the network capacity. At low levels of capacity the expected difference in the nodal prices (after allowing for adjustment to the optimal mix of generation at each node) might be quite large. For large enough levels of capacity the expected difference in the nodal prices (again, after allowing for adjustment in the optimal mix of generation at each node) might be low or zero.

In this simple case of a two-node network the difference in the nodal prices is just equal to the constraint marginal value. In the earlier analysis, we saw that (when network capacity can be added in arbitrary small increments) the optimal level of capacity is where the expected

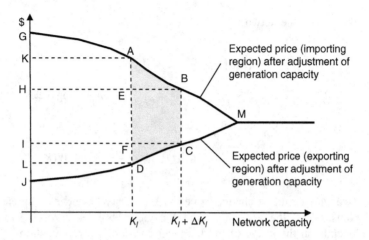

Figure 18.1   Welfare implications of a network augmentation in a two-node network

constraint marginal value is equal to the cost of augmenting the network. In Figure 18.1 this is where the distance between the two lines is equal to the cost of adding network capacity.

Now consider a nontrivial increment in network capacity, from K_l to $K_l + \Delta K_l$. The increase in welfare (after allowing for an adjustment in the mix of generation) is illustrated by the area ABCD in Figure 18.1.

This diagram also allows us to visualise the impact of the changes on market participants. The network augmentation has a different impact on market participants in the exporting node compared to market participants in the importing node.

Let us consider first the impact on the merchandising surplus. As was proven in Section 8.3, the expected merchandising surplus is equal to the expected congestion rents, which is equal to the expected constraint marginal value multiplied by the link capacity (provided each link has a fixed capacity):

$$E[MS] = E[CR] = E\left[\sum_l \mu_l K_l\right] = E[\mu_l] K_l$$

Therefore, the expected merchandising surplus before the augmentation is the area KADL. The expected merchandising surplus after the augmentation (and after adjustment of the mix of generation capacity) is HBCI. The merchandising surplus loses the areas IFDL and KAEH and gains area EBCF.

Now let us explore the impact on market participants in the exporting node. Their welfare increases from the area JLD to the area JIC. The increase in welfare is the area ICDL. Part of this (area IFDL) comes at the expense of the merchandising surplus and the remainder FCD is new welfare gain.

Similarly, market participants in the importing node have their welfare increased by the area KABH. Part of this is due to a reduction in the merchandising surplus by area KAEH and the remainder is overall welfare gain of ABE.

The total welfare gain is the sum of the areas ABE, FCD and EBCF – which is the area ABCD as first noted.

It is important to emphasise once more that the area ABCD is the change in welfare after the mix of generation has changed at both nodes. It is therefore a 'long-run' notion. It cannot be easily determined from the set of assets in the network at any one point in time.

18.2.2 Financial Indicators of the Benefit of Network Expansion

It is sometimes pointed out that if we could determine a financial measure of the economic welfare benefit from a network augmentation (area ABCD in the preceding example) this could be used to develop an incentive on a network operator. In principle, if the network operator received the full benefits from a network augmentation, it could make efficient decisions regarding investment in the network.

For example, let us suppose that we define a financial incentive equal to the difference between the economic welfare arising from a given network configuration (assuming an optimal mix of generation at every node) and the economic welfare arising from some other baseline network configuration (again, assuming an optimal mix of generation at each node).

$$T\left(\theta, K_l^\theta\right) = W\left(\theta, K_l^\theta\right) - W^{\text{baseline}}$$

Amongst the different ways of defining the baseline, two possible ways stand out as representing opposite extremes:

a. The economic welfare that would arise if the network did not exist (and there was an optimal mix of generation at every node); and
b. The economic welfare that would arise if the network was built out to the point where there was no network congestion (and there was an optimal mix of generation at every node);

If we define the baseline as the 'no network' case, the financial indicator before the augmentation is equal to the area GADJ in Figure 18.1. After the augmentation, the financial indicator is equal to the area GBCJ. The increase in the area is equal to the increase in economic welfare ABCD.

Similarly, we could define the baseline as the 'unconstrained network' case. In this case, the financial indicator before the augmentation would be equal to the area AMD. Augmenting the network would reduce the financial indicator to the area BMC. We see that the reduction in financial indicator is equal to the increase in economic welfare ABCD.

It is worth emphasising again that this financial indicator cannot be directly computed using information on the stock of assets in the market at any one point in time. Rather, it requires a counterfactual – that is, a comparison to the stock of assets that would arise if either the network did not exist or the network was never congested. At any point in time, it is very straightforward to compute what the level of economic welfare would be if the network were uncongested (this can be determined simply by removing the network limit constraints in the optimal dispatch task). However, this determines a level of economic welfare given the existing stock of generation and consumption assets. This could be very different from the level of economic welfare that would arise if the stock of generation and consumption assets were allowed to adjust to the absence of network constraints.

18.3   Efficient Investment in a Radial Network

Let us suppose that we have a radial network. The preceding analysis shows that the optimal level of network capacity on any given link depends on the expected constraint marginal value (CMV) and the cost of adding network capacity. As long as the cost of adding network capacity is positive, the expected CMV must be positive at the optimal level of network investment.

However, the CMV is only positive when the flow is equal to the physical limit. In a radial network, the flow on a link is equal to the net injection on the 'exporting' side of the constrained link. The net injection is, of course, equal to the total generation less the total demand. Therefore, we can conclude that the total generation (rate of production) on the exporting side of a constrained link must be larger than or equal to the network capacity.

Mathematically, let us suppose that we have a radial network with one potential constraint that divides the network into two pricing regions – pricing region 1, which is the exporting region, and pricing region 2, which is the importing region. The network capacity must be equal to the maximum injection in the exporting region, which is less than the sum of generation capacity in the exporting region.

$$K_{1 \to 2} = \max_s Z_{1s} = \max_s \sum_t Q^S_{1ts} - Q^B_{1ts} \le \sum_t K_{1t}$$

> *Result*: In an optimally configured radial network (with an optimum level of generation and network capacity) the capacity of generation on the exporting side of a constrained link is always greater than or equal to the network capacity.

Now let us focus on a particular special case. Let us suppose that there is no load on the exporting side of the constrained link and let us suppose that there are no generation cost-shifting factors (i.e. the cost function of the generators is constant over time). Can it be that the total capacity of generation on the exporting side of the constrained link is larger than the network capacity? The answer is no. Since there is no uncertainty in the generation cost functions, if the total generation capacity exceeds the total network capacity there will be some generation capacity that is never utilised. This cannot be efficient. Therefore, we conclude that in the absence of load, and in the absence of uncertainty, the total capacity of generation on the exporting side of a constrained link is equal to the network capacity.

Mathematically, let us suppose that there is no load in the exporting region. There must be a state in which every generator in the exporting region is operating at capacity: $\forall t, Q^S_{1ts} = K_{1t}$. However, the total net injection in the exporting region is less than or equal to the physical flow limit: $Z_{1s} = \sum_t Q^S_{1ts} = \sum_t K_{1t} \le K_{1 \to 2}$.

Putting these last two results together, we see that in a radial network, in a region with no load and no uncertainty in generator cost functions, the total capacity of generation in the region is equal to the total capacity on the network out of the region.

> *Result*: In an optimally configured radial network, in regions where there is no load and no uncertainty in generator cost functions, the total generation capacity in the region is equal to the total network capacity out of that region.

This result allows us to simplify the task of finding the optimal combination of generation and network capacities. Specifically, we find that we can use a simple extension of the screening-curve concept.

Let us consider again the optimisation task presented earlier, but simplified to the case of a two-node radial network, with generation with a constant variable cost and a constant cost of adding generation capacity, and with a constant cost of adding network capacity. Specifically, the variable cost and fixed cost per unit of capacity of generation of type t at location i is assumed to be equal to c_{it} and f_{it}, respectively. The fixed cost per unit of network capacity is assumed to be $f_{1\rightarrow2}$.

The optimisation task can now be written as follows:

$$W = \max_{Z_{1s}, Z_{2s}, K_{1t}, K_{2t}, K_{1\rightarrow2}} E[B_1(Z_{1s}, K_{1t})] + E[B_{2s}(Z_{2s}, K_{2t})] - f_{1\rightarrow2}K_{1\rightarrow2}$$

subject to the following:

$$\forall s, Z_{1s} + Z_{2s} = 0 \leftrightarrow p_s\lambda_s$$

$$\sum_t K_{1t} = K_{1\rightarrow2} \leftrightarrow \gamma$$

From the KKT conditions for this problem, we find the following condition for the optimal level of each type of generation capacity:

$$\forall t, E[(P_1 - c_{1t})I(P_1 \geq c_{1t})] = f_{1t} + f_{1\rightarrow2}$$

In other words, previously we derived the result that at the optimal level of capacity, the area under the price–duration curve and above the variable cost of each generation type was equal to the fixed cost for that generation type. Now we can see that, in the exporting region, the area under the price–duration curve and above the variable cost of each generation type is equal to the fixed cost for that generation type plus the fixed cost of adding network capacity.

This is a useful result. It implies that the simple screening curve analysis that we explored in Chapter 9 can be easily extended to find the optimal mix of generation and network capacity. Specifically, the same screening curve analysis as before applies, but with the fixed cost of additional network capacity added to the fixed cost of each type of generation capacity.

Result: In a simple radial network with only one load node and with no uncertainty in generation cost function at all other nodes, the optimal mix of generation and network capacity can be found using a simple variation on screening curve analysis. In this simple variation, the cost of adding network capacity is added to the fixed cost per unit of capacity of remote generators (i.e. those located away from the load node).

Figure 18.2 A simple two-node one-link network

18.4 Efficient Investment in a Two-Node Network

In order to gain some further insight into the network investment problem, let us apply the analysis in the previous section to a simple network with two nodes and one network link, as illustrated in Figure 18.2.

From this analysis, we know that the optimal level of transmission depends on the expected constraint-marginal-value for this network link at the optimal mix of generation at each end of the link.

Consider the following thought experiment. Let us imagine varying the capacity on the link from zero to a high level, allowing the optimal mix of generation to adjust to the level of capacity at each stage.

The constraint marginal value in this simple network is equal to the price difference between the two nodes: $\mu_{1\to2} = P_2 - P_1$.

Let us consider a plot of the prices P_1 and P_2 against the network capacity. When the network capacity is close to zero we might find that the average prices at each location are quite different (as illustrated in Figure 18.3), so the expected value of the constraint marginal value is quite large. As network capacity is increased, we would expect the average prices to approach each other, so the expected value of the constraint marginal value moderates. Eventually, a point is reached where the network capacity is sufficiently large that the network is no longer constrained under any states of the world and the expected value of the constraint-marginal-value $E(\mu_{1\to2})$ is zero.

The optimal level of capacity is where the expected constraint marginal value is equal to the cost of adding capacity. This is illustrated in Figure 18.4.

As noted in Section 18.3, another way to view the same problem is to consider the constraint-marginal-value–duration curve. We have seen that we have an optimal level of network capacity when the area under the constraint-marginal-value–duration curve is equal to the cost of additional capacity (Figure 18.5).

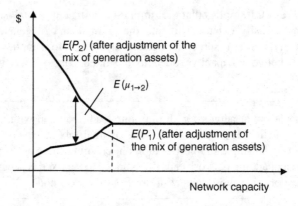

Figure 18.3 Evolution of expected prices with varying network capacity

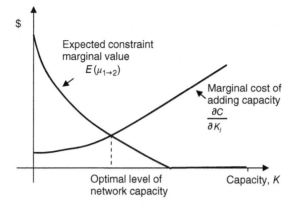

Figure 18.4 Determination of the optimal level of network capacity for a single network link

18.4.1 Example

Let us now consider a simple worked example to illustrate the key points in the preceding sections.

This simple example involves a network with two nodes and one network link. There are three generation types at each node, with constant variable cost and constant cost of adding capacity. The fixed and variable cost of each generation type is shown in Table 18.1.

Table 18.1 Generation cost data for the simple two-node example

Node	Type	Fixed Cost ($/MW)	Variable Cost ($/MWh)
1	1	$100.00	$10.00
1	2	$90.50	$20.00
1	3	$70.00	$50.00
2	1	$140.00	$100.00
2	2	$90.00	$200.00
2	3	$70.00	$500.00

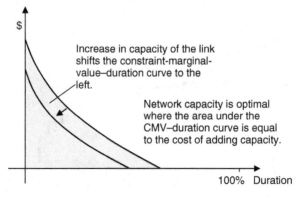

Figure 18.5 Optimal network capacity is where area under CMV–duration curve is equal to the incremental cost of capacity

Table 18.2 Load-duration curve for the optimal network investment example

State	Probability	Load (MW)
1	0.04	500
2	0.06	800
3	0.15	1100
4	0.20	1400
5	0.20	1500
6	0.20	1600
7	0.05	1700
8	0.05	1800
9	0.04	2000
10	0.01	2500

As we have seen, we need to introduce some uncertainty. In this network there are assumed to be 10 demand states, with probability set out in Table 18.2. In each state demand is assumed to be inelastic.

The resulting load–duration curve is set out in Figure 18.6.

For each level of network capacity there is a corresponding optimal mix of generation, and a corresponding optimal dispatch and set of prices. Figure 18.7 shows, for each level of network capacity, the corresponding optimal mix of generation. As is clear from this chart, the total volume of capacity at the exporting node is always equal to the network capacity.

The expected prices for different levels of network capacity are shown in Figure 18.8.

From the analysis, we know that the optimal level of capacity on this network link is where the expected constraint-marginal-value (assuming an optimal mix of generation) is equal to the cost of adding capacity to the network. For example, as Figure 18.9 shows, if the cost of network capacity is constant at $40/MW, the optimal network capacity is 1700 MW.

Now let us consider the impact of expanding the network link from 1000 to 1500 MW. After adjusting the mix of generation, the total cost of generation (the variable dispatch cost plus the fixed cost of capacity) reduces by $54 750/h, so there is a welfare gain of $54 750/h.

However, who gains and who loses from this network expansion? Under the assumptions of constant variable cost and constant fixed cost per unit of capacity, the total profit to generators

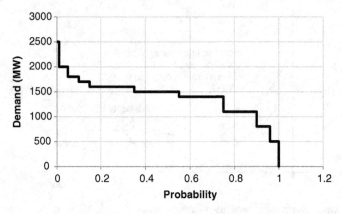

Figure 18.6 Load–duration curve in the optimal-network-planning example

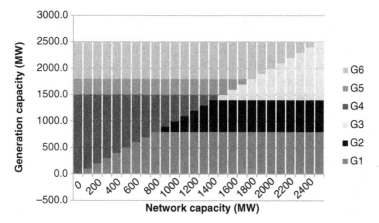

Figure 18.7   Optimal mix of generation at each node as a function of the link capacity

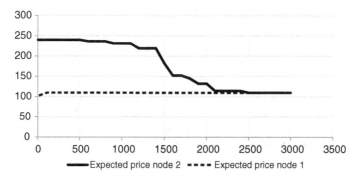

Figure 18.8   Expected price at each node as a function of link capacity

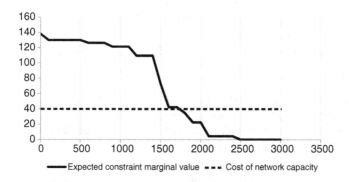

Figure 18.9   Determination of the optimal link capacity

is zero before and after the augmentation. But what about the impact on customers and on the merchandising surplus?

When the network limit is 1000 MW, customers pay an average price of $231.50/MWh, and pay an average of $433 580/h. When the network limit is 1500 MW, the average price paid by customers drops to \$182.50, and customers pay an average of \$366 080/h. Customers benefit by \$67 500/h.

When the network limit is 1000 MW, the expected merchandising surplus is \$121 500/h. With the network limit of 1500 MW, this drops to \$108 750/h, a drop of \$12 750/h. Altogether, customers benefit by more than the drop in the merchandising surplus. The total gain is \$67 500 − \$12 750 = \$54 750/h as expected.

Once again it is important to emphasise that the welfare gain that is measured here is after the change in the mix of generation at each node. In fact, as we have seen, for any given level of network capacity, the capacity of generation at the exporting node is just equal to the network capacity. This means that, if we retain the existing stock of generation, any relaxation of network capacity (such as the increase in capacity from 1000 to 1500 MW discussed earlier) would have no welfare benefits at all. Generation and network investment must be considered in combination.

18.5   Coordination of Generation and Network Investment in Practice

The previous sections have consistently emphasised the point that optimal network investment must be considered in the light of optimal generation investment. It is not possible to separate the task of network investment from the task of generation investment.

Put another way, generation investment is both a substitute and a complement for network investment. Remote generation investment is a complement to network investment (since remote generation investment needs network investment in order to transport its output to the load). Conversely, local generation investment is a substitute for remote generation investment combined with network investment. It is not possible to separate the generation investment task from the network investment task.

However, this raises a serious problem. In liberalised electricity industries, generation investment decisions are primarily the responsibility of independent for-profit generation entrepreneurs. On the other hand, network investment is still almost entirely the responsibility of regulated or government-owned network utilities. This raises important questions about how these investment decisions should be coordinated.

To make matters worse, in practice, both generation and network investment decisions involve substantial sunk long-lived investment. How should such long-lived investments be coordinated?

To an extent there is a chicken-and-egg situation. A generator may not invest in a remote location out of fear that adequate network capacity to that location will not be built. On the other hand the network investor might argue that there is no need to build network capacity to that location until additional generation investment occurs at that location.

Some generation and network investment can be easily coordinated. For example if a generator requires a substantial new network asset to connect its generation plant in a remote location, the generator is usually required to pay for the new network assets (known as connection assets) to the point of connection with the 'shared network'. However, this just raises the question as to the boundary of the shared network. If another generator chooses to locate in the remote region, should it also be required to pay for any network augmentation required? What if the network connection already existed (perhaps to serve a small local load) before either generator arrived?

We can ask the question in this way: Who should move first: The generation investment or the network investment? Should the network investor wait to see where generation investment occurs and then build the network to accommodate? Or should the network investor announce its future plans for expansion of the network and then leave generation entrepreneurs the decision as to where to expand generation capacity?

The case where the network investor waits to observe where generation investment occurs and then builds the network to accommodate it is known as the case of *reactive network investment*. This approach is straightforward to implement in that the network investor does not need any special information on the best location for generation investment. However, this approach would not yield the efficient overall mix of generation and network investment if generators did not face the correct price signals for the use of the network.

The alternative approach is for the network planner to determine the most efficient locations for all the generation investment and then to carry out a joint generation/network co-optimisation to determine the future network expansion plan. This is known as *proactive network investment*. In principle, this approach yields an efficient network expansion path. However, the problem is that the network planner needs a large amount of information on the best location for generation investment. A key benefit of the energy market reforms of the past decades is that they allowed decisions about the operation of and investment in generation to be decentralised to generation entrepreneurs. However, here we see that some of that information needs to be retained by the network planner in order to plan the network – this seems impractical and undesirable.

Ideally, the generation entrepreneur, in making its investment decision, would face (or would expect to face) the long-run marginal cost of network expansion. We have seen that in the optimally configured network the expected constraint marginal value should be equal to the marginal cost of network augmentation (Section 18.1). However, can nodal prices send the correct signals to investors, so as to optimally coordinate generation and network investment decisions?

There are at least two reasons why we might be concerned about the ability of nodal prices to accurately provide a signal as to the long-run marginal cost of network expansion.

The first reason relates to the observation that, due to the inability to optimally dispatch the network in the very short run – in particular, due to the inability to manage network constraints that are binding in the very short run (Section 12.3) – the network is typically built with substantial excess capacity. As a consequence it may not be feasible for congestion to signal the long-run marginal cost of network expansion. This is a deep issue that deserves further research.

The second reason why nodal prices may not accurately signal the long-run marginal cost of network expansion is due to the presence of market power. A single generator connected to the end of a long, dedicated network link (known as a 'congestion asset') essentially has complete control over the nodal price it faces at its local node. The generator will, in effect, never allow congestion to arise on the network link. Therefore, for market power reasons alone, nodal price differences on a dedicated connection asset could never be expected to reflect the long-run marginal cost of network augmentation on that asset. However, in this instance there is a simple resolution that is commonly adopted – it is common to make generators pay for network infrastructure that is dedicated to their use (so-called 'connection assets'). This ensures that generators face the correct price signals at least for the connection assets. However, market power may still disrupt the quality of the nodal price signals in the remainder of the network.

> *Result*: If average nodal prices differences reflected the long-run marginal cost of network augmentation then, in principle, generators would face the correct signals regarding network augmentation, and generation and network investment could be effectively coordinated. However, there are reasons to doubt whether nodal price signals accurately reflect the long-run marginal cost of network augmentation due to (a) the tendency to over-build the network due to the inability to efficiently manage short-run network outages, and (b) the presence of market power.

In practice, network businesses recover the costs of their assets through three sources: (a) charges for connection assets, (b) the merchandising surplus from nodal pricing, and (c) other charges on market participants. It is worth emphasising that, in practice, for the reasons just mentioned, nodal pricing is unlikely to recover the full costs of the transmission network. Some additional charges will be necessary. Ideally, these residual charges should be levied on network customers in a manner which:

a. Reflects network usage – network users (generators and loads) should not be required to pay for the residual costs of network elements that they do not use;
b. Is stable over time – network users (generators and loads) must make a sunk investment in reliance on continuing access to the network. These investments will be threatened if network charges increase due to factors outside their control; and
c. Does not distort usage or investment decisions.

In practice, additional charges to recover the residual costs of network businesses are typically levied on customers.

18.6 Summary

The problem of choosing the efficient level of investment in an AC network cannot be separated from the problem of choosing the optimal mix of generation investment.

The general principle is that for each network element the average constraint-marginal-value for the flow limit constraint through that network element should be equal to the long-run marginal cost of adding capacity to that network element. However, this constraint-marginal-value is calculated at the optimal mix of generation.

In the case of a two-node network there is a simple graphical representation of the effects of a network expansion. A network expansion (combined with a change in the optimal mix of generation) has the effect of raising prices in the exporting region, reducing prices in the importing region, and has an ambiguous effect on the merchandising surplus.

We can construct a measure of the economic welfare gain from a network expansion by constructing a measure of the difference between the economic welfare of the network relative to a baseline. That baseline could be, for example, the network with all constraints removed, or the situation with no network at all. In each case however, the mix of generation and consumption assets must adjust to the new optimal level. It is not possible to find a measure of the economic welfare of a network augmentation by comparing the economic welfare of the network to a baseline using the existing mix of generation.

In a simple radial network with a single load node, the level of generation capacity on the remote side of a network link is always exactly equal to the capacity of the network link, further emphasising the importance of close coordination between network and generation investment.

Coordination between generation and network investment requires that generators face (or expect to face) a signal of the long-run marginal cost of network expansion. In principle, nodal prices provide such a signal. However, there are at least a couple of reasons to be concerned that, in practice, nodal prices may not provide such a signal. The first reason relates to the tendency to over-build capacity in the network due to the inability to efficiently handle short-run network outages. The second reason relates to the presence of market power that almost certainly distorts price signals where the number of market participants connected to a network asset is small (such as the case of a generation connection asset).

In practice, nodal prices do not provide sufficient revenue to cover the total costs of network businesses (partly for the reasons just mentioned). Some other charges are likely to be required. These charges should not distort the operational and investment decisions of the market participants, should cover attributable costs and should be stable over time.

Questions

18.1 Is it the case that, in a radial network with a single load node, the optimal level of network capacity on any given network element is equal to the generation capacity 'behind' that network element? Why or why not?

18.2 Is it possible to use the merchandising surplus as a measure of the economic benefit from a network augmentation? Should the network business be allowed to augment the network if doing so increases the merchandising surplus by more than the cost of the augmentation?

18.3 What are the conditions for efficient investment in a DC link? When should a DC link be augmented?

18.4 If demand is dropping, so that there is excess capacity on the network, would you expect the revenue from nodal price differences (the merchandising surplus) to be equal to the long-run marginal cost of network expansion? Why or why not?

Further Reading

See Sauma and Oren (2006, 2007).

Part IX

Contemporary Issues

Laying cables for the Auckland City Council's power distribution system, ca 1908 (Source: Neil Rennie)

19

Regional Pricing and Its Problems

We have seen throughout this text that, in principle, the twin goals of efficient use of existing assets and efficient investment in production and consumption assets can be achieved through prices that are differentiated by time and by location known as nodal prices or locational marginal prices. Nodal pricing is used by many electricity markets around the world. However, many markets approximate or vary nodal prices in some way by, for example, failing to differentiate prices adequately by location, or by differentiating prices for producers but not for consumers. In this chapter, we focus on the consequences of failing to differentiate prices adequately by location. Specifically, we will focus on pricing schemes known as regional or zonal pricing.

19.1 An Introduction to Regional Pricing

As we have seen, nodal pricing requires prices that are differentiated by geographic location. Depending on the nature of congestion on the underlying network, nodal pricing may require prices that differ at every node.

Let us consider an alternative approach in which (a) we group nodes into regions, and (b) we set the wholesale price for each node in a region to be the same. This is known as *regional* or *zonal pricing*. Figure 19.1 illustrates how different nodes in a network might be grouped into pricing regions. We will be more precise about exactly how the regional price is related (if at all) to the underlying nodal prices shortly.

Regional pricing is, at best, an approximation, to nodal pricing. One question we can ask is: Under what conditions is regional pricing a good approximation to nodal pricing?

The extent to which regional pricing approximates nodal pricing depends on the number of regions and on how well the regions reflect the pricing patterns that would emerge under nodal pricing. If there were a very large number of regions (in fact if there were the same number of regions as nodes), regional pricing and nodal pricing would be equivalent. At the other extreme, if there were just a single region, nodal pricing and regional pricing could potentially lead to very different outcomes.

The Economics of Electricity Markets, First Edition. Darryl R. Biggar and Mohammad Reza Hesamzadeh.
© 2014 John Wiley & Sons, Ltd. Published 2014 by John Wiley & Sons, Ltd.

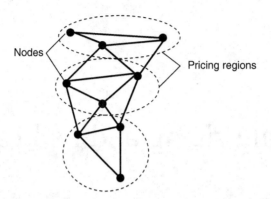

Figure 19.1 Illustration of grouping nodes into regions

We have seen that in simple radial networks a single network constraint divides the network up into two separate pricing regions. In principle, therefore, if the network constraints were clearly identifiable in advance and persistent, in a radial network it would be possible to replicate nodal pricing outcomes with carefully chosen regions. Specifically, the region boundaries should be located across network elements that have persistent binding constraints.

However, real-world networks are not radial. In practice, dividing up nodes into pricing regions is equivalent to nodal pricing only when the following conditions hold:

a. Congestion only occurs on network elements that cross the boundary of regions (there is no congestion on network elements that lie wholly within a region); and
b. For any two nodes in the same region there is no path between those two nodes that passes through a potentially congested network element.

However, these conditions fail to hold in almost all real-world networks. In fact, in most real networks, changing physical conditions (such as changing demand and supply patterns or periodic network outages) are such that for any given node the set of nodes that *always* share the same nodal price in optimal dispatch turns out to be relatively small. In other words, in most real networks, the number of regions you would need to accurately replicate nodal pricing is very close to the number of nodes.

In addition, we saw in Part IX that in an optimally sized network there must be congestion on every single network element with some positive probability. Therefore, even in a radial network, if that network is optimally sized, it is only possible to achieve efficient price signals with nodal pricing.

In practice, therefore, regional pricing is not the same as nodal pricing. Regional pricing therefore has the potential to result in inefficient operational decisions and inefficient investment decisions. As we will see, in some cases associated measures reduce the inefficiency in operational decisions. However, we know of no mechanism that completely restores the efficiency of nodal pricing short of nodal pricing itself.

> *Result*: In simple network configurations (such as radial networks) the presence of binding constraints gives rise to a small number of clearly identifiable pricing regions. However, theory shows that if the network is to be optimally sized, every network element must experience congestion with some probability so that the number of regions must match the number of nodes. Furthermore, in practice, networks are highly meshed. In a meshed network even a single binding constraint can give rise to multiple pricing regions. In practice, regional pricing always results in the mispricing of some nodes at some times.

Approaches to regional or zonal pricing differ in three main ways:

a. How the regional or zonal price is calculated. For example, the uniform regional price could be a simple average of the nodal prices, a weighted average of the nodal prices, or equal to a specific nodal price (such as the price at a designated *reference node*). In other cases, the uniform regional price is calculated through a separate mechanism, such as a hypothetical dispatch with no intraregional network constraints, and therefore may not be directly linked to the underlying nodal prices at all.
b. Whether or not market participants are paid *constrained-on* and *constrained-off* payments. In some cases, market participants are paid the difference between the local nodal price and the notional regional price. This approach is arguably not regional pricing but rather a form of automatic allocation of financial transmission rights. This approach is discussed further later. In other cases (such as in Australia) there are no additional payments to or from market participants.
c. Which market participants are exposed to the regional price. For example, do generators face a nodal price and consumers a regional price? Or does the same regional price apply to all market participants? For example, in the PJM market in the United States, consumers face a regional price even though generators receive the nodal price. In Australia, both generators and consumers face the regional price.

We will look first at the case where there are no constrained-on and constrained-off payments. This is the case where the most significant problems arise, as we will see later.

One final point about regional pricing is worth emphasising. In Part III, we observed that, in a radial network power always flows from low-priced nodes to high-priced nodes. This result does not carry over to meshed networks (in a meshed network power will often flow from higher-priced nodes to lower-priced nodes).

In addition, in the context of regional pricing, we can observe that, even if the underlying dispatch is perfectly efficient, and even if the underlying network is radial, power will not always flow from low-priced regions to high-priced regions. (Furthermore, as we will see later, the regional pricing can further distort the dispatch to increase the likelihood of counter-price flows).

19.2   Regional Pricing Without Constrained-on and Constrained-off Payments

Let us suppose that the system operator solves an optimal dispatch problem based on the bids and offers of the market participants, taking into account the physical network limits, in exactly the way we have discussed throughout this text. However, now let us assume that, rather than market participants facing the local nodal price, market participants will be assumed to face a regional price, calculated via some alternative mechanism. For example, as noted earlier, the regional price could be the price at a designated node in the region, or the average of the prices of the nodes in a region. In this section, we will focus on scheduled market participants – that is, those market participants that submit a bid or offer curve to the system operator and receive a dispatch target in return.

Note that the dispatch task carried out by the system operator *must* take into account the actual physical limits of the network. Even if the regional pricing arrangements ignore or overrule price differences between regions, the physical constraints that might give rise to price differences within regions cannot be ignored if the power system is to be operated safely. We will assume that the optimal dispatch problem does not ignore any relevant physical limits. It could be said that 'dispatch is nodal' even though 'pricing is regional'.

However, this gives rise to a problem. In Section 5.2, we saw that provided the nodal prices are set correctly, a market participant is always dispatched to a price–quantity combination that is on its bid or offer curve. As a result, a market participant always has an incentive to voluntarily comply with the dispatch instructions. There could be penalties for noncompliance with dispatch instructions, but those penalties should rarely, if ever, need to be invoked.

However, this no longer holds under regional pricing (at least without additional constrained-on or constrained-off payments). If the price paid to a generator (or the price paid by a customer) is not the correct nodal price for the location of that generator, then there can arise a *mismatch between pricing and dispatch*. Specifically, the price paid to a generator (or paid by a customer) and the dispatch target may not fall on that participant's bid or offer curve. Put another way, there is a problem of *mispricing*: The market participant faces the incorrect price for production or consumption at its location.

Importantly, when this happens, the market participant no longer has an incentive to comply with the dispatch target. In fact, the market participant has an incentive to take whatever actions it can in order to move either its actual production or consumption or its dispatch target closer in line with the price it faces. The participant may choose not to comply with the dispatch instructions or, if the penalty for noncompliance is severe, will seek to change its bid in an attempt to have its dispatch target move in line with the price it faces.

This can be illustrated in a diagram, such as Figure 19.2. Let us suppose that, initially, a generator submits an offer curve that reflects its marginal cost curve. The system operator

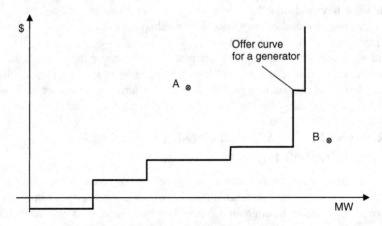

Figure 19.2   Mismatch between pricing and dispatch

computes the optimal dispatch outcome given the bid and offer curves. However, let us suppose that the price paid to this generator does not reflect the local nodal price. As a consequence, the price–quantity combination facing the generator (the price it is paid and the quantity for which it is dispatched) does not lie on its offer curve.

Instead there are two cases to consider: Where the price–quantity combination lies above the offer curve and where the price–quantity combination lies below the offer curve.

Let us consider first the case where the price–quantity combination lies above the offer curve, as at point A in Figure 19.2. In this case, given the price the generator is paid, the generator would like to be dispatched to a higher quantity. In this case, the generator is said to be *constrained off*.

Similarly, in the case where the price–quantity combination lies below the offer curve, as in point B in Figure 19.2, the generator would like to be dispatched for a lower quantity, given the price it is paid. In this case, the generator is said to be *constrained on*.

A situation where a generator is constrained off or constrained on is not an equilibrium. The generator would like to take whatever actions are within its power to bring its output back into line with the price it is paid.

For example, the generator may simply not comply with the dispatch instruction. However, there could be serious penalties for noncompliance – and for good reason: noncompliance of generators could place the power system under physical threat. Having noted the incentive for noncompliance, let us set this possibility aside and focus on other possible responses of generators.

Another way a generator can seek to bring its output back into line with the price it is paid is to alter its offer curve. If the generator is constrained off, it would like to increase its dispatch target. It can attempt to do this by lowering the price at which it offers its output to the market. Similarly, if a generator is constrained on, it would like to reduce its dispatch target. It can attempt to do this by raising the price at which it offers its output to the market.

In fact, if there is more than one generator that is constrained on or off, they will attempt to 'outbid' each other in an attempt to increase or decrease their dispatch target. In markets with price caps, generators that are constrained on are likely to offer its output to the market at the price ceiling. Conversely, generators that are constrained off are likely to offer their output to the market at the price floor. In the Australian National Electricity Market (NEM), this is known as *disorderly bidding*.

This is important because, as we have seen, a primary objective of the dispatch task is to achieve short-run efficiency in the use of a set of production or consumption assets. However, this is not possible if the offer curves of market participants do not reflect their true marginal cost curves. Mispricing induces market participants to distort their bids and offer curves away from their true marginal cost or demand, reducing the short-run efficiency of dispatch outcomes.

In addition to distorting the bid or offer curves, there may be other bid or offer parameters that are affected by the mispricing implicit in regional pricing. For example, in the Australian market, generators can alter the rate at which they are able to ramp up or down, which can limit the extent to which the system operator can change their dispatch target. Again, this results in inefficient short-run economic outcomes. Such practices can also threaten system security and reliability. If generators are not able to respond to changes in demand, the system operator may not be able to balance supply and demand, resulting in very high prices or (in the presence of price caps) involuntary load shedding.

There is one more consequence of regional pricing without constrained-on or constrained-off payments that should be mentioned. As we have seen, mispriced generators and loads within a region have an incentive to distort their bids and offers. However, generators in other regions will typically may not have an incentive to distort their bids and offers. This can lead to flows between regions that are counter-intuitive. For example, flows between regions can be in the direction from high-priced regions to low-priced regions, resulting in negative settlement residues on the network links between regions. This is a problem where (as in the Australian market) the residues on inter-regional network links are provided to the market as a hedging instrument. The presence of negative settlement residues reduces the value of the inter-regional settlement residues as a hedging device, and undermines the financial viability of the system operator. There are illustrations of this effect later.

Result: Without any constrained-on or constrained-off payments, regional pricing results in the mispricing of at least some market participants some of the time. Mispriced market participants are either constrained on or constrained off. Mispriced market participants have an incentive to not comply with dispatch instructions and/or to distort their bids and offer curves away from their true cost or demand. This can threaten system security and result in inefficient short-run dispatch outcomes.

19.2.1 Short-Run Effects of Regional Pricing in a Simple Network

Let us suppose we have the following simple two-node network. Let us suppose that there is a binding network constraint on the link in the direction from A to B. Under nodal pricing the price at node B must be higher than the price at node A (Figure 19.3).

Let us consider first the case where the regional price lies below the nodal price at node B. This would be the case, for example, if the regional price were the price at node A or if the regional price were the average of the two prices.

In this case, generators at node B would be dispatched according to the local nodal price but only paid the lower regional reference price. Such generators would be likely to be constrained on. As we have noted earlier, such generators would like to reduce their dispatch target. They can do this by, for example, pretending to be high-cost generators, or pretending to be unavailable.

For example, let us suppose that the nodal price at node B is $100/MWh, but the regional price is $20/MWh. Consider the position of a generator at node B with a variable cost of, say, $50/MWh. This generator will be paid the regional price of $20/MWh, but will be dispatched

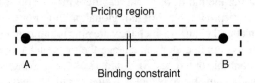

Figure 19.3 Regional pricing in a simple network

as though the local nodal price is $100/MWh. At the local nodal price of $100/MWh, the generator will be dispatched for its full capacity and will be forced to incur a cost of $50/MWh for every MW of production, but will only be paid the regional price of $20/MWh. To avoid being dispatched at this time the generator could offer its output to the market at a much higher price (as high as the price ceiling) or could perhaps pretend to be unavailable.

If many of the generators at node B pretend to be unavailable, system security and reliability could be threatened – in fact, there is a risk that some customers at B will be left unserved. In any case, the misrepresentation of the true costs of these generators means the system operator is unable to achieve a short-run efficient dispatch outcome.

Now let us consider the case where the regional price is higher than the nodal price at node A. In this case, generators at node A are likely to be constrained on – at the price they are paid they would prefer to be dispatched to a higher level of output. They may be able to achieve this by pretending to be much lower cost generators.

For example, let us suppose that the regional price is $50/MWh, but the local nodal price at A is $10/MWh. A generator at A with a variable cost of $20/MWh would not be dispatched but would like to be dispatched given that it will receive the regional price of $50/MWh. Such a generator has a strong incentive to offer its output to the market at a lower price, as low as the price floor in the market.

In the Australian NEM, market participants can offer their output to the market at a price as low as $-1000/MWh. When generators are constrained off, it is common to find generators offering their output to the market at this price floor. Again we find that the misrepresentation of the true costs of these generators means the system operator is unable to achieve a short-run efficient dispatch outcome.

19.2.2 Effects of Regional Pricing on the Balance Sheet of the System Operator

From Part III, we know that under nodal pricing the system operator receives a positive flow of funds known as the merchandising surplus. Let us now explore the situation under regional pricing. Does the system operator always receive a positive flow of funds? Does it depend on how the regional prices are defined?

Let us suppose that we have a mapping from nodes to regions given by the function $R(i)$. Let us suppose that the regional price in region r is P^r. Similarly, let us define the total net injection in a region as the sum over the net injection of the nodes in that region:

$$Z^r = \sum_{i:R(i)=r} Z_i$$

Let us define the merchandising surplus received by the system operator under regional pricing as follows:

$$MS^R = -\sum_r P^r Z^r$$

Is this value always positive (or at least non-negative)? In the case where there is just one region, this value is always zero. But what if there are two or more regions?

Regional pricing can be thought of as nodal pricing combined with a system of transfers to every market participant. Under regional pricing, a producer at node i that produces at the rate Q_i^S can be thought of as receiving the nodal price P_i plus a payment equal to the difference between the regional price and the nodal price multiplied by the rate of production: $(P^{R(i)} - P_i)Q_i^S$ (and similarly for consumers). Here $R(i)$ is the region in which node i is located. Let us define the total payments to (and from) market participants to achieve regional pricing (the 'regional pricing payments') as follows:

$$\text{RPP} = \sum_i \left(P^{R(i)} - P_i\right)Z_i$$

It is straightforward to verify that the merchandising surplus under regional pricing is equal to the merchandising surplus under nodal pricing less these regional pricing payments:

$$\text{MS}^R = -\sum_r P^r Z^r = -\sum_i P^{R(i)} Z_i = -\sum_i P_i Z_i - \sum_i \left(P^{R(i)} - P_i\right)Z_i = \text{MS} - \text{RPP}$$

Although (as we have seen) the merchandising surplus is always positive, the merchandising surplus less the regional pricing payments is not always positive. Importantly, if the regional price is above the nodal price for some generators (i.e. if some generators are constrained off), then the regional pricing payments may easily exceed the merchandising surplus, so that the system operator is losing money.

In the special case of just one region, the regional pricing payments are equal to the merchandising surplus – in effect the merchandising surplus is paid out in full to market participants. However, as we have seen, the way the merchandising surplus is paid out distorts the production and consumption decisions.

It is easy to see this in a simple network. Let us extend the simple network to include two pricing regions as illustrated in Figure 19.4. The underlying network, as before, is a simple radial network. From the analysis in Part III, we know that in a simple radial network power always flows from low-priced nodes to higher-priced nodes, but this no longer holds under regional pricing.

Let us suppose that there is a network constraint in the direction from node B to node C, but the regional price in region 2 is larger than the nodal price at node B. Let us also assume the regional price in region 2 is larger than the regional price in region 1 so that we would expect that power would flow from region 1 to region 2. However, in this case, generators at node B are constrained off. They are likely to respond by pretending to be low-cost generators – perhaps even offering their output to the market at the price floor. Faced with a large amount of low-cost generation at node B, the optimal dispatch task is likely to result in this generation

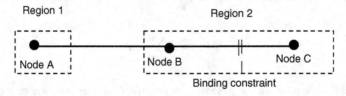

Figure 19.4 Impact of regional pricing on inter-regional settlement residues

Table 19.1 Illustration of possible prices, injections and flows when generators at node B are constrained off in the simple network of Figure 19.4

	Nodal		Regional	
Prices		Injections	Prices	Injections
$P_A = \$20$		$Z_A = -200$	$P^1 = \$20$	$Z^1 = -200$
$P_B = \$ - 10$		$Z_B = 1000$	$P^2 = \$100$	$Z^2 = 200$
$P_C = \$100$		$Z_C = -800$		

being dispatched ahead of generation at node A. As a result it is likely that power will flow from node B to node A, even though the regional price in region 2 is larger than the regional price in region 1.

The reverse can also happen. The regional price in region 2 may be below the nodal price at node B, in which case the generators at node B may pretend to be very high cost generators or pretend to be unavailable. In this case, power may flow from node A to node B even though the regional price in region 2 is below the regional price in region 1.

Let us work out the merchandising surplus and payment flows in a specific case. Suppose that the regional price in region 2 is above the nodal price at node B, so that generators at B are constrained off. They therefore pretend to be low-cost generators, reducing the nodal price at node B. The nodal prices and injections and regional prices and injections are as set out in the following Table 19.1.

Notice that the flow between the regions is in the direction of region 1 even though region 2 has a higher price.

Under nodal pricing, the merchandising surplus would be $94 000/h, which is positive as expected. However, the regional pricing payments are $110 000/h, so the merchandising surplus under regional pricing is $ - 16 000/h. The system operator incurs a loss.

> *Result*: Regional pricing can result in counter-price flows on network links between regions, even in radial networks. This can undermine the financial viability of the system operator and can undermine the value of the inter-regional settlement residues as a hedging device.

19.2.3 Long-Run Effects of Regional Pricing on Investment

In Part IV, we explored how nodal pricing in theory provides the correct investment signals for generation and load assets. Unsurprisingly, regional pricing does not result in correct investment incentives for market participants.

Untangling the investment incentives that arise under regional pricing is tricky. On the surface, regional pricing results in higher prices for some market participants under some circumstances and lower prices for some market participants in other circumstances. As a first cut, therefore, we may suggest that regional pricing reduces incentives for investment in locations that are routinely constrained-on and increases incentives for investment in locations that are routinely constrained-off.

More specifically, we have seen that regional pricing is equivalent to nodal pricing combined with payments to market participants. We noted earlier that the payment to a generator takes the form: $(P^{R(i)} - P_i)Q_i^S$. Where the regional price is above the nodal price (i.e. where the generator is constrained off), the generator receives a positive payment from the system operator. Other things being equal, this would enhance the incentive for the generator to invest in that location (and reduce the incentive for a load to invest in that location).

However, this analysis is not entirely correct. A generator in a constrained-off location receives a higher price under regional pricing than under nodal pricing, but may not be able to be sure that it will be able to produce as much as it desires – after all, if there are many generators that are each offering their output to the market at the price floor, the generator may have to share the total output with many other generators. Even though the regional price is higher than the nodal price, the generator may not be able to be assured that it will be able to take advantage of that higher price.

This lack of assurance is important. Under normal market operation, a low-cost generator has an assurance that it will be able to be dispatched at most times, including high-price times. However, under regional pricing, a low-cost generator has no such assurance – every constrained-off generator has an incentive to pretend to be a low-cost generator. Under nodal pricing, a low-cost generator can locate in a region with limited export network capacity, confident that it will be able to undercut other generators in the event of an export constraint limit. However, under regional pricing, a low-cost generator has no such assurance – another higher-cost generator can locate in the same region and still have a chance of being dispatched.

The analysis in Part IV showed how the price–duration curve at a location provides important signals as to the type and mix of investment at that location. These signals would seem to be at best muted or completely absent under regional pricing.

The overall impact on investment incentives is difficult to predict. It seems almost certain that investment incentives will be distorted, but further work is required to establish the exact nature of those distortions.

Result: Regional pricing distorts investment incentives. Generators that are constrained off receive a higher price, but cannot be assured of being dispatched. The overall impact on investment incentives is complex.

Regional pricing may also have an impact on network investment decisions. Some electricity markets have a policy of building the network in such a way as to eliminate the risk of congestion within each region. To the extent that intraregional congestion is eliminated, the pricing and investment distortions just noted are no longer relevant. However, this comes at the cost of inefficient network augmentation decisions.

19.3   Regional Pricing with Constrained-on and Constrained-off Payments

Some liberalised electricity markets use a form of regional pricing but combine regional pricing with payments to and from market participants to ensure that market participants face the correct price at the margin.

We saw in the previous section that regional pricing can be thought of as nodal pricing combined with payments to producers of the form $(P^{R(i)} - P_i)Q_i^S$, and payments to consumers of the form $-(P^{R(i)} - P_i)Q_i^B$.

The distortions from regional pricing arise from the fact that these payments depend on the production or consumption decisions of the market participants themselves. This gives market participants a strong incentive to distort their production or consumption decisions in order to achieve a higher payment from the system operator.

This incentive can be eliminated by making the payment to each market participant independent of the actual production or consumption decision. Let us suppose that for each market participant we have a predetermined value $\overline{Q}_i^S$ or $\overline{Q}_i^B$. We now view regional pricing as a nodal pricing combined with payments to producers equal to: $(P^{R(i)} - P_i)\overline{Q}_i^S$, and payments to consumers equal to $-(P^{R(i)} - P_i)\overline{Q}_i^B$.

In this framework the amounts $(P^{R(i)} - P_i)(\overline{Q}_i^S - Q_i^S)$ are the *constrained-on and constrained off payments*.

To make this clearer, consider the revenue received by a generator. In a regional pricing framework the revenue received by a generator is equal to $P^{R(i)}Q_i^S$. Combined with the constrained-on/constrained-off payments, we find the revenue of the generator as follows:

$$P^{R(i)}Q_i^S + \left(P^{R(i)} - P_i\right)\left(\overline{Q}_i^S - Q_i^S\right) = P_iQ_i^S + \left(P^{R(i)} - P_i\right)\overline{Q}_i^S = P^{R(i)}\overline{Q}_i^S + P_i\left(Q_i^S - \overline{Q}_i^S\right)$$

From this expression we see that the generator receives the regional price on a proportion of its production and the nodal price on the remainder.

The key observation here is that since the generator faces the nodal price at the margin, the generator faces the correct local price; there is no longer a mismatch between pricing and dispatch; the mispricing problem has been solved. The generator no longer has an incentive to distort its offer curve.

The values $\overline{Q}_i^S$ and $\overline{Q}_i^B$ for each generator and load are known as the *entitlement* for each generator and load. A key question, of course, is the definition of the entitlement. If the entitlement is set equal to zero, this form of pricing just reduces to nodal pricing.

Result: By making constrained-on/constrained-off payments to generators and loads the system operator can ensure that, even under a regional pricing framework, generators and loads face the correct nodal prices, eliminating the incentive to distort bids and offers. A key question, however, is the definition of the entitlement.

It is important to note that the use of constrained-on/constrained-off payments to market participants may distort their investment incentives. Let us suppose that, as just suggested, in addition to the revenue from nodal pricing each new generator in the market receives a payment equal to $(P^{R(i)} - P_i)\overline{Q}_i^S$. Then (as long as the entitlement for the new generator is positive), a generator has an incentive to invest in a location where the regional price is above the nodal price $P^{R(i)} > P_i$. In effect a new generator receives a subsidy for investing in such a location. Conversely, the new generator has a dis-incentive to invest in locations where the regional price is below the nodal price $P^{R(i)} < P_i$.

Note that this investment incentive is likely to be quite perverse: Generators are given an enhanced incentive to locate in locations where the nodal price is low relative to the regional price – which is likely to be precisely those locations where there are physical limits on exports. Conversely, generators have too little incentive to invest in locations where the nodal price is high relative to the regional price – precisely those locations that face a shortage of supply.

This problem could, in principle, be solved by providing no entitlements to new generation or load. Any new generation would then merely face the correct nodal price. However, there would still be a distortion to the (dis)investment decision of an existing plant – the existing plant would have a reduced incentive to retire in locations where the regional price is above the nodal price and an enhanced incentive to retire in locations where the regional price is below the nodal price.

Result: The use of constrained-on/constrained-off payments may affect the incentive to invest. As long as new generation or load automatically receives a nonzero entitlement, it will have a distorted incentive to invest – specifically, an enhanced incentive to invest in locations where the regional price is above the nodal price and a reduced incentive to invest in locations where the regional price is below the nodal price (and vice versa for load). This problem can be partly addressed by ensuring that any new generation or load receives no entitlement automatically (and thereby faces the nodal price).

How should we set the entitlement? There are several possibilities.

First, if the entitlement is set equal to the actual or out-turn rate of production or consumption (as we discussed earlier in Part VI in the discussion of optimal hedging), then each generator and load is left in exactly the same position as if he/she faced the regional price for all of his/her output. The generator or load could, in principle, then design hedging instruments based on that regional price.

This approach would still, of course, lead to distortions in generation and load investment decisions. Furthermore, as we noted earlier, the system operator may still face situations where it is forced to pay out more than it takes in through the merchandising surplus.

Another possibility for setting the entitlement arises when we are currently using regional pricing (without constrained-on/constrained-off payments) and we want to introduce such payments. In this case, provided we set the entitlement equal to the rate of production or consumption that would arise under the status quo, then no generator or consumer would be left worse off as a result of the change.

Intuitively, the reason is as follows: As long as each market participant can choose exactly the same rate of production or consumption as under the status quo and not be left worse off than under the status quo, then each market participant can choose another rate of production or consumption and still be left not worse off than under the status quo.

A third possibility is to set the entitlement on some other basis, such as the dispatch that would arise under a hypothetical network with no intraregional congestion. This is, we understand, the approach used in the England and Wales market.

19.4 Nodal Pricing for Generators/Regional Pricing for Consumers

Some liberalised electricity markets differentiate between the prices paid to generators and the prices paid by loads. Specifically, it is sometimes the case that generators are paid the correct nodal price for their output while consumers pay a regional price.

It needs to be emphasised that, of course, such an arrangement cannot achieve an efficient overall outcome. As long as there is some scope for consumers to adjust their consumption in the light of the local wholesale market conditions, an arrangement under which consumers face a price that is different from the local nodal price will lead to inefficient market outcomes.

For example, under a system of regional pricing for consumers, consumers will not have an incentive to curtail their demand at times when local network conditions cause the local wholesale price to rise to high levels. Similarly, consumers will not be able to respond (perhaps by charging electric vehicles or reducing the output of local generation) at times when the local wholesale price is very low. As we have emphasised throughout this text, it is expected that in the future small customers will have the potential to become more responsive to wholesale market conditions. Regional pricing for customers limits the benefits that can be obtained from such responsiveness.

Furthermore, regional pricing for customers distorts investment decisions for customers. Customers have too little incentive to invest in expanding their consumption in locations where the nodal price is low, and too much incentive for investing in locations where the nodal price is high.

19.4.1 Side Deals and Net Metering

As long as the price paid by consumers at a given location differs from the price paid by generators at the same location there arises the possibility of a 'side-deal' between generators and consumers in which they agree to report a different total generation and consumption to the system operator, while keeping their actual net injection the same.

For example, if generators are paid a nodal price while consumers pay a regional price, a situation might arise where the price paid to a generator at a particular location is less than the price paid by the customers at that location. In principle, a generator could approach a consumer and offer to 'sell' electricity to the consumer at a rate that is below the consumer price and above the generator price. The sale occurs by the generator reporting less production and the consumer reporting less consumption to the centralised market. The total net injection (generation less consumption) remains the same.

For example, let us suppose a customer consuming 100 MW is located at the same node as a generator producing 50 MW. Let us suppose the generator is currently receiving $10/h for its production while the customer must pay $100/h. Let us suppose that the customer and the generator both agree to report 50 MW less production and consumption to the system operator (perhaps through 'net metering'). The customer pays $5000/h less than before. If he/she shares at least $500/h with the generator, both parties are better off.

The same applies if the generator is producing more than the customer. Let us suppose that the generator is producing 150 MW (as before the customer is consuming 100 MW). Now both parties can agree to report 100 MW less to the system operator. The customer is now better off by $10 000/h. If the customer pays the generator more than $1500/h both parties are better off.

Side deals of this kind result in both parties in effect paying a price that lies between the generator price and the consumer price.

Now let us consider the case where the generator price is above the consumer price. In this case both parties have an incentive to report a higher production and consumption.

For example, suppose, as before, a customer consuming 100 MW is located at the same node as a generator producing 50 MW. Let us suppose the customer is paying $10/MWh, but the generator is paid $80/MWh. If both parties agree to increase their reported production and consumption by 1 MW, the system operator must pay an additional $80 − $10 = $70/MWh. The generator can take the extra $80/h it receives and pay the consumer something more than $10/h and both parties will be better off.

There must be a limit on the reported production and consumption. Otherwise, this scheme would make the parties arbitrarily rich. For example, the generator may not be allowed to report higher production than its officially nominated capacity.

Result: Whenever generators and loads at the same physical network location are paid different prices, there arises an incentive for the generators and loads to co-operate to jointly report higher or lower production and consumption to the system operator, while holding the net injection of the parties constant.

In the case where the generation price is below the consumption price, there is an incentive to reduce the reported production and consumption. This can be achieved through net metering, where the generation production is used to offset local consumption or vice versa.

In the case where the generation price is above the consumption price, there is an incentive to increase the reported production and consumption. There must be a limit on the amount the parties can over-report or they will be able to extract arbitrarily large amounts of money from the system operator.

If we adopt a system with nodal pricing for generators and regional pricing for consumers, how should the regional price for consumers be set?

One possible objective is to set the regional price for customers in such a way that the surplus received by the system operator is equal to the surplus that would arise if each region were a single-priced region with a price equal to the consumer price.

With regional pricing for consumers, the merchandising surplus is as follows:

$$\text{MS} = \sum_r P^r \sum_{i:R(i)=r} Q_i^B - \sum_i P_i Q_i^S = \sum_r P^r \sum_{i:R(i)=r} \left(Q_i^B - Q_i^S\right) - \sum_i \left(P_i - P^{R(i)}\right)Q_i^S$$

$$= \sum_r P^r Z^r - \sum_i \left(P_i - P^{R(i)}\right)Q_i^S$$

Therefore, we can set the merchandising surplus equal to the surplus that would arise if each region were a single-priced region ($\sum_r P^r Z^r$), by setting $\sum_i (P_i - P^{R(i)})Q_i^S = 0$. This implies that we should choose the regional price equal to the production-weighted nodal price in each region:

$$P^r = \frac{\sum_{i:R(i)=r} P_i Q_i^S}{\sum_{i:R(i)=r} Q_i^S}$$

Result: Under a system of nodal pricing for generators and regional pricing for customers, the system operator can assure itself of obtaining the same merchandising surplus as it would under a system of regional pricing by choosing the regional price equal to production-weighted nodal price.

19.5 Summary

We have seen that nodal pricing provides efficient signals for operational and investment decisions. However, many liberalised electricity markets do not use nodal pricing. Some liberalised electricity markets use a form of pricing in which all the nodes in a region face the same price (in some cases this applies only to one side of the market, such as consumers). This is known as regional or zonal pricing.

In certain special circumstances (such as when the number of regions is as large as the number of nodes), grouping nodes into pricing regions results in no distortions to the pricing signals. Theory shows that in an optimal-sized network congestion will arise on every network element with some positive probability. Therefore, as long as the number of regions is less than the number of nodes, there will be some distortion to the pricing signals in at least some circumstances.

In some markets with regional pricing the local pricing signals are restored through constrained-on/constrained-off payments. In the absence of these payments, a situation may arise where a generator or load is paid a price and dispatched for a quantity that does not fall on its bid or offer curve. In this case the market participant has an incentive to not follow its dispatch instructions and/or to distort its bid or offer in order to increase or decrease the amount that it is dispatched. A generator that is dispatched to a price–quantity combination that lies above its offer curve is said to be constrained off. Such a generator has an incentive to submit a lower offer curve – as low as the price floor. A generator that is dispatched to a price–quantity combination that lies above its offer curve is said to be constrained on. Such a generator has an incentive to submit a higher offer curve – as high as the price ceiling.

The distortion in the offer curves results in inefficient short-run dispatch outcomes. It also may lead to counter-price flows on the interconnectors between regions. In addition, we have seen that under nodal pricing the system operator receives a positive flow of funds known as the merchandising surplus. Under regional pricing the merchandising surplus may be negative.

The mispricing that arises from regional pricing without constrained-on/constrained-off payments also distorts investment incentives. A location that is routinely constrained off receives a higher price under regional pricing than under nodal pricing which, other things being equal, would encourage investment. However, a generator investing at that location has no assurance that it will be able to be dispatched. Overall, the implications for investment under mispricing are complex.

The correct local operational decisions under regional pricing can be restored using a system of constrained-on/constrained-off payments. These payments ensure that the generator or load faces the correct local nodal price at the margin. A key parameter in these payments is the participant's entitlement, which determines the size of the payment when the market participant produces or consumes nothing. Setting the entitlement equal to zero restores nodal pricing. A nonzero entitlement will have distortionary implications for investment decisions.

Some markets make use of a partial regional pricing arrangement. For example it may be that generators are exposed to nodal pricing while customers are exposed to a regional price. Differences in prices between generators and customers gives rise to an incentive for generators and customers to co-operate to report a higher or lower production and consumption to the system operator (while holding the net injection constant). If the generation price is below the consumption price, the parties have an incentive to make the reported generation and consumption appear as low as possible. If the generation price is above the consumption price, the parties have an incentive to make the reported generation and consumption as large as possible.

Questions

19.1 The discussion in this chapter has focused on the impact of mispricing on scheduled generators (which must submit an offer curve to the dispatch engine and which must follow dispatch instructions). What is the impact of mispricing on smaller generators that are unscheduled?

19.2 In the Australian NEM, mispricing of generators gives rise to what is known as 'disorderly bidding'. Can you describe what disorderly bidding looks like and explain how it is different from the exercise of market power?

19.3 We have seen that regional pricing (in the absence of constrained-on/constrained-off payments) results in inefficient operational decisions, negative merchandising surplus and negative settlement residues, and inefficient investment decisions. Can you think of an argument in its favour?

19.4 Do generators have an incentive to locate at a node that is routinely constrained off? Why or why not?

19.5 Does the use of constrained-on/constrained-off payments solve the problem of inefficient operational decisions and inefficient investment decisions? Why or why not?

19.6 What problems arise when we pay a different price to generators and loads at the same location?

19.7 What is the effect of nodal pricing for generators/regional pricing for consumers on generator market power?

Further Reading

See Brunekreeft, Neuhoff and Newbery (2005) and Ding and Fuller (2005).

20

The Smart Grid and Efficient Pricing of Distribution Networks

Chapter 19 dealt with pricing issues at the transmission network level. Many liberalised wholesale electricity markets use nodal pricing (at least for generators) at the transmission network level. However, even the most advanced markets today do not seek to determine the correct price signals for customers connected to the distribution network. This remains one of the most important areas for further reform.

20.1 Efficient Pricing of Distribution Networks

How can we achieve efficient use of a given set of production and consumption assets connected to a distribution network?

In principle, a distribution network is no different to any other form of network. The question of the efficient use of production, consumption and network assets has been addressed throughout this text. We are now very familiar with the mechanism for achieving efficient use of a given set of assets. As we have seen, this mechanism involves spot prices for electricity, which are differentiated by time and by location – known as nodal prices or locational marginal prices. Efficient pricing of a distribution network requires some form of nodal pricing.

However, although some liberalised electricity markets around the world make use of nodal pricing at the transmission network level, to our knowledge, nodal pricing of the distribution network has not been implemented anywhere to date. Instead, distribution network charges have tended to be highly averaged over time and often over geographic locations. To our knowledge, distribution networks have not been efficiently priced anywhere in the world.

This distortion in pricing of distribution networks has significant consequences. Specifically, time-averaged distribution network charges have the following consequences:

- Consumers of electricity do not have an incentive to operate their local consumption assets in a manner that curtails demand (or increases local production or reduces the amount being injected into storage) at times of local (importing) network congestion.

The Economics of Electricity Markets, First Edition. Darryl R. Biggar and Mohammad Reza Hesamzadeh.
© 2014 John Wiley & Sons, Ltd. Published 2014 by John Wiley & Sons, Ltd.

- Embedded generation (such as solar PV generation or other local generation) does not have an incentive to increase production at times of local (importing) network congestion, or to reduce production at times of local (exporting) network congestion.
- As a consequence of the lack of responsiveness to local supply and demand conditions, the distribution network must be over-built (that is, built to a larger scale than is economically efficient) in order to maintain reliability.
- Consumers of electricity do not make efficient decisions regarding investment in local production and consumption assets. Specifically, consumers face too little disincentive (too strong an incentive) to invest in assets that are disproportionately used at times of network congestion, such as air-conditioning loads. In addition, consumers have too little incentive to invest in devices that allow consuming assets to be switched off at times of network congestion.
- At the same time, consumers have too little incentive to invest in local generation, which is a substitute for the network at times of local network congestion, and too much incentive to invest in local generation, which produces output at all times, to reduce average network usage. Later, we will see that current network charging arrangements distort the decision to invest in, say, solar PV assets.

Result: Efficient pricing of distribution networks involves an application of the same principles that have been used throughout this text. Specifically, efficient pricing of a distribution network requires some form of nodal pricing of the distribution network. The absence of nodal pricing of distribution networks leads to a variety of distortions including distorted incentives to use local production and consumption assets, and distorted incentives to invest in local production and consumption assets, such as too much incentive to invest in air conditioning loads and solar PV. One consequence is that the distribution network must be built to an inefficiently high level of capacity to maintain reliability.

Efficient pricing of distribution networks requires that certain infrastructure be put in place. The required infrastructure includes (a) some form of mechanism that can compute the efficient nodal prices on the distribution network and (b) devices capable of metering electricity consumption at each node on the distribution network at different points in time.

Historically, the vast majority of smaller electricity customers (particularly customers connected to distribution networks) have not had meters capable of monitoring consumption at each point in time. For the majority of customers, therefore, efficient pricing of electricity at all was simply impossible. In any case, customers typically had very few devices capable of responding to local market conditions. (One common exception is electric hot water heating which, in Australia and New Zealand, is often centrally controlled by the distribution network and is activated at off-peak times.)

However, this situation is changing rapidly. Many countries and regions have made a decision to roll out new metering technology (known as interval meters or smart meters), which are capable of recording the customer's rate of consumption at different points in time.

In addition, a number of new devices are becoming available that allow electricity consuming assets to respond to wholesale market conditions. A number of manufacturers of appliances have committed to making their appliances capable of responding to the wholesale market. The future growth of electric vehicles has the potential to add substantial new demand to the grid – but demand that can be shifted in time.

To an extent, we can view the historical decision to not efficiently price distribution networks as an internally consistent equilibrium. As long as small customers had no possibility to respond to local market conditions, no local generation and no devices that would allow them to do so, there was little economic value in investing in a market infrastructure (including smart meters and a distribution dispatch process) in which customers would be exposed to wholesale market conditions through nodal prices. At the same time, given the absence of price signals, customers have no incentive to develop or invest in devices that allow them to respond to local market conditions. The network is over-built to cater for the peak demand without load shedding. This is an internally consistent equilibrium.

On the other hand, as long as there is some scope for customers to respond to local network conditions, and local generation, appliances and devices that can do so, the benefits of efficient pricing of the distribution network are substantially larger. In this case, it can make sense to establish the required infrastructure (smart meters and the distribution dispatch process) that exposes customers to efficient local marginal prices. At the same time, given the price signals to which the customer is exposed, the customer has strong incentives to invest in devices and appliances that can respond to those signals. In addition, the network capacity can be adjusted to the efficient level. Again, this is an internally consistent equilibrium.

It seems likely to us that the increasing integration of small customers into the wholesale electricity market will be one of the most important transformations of the electricity industry over the next few decades.

20.1.1 The Smart Grid and Distribution Pricing

Advocates of the so-called Smart Grid envisage a future in which even small customers are responsive to local market conditions with devices that reduce or defer electricity consumption at times of high prices and increase consumption at times of low prices. This might occur by, for example, turning off air conditioners, pool pumps or refrigerators, or by drawing on local storage temporarily during price spikes. At the same time, electric vehicle charging, or injections into local storage can be deferred to off-peak or low-price times. Local devices could also, in principle, respond to short-term supply/demand imbalances, for example, drawing on local storage temporarily, following the outage of, say, a local generator, or cutting solar production following a local network outage.

Achieving the Smart Grid vision will require efficient network pricing. Smart meters are of little value with dumb pricing. In our view, achieving the full benefits of the Smart Grid vision depends on efficient pricing of the distribution network. The Smart Grid requires smart pricing.

Result: The increasing availability of devices capable of responding to wholesale market conditions increases the value of establishing efficient distribution price signals. At the same time, establishing efficient distribution price signals increases the value of developing devices capable of responding to wholesale market conditions. Achieving the full benefits of the Smart Grid vision will require efficient distribution pricing. The Smart Grid requires smart pricing.

20.2   Decentralisation of the Dispatch Task

As we have just emphasised, efficient pricing of distribution networks requires some form of nodal pricing – that is, pricing that reflects the local network congestion. But how exactly should we go about computing nodal prices for distribution networks? Are there any new issues that arise from nodal pricing at the distribution network level?

In principle, the answer is no. Nodal pricing at the distribution network level is merely an application of the principles that have been discussed earlier.

However, in practice, there could be issues that need further consideration. For example, it may be that, with the increasing geographic differentiation of nodes, the number of customers exposed to any particular nodal price may be quite small (in the extreme, each customer could have its own nodal price). In this context, market power issues may become significant.

Another issue involves the practicality of computation of nodal prices and the complexity of the dispatch task. We have seen that the computation of nodal prices involves the carrying out of a constrained optimisation problem in which the system operator seeks to maximise economic welfare subject to the physical constraints of the network. The system operator must therefore have a detailed knowledge of the network and its physical limits. The volume of data involved in creating an accurate short-term representation of the transmission network is typically already large. The amount of additional data involved in creating an accurate short-term representation of each distribution network could be much, much larger. This information would have to be communicated to the system operator and updated on a frequency at least equal to the dispatch cycle.

Although in principle this approach is feasible, it seems more sensible to look for mechanisms to decentralise the dispatch task. Ideally, it seems desirable to look for ways to create a *hierarchy of dispatch mechanisms*, with each dispatch mechanism responsible for its own geographic area. The benefit of this approach is that each dispatch mechanism need only be concerned about modelling the network within its own region – it does not need to communicate detailed information to a distant or remote system operator.

For example, we could imagine, say, a three-level structure. At the bottom level, large customers with their own networks (such as a university campus) might carry out their own local optimisation, taking into account their own local network constraints. The result of this optimisation could then be communicated in some way to the next level of the hierarchy, which could be, say, a local city-wide distribution company. The local distribution system operator would then carry out another optimisation, taking into account the supply and demand of any subnetworks, and taking into account the local distribution network constraints. The result of this optimisation could then be communicated in some way to the next level of the hierarchy. The top level of the hierarchy could be a transmission network system operator at the national or regional level. This system operator would then carry out its own optimisation, taking into account the supply and demand at each connection point and the transmission network constraints. Once the top level of the hierarchy is completed, the supply and demand conditions are determined, and the resulting nodal prices can be computed and communicated back down the hierarchy. This is illustrated in Figure 20.1.

20.2.1   Decentralisation in Theory

Let us imagine that we have a subnetwork that connects to a larger network at two points of interconnection, labelled A and B. We have chosen two points of interconnection as the case of

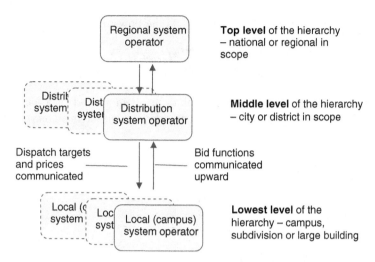

Figure 20.1   A hierarchical dispatch structure

just one point of interconnection is trivial, and the case of two points of interconnection can be generalised to more points of interconnection.

We can imagine that at each point of interconnection there is a hypothetical interconnector representing the flows across the boundary of the networks at that point of interconnection. This is illustrated in Figure 20.2. For each node connected to a link that crosses the boundary of the region, we imagine a matching node in the other region, and a new hypothetical network link, in this case the links $A' \to A$ and $B \to B'$. Without loss of generality, we can take the links to be in the direction from the main network to the subnetwork. Let us denote the flow on the interconnectors (in the direction of the subnetwork) as $Z_{A' \to A}$ and $Z_{B \to B'}$. Let us use S to denote the nodes in the subnetwork.

Throughout this text, we have seen how we can express the flow on any link in the combined network as a function of the set of net injections over the nodes of the network (see Chapter 7). Here, we observe that we can express the flow on any link of the subnetwork as a function of the

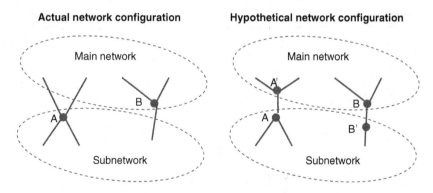

Figure 20.2   Modelling of connections between networks

net injections at the nodes in the subnetwork and the flows on the interconnectors:

$$\forall l \in S, \quad F_l = \sum_i H_{li} Z_i = \sum_{i \in S} H_{li} Z_i + H_{lA} Z_{A' \to A} + H_{lB} Z_{B \to B'}$$

Similarly, we can express the flow on any link of the main network as a function of the net injections at the nodes in the main network and the flows on the interconnectors:

$$\forall l \notin S, \quad F_l = \sum_i H_{li} Z_i = \sum_{i \notin S} H_{li} Z_i - H_{lA} Z_{A' \to A} - H_{lB} Z_{B \to B'}$$

The energy-balance equation in the subnetwork is simply

$$\sum_{i \in S} Z_i + Z_{A' \to A} + Z_{B \to B'} = 0$$

Similarly, the energy-balance equation in the main network is

$$\sum_{i \notin S} Z_i - Z_{A' \to A} - Z_{B \to B'} = 0$$

We can now see that we can separate the dispatch task into two stages. In the first or lower stage, the subnetwork operator carries out the following optimal dispatch task:

$$B^{SUB}(Z_{A' \to A}, Z_{B \to B'}) = \max \sum_{i \in S} B_i(Z_i)$$

subject to the following:

$$\sum_{i \in S} Z_i + Z_{A' \to A} + Z_{B \to B'} = 0$$

$$\forall l \in S, \sum_{i \in S} H_{li} Z_i + H_{lA} Z_{A' \to A} + H_{lB} Z_{B \to B'} \le K_l$$

In the second or upper stage, the main network operator computes the optimal dispatch taking into account the benefit function $B^{SUB}(Z_{A' \to A}, Z_{B \to B'})$, which contains all the relevant information about the subnetwork. The main network operator carries out the following optimisation:

$$\max \sum_{i \notin S} B_i(Z_i) + B^{SUB}(Z_{A' \to A}, Z_{B \to B'})$$

subject to the following:

$$\sum_{i \notin S} Z_i - Z_{A' \to A} - Z_{B \to B'} = 0$$

$$\forall l \notin S, \sum_{i \notin S} H_{li} Z_i - H_{lA} Z_{A' \to A} - H_{lB} Z_{B \to B'} \le K_l$$

The overall dispatch task would now operate as follows: The subnetwork operator would collect information on the bids, offers and network conditions within the subnetwork. It would then carry-out an optimisation (or, more accurately, a series of optimisations, each contingent on the flows on the hypothetical interconnectors) to form the subnetwork benefit function $B^{\mathrm{SUB}}(Z_{A' \to A}, Z_{B \to B'})$. It would then communicate this (in the form of sophisticated bid/offer function) to the main network operator. Any other subnetwork operators could also do the same.

The main network operator carries out its own optimal dispatch task, determining the optimal prices, dispatch and flows, including the flows on the hypothetical interconnectors to the subnetwork. With this information the subnetwork can determine the optimal prices, dispatch and flows within its own network.

It is clear from this presentation that the only extension in this analysis from the material presented earlier is that the benefit function has been generalised to be a function of injections at more than one location. This implies that the way in which bids are communicated to the system operator must also be generalised to allow for bids that are a function of net injections at two or more locations.

Provided we can efficiently determine the subnetwork function and then communicate this function to the main network operator, this approach can allow for effective decentralisation of the optimal dispatch process. Each network operator need only be concerned about demand, supply and network conditions at its own level of the network.

Result: The optimal dispatch task can, in principle, be decentralised. Each network layer must determine a benefit function as a function of the net flows into the next-higher layer of the network. Provided this function can be estimated efficiently and communicated to the next-higher layer of the network, the optimal dispatch task could be carried out as a hierarchy, with each network layer only responsible for information about demand, supply and network conditions at its own level of the network.

20.3 Retail Tariff Structures and the Incentive to Misrepresent Local Production and Consumption

In Section 20.1, we observed that efficient distribution pricing requires some form of nodal pricing of distribution networks. However, in practice, distribution network tariffs are typically far from efficient. As noted in Section in 20.1, this causes a wide range of inefficient outcomes including the following:

a. Inefficient usage decisions at times of network congestion (customers have too much incentive to continue to draw from the grid and too little incentive to make-use of local generation or storage);

b. Inefficient network investment decisions, as the network has to be sized to match peak load even though it would be cheaper for customers to respond by reducing consumption at peak times; and

c. Inefficient customer investment decisions, including decision to invest to consuming appliances (such as air conditioners) and decisions to invest in local generation.

In the next few sections, we will focus on this last question. To what extent do inefficient retail tariffs give rise to inefficient incentives to invest in on-site or embedded generation?

These questions are important. In recent years there has been very substantial investment in embedded generation in many countries. In particular, there has been rapid take-up of rooftop solar PV generation. This generation has been driven, in part, by government subsidies. In particular, many governments have established schemes under which local solar generation is paid a generous price for its output, known as a feed-in-tariff. In recent years in Australia the feed-in-tariff has been substantially reduced. Nevertheless, the question remains: Is it the case that existing tariff structures are giving rise to inefficient incentives for smaller customers to invest in on-site generation?

20.3.1 Incentives for Net Metering and the Effective Price

In the analysis that follows, we will assume that the price paid for the output of an embedded generator (also known as the feed-in-tariff) P^S is different to the price paid for local electricity consumption P^B.

As before, we will assume that the customer is a price-taker in both the buying and selling of electricity.

The first point that we can observe is that, as we noted in Chapter 19, when a different price is paid to generation and consumption at the same location, there arises an incentive for the local generation and consumption to co-operate to misreport the amount of local generation and consumption, while holding the net injection the same. This is particularly the case when the generation and consumption assets are owned by the same entity.

Let us suppose that the customer chooses to invest in K (MW) units of capacity of local generation. As in Chapter 9, let us focus on the case of a controllable generator with constant variable cost c ($/MWh) and a constant capacity cost f ($/MW).

If the customer consumes at the rate Q^B and produces at the rate Q^S, the net utility of the customer is as follows:

$$\varphi(Q_B, Q_S) = U(Q_B) - P_B Q_B + P_S Q_S - c Q_S - fK = U(Q_B) - P_B Q_B + (P_S - c)Q_S - fK$$

Let us suppose that the price paid for production and consumption is P_S and P_B, respectively. Let us suppose that the reported local production and consumption is $\hat{Q}_S$ and $\hat{Q}_B$, respectively. The system operator is assumed to be able to measure the net injection to the network Z but not the individual local production and consumption. The reported net injection to the network must equal the actual net injection, so $Z = \hat{Q}_S - \hat{Q}_B$.

The customer is then paid

$$P_S \hat{Q}_S - P_B \hat{Q}_B = (P_S - P_B)\hat{Q}_S + P_B (\hat{Q}_S - \hat{Q}_B) = (P_S - P_B)\hat{Q}_S + P_B Z = (P_S - P_B)\hat{Q}_B + P_S Z$$

From this expression, we can see that when the price paid to generation exceeds the price paid for consumption (when $P_S > P_B$), the customer has an incentive to make the local production (and, at the same time, the local consumption) appear as large as possible – in fact, as large as the market will allow.

Conversely, when the price paid to generation is less than the price paid for consumption (when $P_S < P_B$), the customer has an incentive to make both the local production and local consumption appear as small as possible. The system operator is not likely to allow either local production or local consumption to go negative, so the best the customer can do is to report the smaller of the two equal to zero.

When the net injection is negative (i.e. when the customer is a net buyer from the network), the best the customer can do is to report no local production and to use local production to offset local consumption (known as net metering). In this case the customer is, in effect, paid the buying price for both its local production and consumption:

$$P_S \hat{Q}_S - P_B \hat{Q}_B = P_B Z$$

When the net injection is positive (i.e. when the customer is a net seller to the network), the best the customer can do is to report no local consumption. In this case the customer is, in effect, paid the selling price for both its local production and consumption:

$$P_S \hat{Q}_S - P_B \hat{Q}_B = P_S Z$$

In effect, when the local price for production is less than the local price for consumption, the effective price paid by the customer for both production and consumption depends on whether the customer is a net buyer or seller:

$$P^{\text{eff}} = \begin{cases} P^B, & \text{if } Q^B > Q^S \\ P^S, & \text{if } Q^B < Q^S \end{cases}$$

For example, let us suppose that a customer is paid 80 c/kWh for its reported local generation, but pays 20 c/kWh for local consumption. The buying price is less than the selling price. Let us suppose that at a given point in time the customer is producing 1 kWh and consuming 2 kWh, for a net withdrawal from the network of 1 kWh. If production and consumption were metered accurately and separately, the customer would be paid $1*80-2*20 = 40$ c/h. However, the customer has an incentive to misrepresent both production and consumption to make them appear larger. For example, the customer could report local production of 4 kWh and local consumption of 5 kWh. In this case the customer would be paid $4*80-5*20 = 220c/h$.

Now let us suppose that the customer pays 80 c/kWh for its local consumption but is only paid 20 c/kWh for local production. As before the customer is producing 1 kWh and consuming 2 kWh. If production and consumption were metered accurately, the customer would pay $2*80-1*20 = 140$ c/h. However, using net-metering, the customer can report no local production and local consumption of 1 kWh, and would pay only 80 c/h. In this scenario the customer is in effect paid the local buying price for both its production and consumption.

Now let us consider the case where the customer is producing 2 kWh and consuming 1 kWh. With accurate separate metering the customer would pay $1*80-2*20 = 40$ c/h. However, using net-metering, the customer can report no local consumption and local production of 1 kWh, and would be paid 20 c/h. In this scenario the customer is in effect paid the local selling price for both its production and consumption.

Result: When a different price is paid to local generation compared to the price paid for local consumption, a customer has an incentive to manipulate apparent production and consumption volumes in the manner we saw in Chapter 19. When the customer is a net buyer from the network, the local buying price is the relevant price for both local production and consumption. When the customer is a net seller to the network, the local selling price is the relevant price for both local production and consumption.

20.4 Incentives for Investment in Controllable Embedded Generation

Recall that the problem we are interested in is whether or not a small customer has efficient incentives to invest in local generation. Let us focus on the case where the price paid for local generation is less than the price paid by local consumption $P^B > P^S$.

We can write the net utility of the customer as comprising two parts: A part that relates to the utility from consumption and a part that relates to the profit from production:

$$\varphi\left(Q^B, Q^S\right) = U\left(Q^B\right) - P^B Q^B + \pi\left(Q^S | Q^B\right)$$

where the profit from production is given as follows:

$$\pi\left(Q^S | Q^B\right) = \begin{cases} \left(P^B - c\right)Q^S - fK, & \text{if } Q^S < Q^B \\ \left(P^B - c\right)Q^B + \left(P^S - c\right)\left(Q^S - Q^B\right) - fK & \text{if } Q^S > Q^B \end{cases}$$

In effect, this generator can be thought of as facing a downward-sloping demand curve, and a marginal cost curve, as illustrated in Figure 20.3.

The profit-maximising level of output for this generator is where the demand curve and the marginal cost curve intersect (there is no problem of price–volume trade-off in this problem) as follows:

$$Q^S = \begin{cases} K, & c < P^S \\ Q^B, & P^S < c < P^B \\ 0, & P^B < c \end{cases}$$

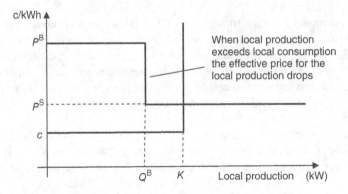

Figure 20.3 Demand curve and marginal cost curve for an embedded generator

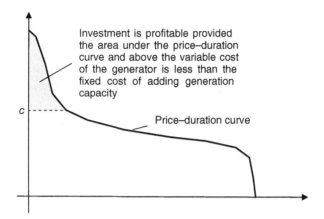

Figure 20.4 Review of investment principles from Chapter 9

Now let us consider the incentive for investment in embedded generation. Let us consider the case where there is always some local consumption (i.e. $Q^B > 0$). In this case, we can start by adding a small amount of local generation. For a sufficiently small local generation capacity, that local generation will always be less than the local consumption, so the effective price paid is the price for local consumption.

However, does the price paid for local consumption provide the right incentives for investment?

We know from Chapter 9 that the incentives for investment in generation depend on the shape of the price–duration curve. The customer has an incentive to invest in additional capacity of this generator as long as the area under the price–duration curve and above the variable cost is less than the fixed cost of adding capacity. This is illustrated in Figure 20.4.

Therefore, as long as the price paid by local consumption is exactly equal to the efficient local nodal price, the customer has the right incentives for investment in local embedded generation. (In addition, if the price paid by local consumption is exactly equal to the efficient local nodal price, there is no point in distinguishing between the price paid by local consumption and the price paid to local generation, so both can be equal to the local nodal price – and the customer has efficient incentives for investment in local generation).

However, as we noted earlier, around the world distribution networks are not efficiently priced. Customers connected to distribution networks typically face a price which is only loosely related to the local nodal price (if there is any connection at all).

Let us start by considering the case where the customer faces a simple time-averaged price – a fixed price for the consumption of electricity. Specifically, let us suppose that the customer faces a fixed price equal to the simple average of the underlying nodal spot price. In this case the price–duration curve is essentially flat, as illustrated in Figure 20.5.

What effect does the time-averaging of the price have on the incentives for investment? Let us consider a small investment of capacity K in embedded generation. Here, the installed capacity is small so that the local generation never exceeds the local consumption. If the customer faced a time-varying local nodal price P, the expected profit from the investment is as follows:

$$\pi^{\text{nodal}} = E(P - c \,|\, P \geq c)\Pr(P \geq c)K - fK$$

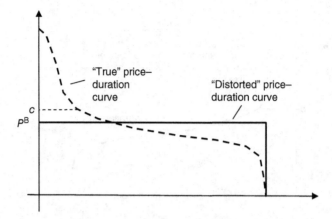

Figure 20.5 Price–duration curve with a flat time-averaged price

On the other hand, if the local time-averaged consumption tariff is P^B, the expected profit from the investment is as follows:

$$\pi^{\text{averaged}} = \left(P^B - c\right)I\left(P^B \geq c\right)K - fK$$

It is clear that for embedded generation with a variable cost above the time-averaged price, there is no reward from investment at all (it will never be efficient to operate such generation). On the other hand, for embedded generation with a variable cost below the time-averaged price, the embedded generation will operate at capacity all the time. This does not correspond to an efficient operational decision. But what about the incentives for investment?

It is possible to show that the time-averaged profit per unit of capacity is related to the efficient profit under nodal pricing as follows:

$$\frac{\pi^{\text{averaged}} - \pi^{\text{nodal}}}{K} = \left[c - E\left(P|P < c\right)\right]\Pr(P < c) + \left(P^B - E(P)\right)$$

The first term on the right-hand side is always positive and goes to zero as the variable cost of the generator becomes arbitrarily small. We can conclude that a customer facing a fixed time-averaged price equal to the average of the local nodal price ($P^B = E(P)$) will face too little incentive to invest in controllable embedded generation except in the special case where the variable cost of that generation is below every possible realisation of the nodal price (i.e. $\Pr(P < c) = 0$).

More generally, if the flat retail tariff is above the average of the local nodal price ($P^B > E(P)$), the customer can have too much incentive to invest in generation with a very low variable cost.

We can gain a little more insight into what the earlier expression is telling us with the aid of a graph. Figure 20.6 illustrates the expected profit per unit of capacity for generating types with different variable cost under nodal pricing and under time-averaged pricing. The dotted line illustrates the expected profit per unit of capacity under nodal pricing. In this example the average price $E(P)$ is chosen to be \$44/MWh and the time-averaged price P^B is \$50/MWh. As

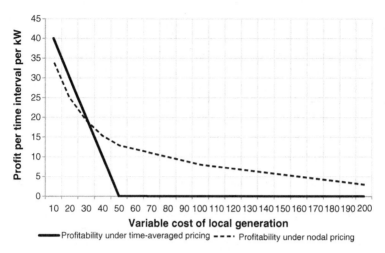

Figure 20.6 The expected profit from a local generation investment under nodal pricing and under time-averaged pricing

can be seen, the time-averaged charge results in too little incentive to invest in generation with a variable cost above around \$36/MWh. It also results in too large an incentive to invest in generation with a very low variable cost.

Finally, we can make the following observation: In order for the time-averaged tariff to yield the same revenue as the nodal price, it must satisfy the following condition:

$$P^B = \frac{E(PZ)}{E(Z)}$$

Here, Z is the net injection of the customer at each point in time (we are assuming that the local consumption always exceeds the local production, so this amount is negative). It seems likely that the local nodal price is correlated with the net withdrawal of the customer from the network. As a consequence, the time-averaged tariff must be above the simple average of the nodal prices $P^B > E(P)$. As a result, the customer is likely to have too much incentive to invest in local 'baseload' generation.

This analysis was carried out for the case of a very small amount of embedded generation so that the local production is always less than the local consumption. If the local production exceeds the local consumption, the customer is paid the lower 'feed-in-tariff' price for local production. This further reduces the incentive for investment.

Result: A customer considering an investment in a local generator facing a flat time-averaged retail tariff faces too little incentive to invest in generation with a high variable cost and, where the investment is nevertheless undertaken, has inefficient incentives for the operation of the generator (i.e. will choose to operate the generator when the underlying local nodal price is below the variable cost of the generator).

If the flat time-averaged retail tariff is equal to the average of the underlying nodal price, the customer will have too little incentive to invest in all generation types except the generation type with a variable cost below all possible realisations of the nodal price. Furthermore, if the flat time-averaged retail tariff is *above* the average of the underlying nodal price, as is likely if the times of high local nodal prices are correlated with the net withdrawal of the customer, the customer may have too much incentive to invest in the generation with the lowest variable cost.

20.4.1 Incentives for Investment in Intermittent Solar PV Embedded Generation

In recent years, there has been very substantial investment in rooftop solar PV. As noted earlier, much of this investment has been driven by generous government subsidies including, in particular, high feed-in-tariffs. However, for the most part, these high feed-in-tariffs have been unwound. Is there too much incentive to invest in solar PV, even in the absence of a feed-in-tariff?

Solar PV is not a controllable generation technology. Nevertheless, much of the previous analysis can still be applied. In particular, at least for small volumes of solar generation, we can assume that the solar generation is exclusively used to offset local consumption. In this case the relevant price received for the generation is the local consumption tariff (the 'buying price').

Let us suppose the local consumption tariff is P^B. Let us suppose that the production of a solar installation as a proportion of total capacity at a given point in time is ρ (a random variable) and the local nodal price at the same point in time is P (also a random variable). The variable cost of solar production is assumed to be zero. Let us assume that the local nodal price always exceeds the variable cost of local production (which is zero) so that it is always efficient to produce as much solar output as possible. If the customer installs a capacity K of solar generation, the expected profit under time-averaged pricing is as follows:

$$P^B E(\rho)K - fK$$

Similarly, the expected profit under nodal pricing is

$$E(\rho P)K - fK$$

Clearly, the customer will have an efficient incentive to invest in solar if and only if local consumption tariff is equal to a weighted-average nodal price, with the weights in the weighted average price determined by solar output:

$$P^B = \frac{E(\rho P)}{E(\rho)}$$

If the total solar production is very small and/or uncorrelated with the local nodal price, and if the local consumption tariff is equal to the average local nodal price ($P^B = E(P)$), the customer will have efficient incentives to invest in solar generation.

However, this is a special case. It is more likely that solar generation will be somewhat inversely correlated with the local nodal spot price (prices are likely to be lower at times of high solar production). Furthermore, as noted earlier, the time-averaged consumption tariff is likely to be above the average of the local nodal prices. We can conclude that customers will have too much incentive to invest in solar PV generation, even in the absence of high feed-in-tariffs.

20.4.2 Retail Tariff Structures and the Death Spiral

It has often been suggested that the structure of electricity distribution tariffs results in an over-incentive to invest in embedded generation. The argument is as follows:

Most small electricity consumers pay a retail tariff for electricity (a per kWh charge), which reflects both wholesale energy generation costs and network costs.

In contrast, it is observed that most network costs are fixed in the short run. By investing in local generation, customers are able to avoid the generation costs and the network charges. Yet, in practice, the network costs are fixed. Customers are over-rewarded for installing local generation.

Furthermore, depending on how network charges are set, there may be a self-reinforcing loop: As more customers install local generation, the demand for network services reduces, reducing the revenue received by the network business. In order for the network business to be able to cover its costs, it is argued that charges must be raised to the remaining customers. This may increase the incentive for customers to install local generation, further reducing demand, resulting in what is known as a downward spiral or *death spiral*.

It has been argued that one way to address this problem is to rebalance the network tariffs, raising the fixed charges and reducing the variable charges. In this way, it is argued that customers will not have an incentive to invest in local generation unless that local generation is cheaper than generation supplied over the network. This rebalancing could, in principle, be limited to those customers that install local generation.

But is this line of argument correct?

There is a grain of truth in this line of argument. However, the reasoning is incorrect.

From the analysis in previous chapters, we know that provided the customer faces the correct local marginal price for electricity (the local nodal price), the customer has the correct incentive to use and invest in local consumption and production assets. The local nodal price reflects both network congestion and generation costs. The local nodal price is also a per kWh charge. Those charges, as we have seen, give rise to a source of funds known as congestion rents, which can be used to fund network expansion. It is not, in itself, wrong to recover (some) network costs through per kWh charges.

If there were no congestion on the local (distribution) network, it is correct that the charges paid by customers should only reflect the generation costs and should not (ignoring losses) include a component designed to recover network costs.

However, it is not efficient to build a network that is never congested. At times of network congestion the local nodal price will be higher (perhaps much higher) than the generation costs. At these times, as we have noted, the network receives a stream of revenue that can be used to fund network expansion. To repeat the point we just made, it is not, in itself, incorrect to recover network costs through per kWh charges.

The problem with existing network tariff structures is not that network costs are (partly) recovered through usage charges. Rather, the primary problem with the existing tariff structures is the way those charges are structured. As the preceding sections have highlighted, if those charges were structured in such a way as to mimic locational marginal pricing, customers would have the correct incentive to invest in local generation.

Therefore, the primary problem with existing network tariff structures is the time-averaging of charges. As we have seen earlier, time-averaged charges will often result in *too little* incentive to invest in local generation with a high variable cost (at the same time, time-averaged charges result in too much incentive to invest in local consumption assets, such as air-conditioners, which tend to be used at times of network congestion).

At the same time, we observed that time-averaged charges will typically result in *too much* incentive to invest in local generation with a low variable cost, especially for generation that is likely to receive less, on average, for its output than the average local nodal price.

> *Result*: Some commentators have argued that the practice of recovering network charges partly through electricity consumption (kWh) tariffs results in too much incentive to invest in local generation, and have argued that network tariffs should be rebalanced, with higher fixed charges and lower variable charges. This conclusion is not quite right. It is efficient to recover at least some network costs through electricity consumption (kWh) tariffs – but those prices should reflect episodes of congestion on the network. The practice of time-averaging of small customer charges distorts incentives for investment – increasing the incentive for investment in low variable cost or 'baseload' embedded generation and reducing the incentive for investment in embedded generation that can respond at peak times. The rebalancing that is required is not between fixed and variable charges but between congested and noncongested times.

20.4.3 An Illustration of the Death Spiral

In the previous sections, we have emphasised that time-averaging of network charges can result in inefficient incentives for investment in embedded generation. However, the outcome may be even worse. The time-averaging of network charges may result in a situation where there is no combination of charges that allows the network to cover its costs.

This can be illustrated with a simple example. Let us suppose that we have a customer that faces 10 possible future demand states (labelled 1–10). The demand of the customer in each state and the probability of each state is set out in Table 20.1.

Table 20.1 Demand-duration data for the simple death spiral illustration

State	Probability	Load (MW)
1	0.01	2500
2	0.04	2000
3	0.05	1800
4	0.05	1700
5	0.2	1600
6	0.2	1500
7	0.2	1400
8	0.15	1100
9	0.06	800
10	0.04	500

Table 20.2   Cost data for available generation technologies in the simple death spiral illustration

Technology	Fixed Cost ($/MW/h)	Variable Cost ($/MWh)
A	$80	$1
B	$23	$100
C	$0	$300

Let us suppose that this customer has access to three different local generation technologies. Technology A is a 'baseload' technology with a moderately high fixed cost of $80/MW/h and a low variable cost of $1/MWh. Technology B is a 'mid-merit' technology with a fixed cost of $23/MW/h and a variable cost of $100/MWh. Technology C is a 'peaking' technology with zero fixed cost and a high variable cost of $300/MWh (this could be some form of demand response). All three generation technologies are assumed to be fully controllable (i.e. not intermittent). The key cost data is summarised in Table 20.2.

Let us suppose that the network operator can build a link to connect this customer to the broader network at a cost of $30/MW of capacity. In the broader network there is assumed to be an unlimited supply of generation with a variable cost of $50/MWh.

First, let us ask what is the socially efficient mix of generation and network? It turns out that the socially efficient least-cost mix of generation and network capacity in this example involves 1700 MW of network being built to this customer, and 800 MW of local high-cost generation (demand response). When demand exceeds 1700 MW, the local generation (technology C) makes up the remainder of the load. The total cost of meeting load (including the fixed cost of the network) is $127 900/h.

We could achieve this outcome with efficient nodal prices. In this case, the nodal prices would set a price of $300/MWh when demand is larger than 1700 MW and $50/MWh when demand is less than 1700 MW. The expected revenue received by the network operator (after paying for the $50/MWh generation) is $51 000/h which, since we are assuming constant returns to scale, is precisely equal to the cost of the network ($30/MW/h times 1700 MW).

Now let us suppose that we insist that prices be uniform across time. If we try to start with electricity prices fixed at, say, $50/MWh we find that it is no longer privately worthwhile for the customer to install the high-cost generation. The customer can obtain all the supply he/she wants at $50/MWh, so there is no incentive to install demand response. However, since the customer has no incentive to curtail demand, the outcome is that all 2500 MW of the load must be met through network generation and network capacity. The network is built 47% larger than is efficient (and the network costs 47% higher than is efficient).

However, this outcome is inconsistent with allowing the network cost recovery. If the network operator can only charge a flat (time-averaged) fee, that charge must be above the cost of generation elsewhere in the network so as to provide a contribution to the network. It turns out in this example that this means that we have to charge $103.08/MWh ($50 generation cost plus $53.08 network cost) in order for the network to break even, assuming that consumption patterns remain the same when we increase the price.

However, consumption patterns do not stay the same. When we raise the electricity price to $103.08/MWh, it becomes profitable for the customer to invest in technology A. In fact, it is now profitable for the customer to install 1100 MW of technology A. Since this is a 'baseload' technology, the customer runs this technology almost all the time. This reduces the usage of the network, so that average demand for electricity from the network drops from 1388 MW in the socially efficient mix to 355 MW. Now, in order for the network to break even (assuming that consumption patterns remain the same), the flat time-averaged price for electricity must be increased to $168.31/MWh.

However, this higher electricity price further reinforces the value of the local generation. Now it is profitable for the customer to invest in 1400 MW of technology A and 200 MW of technology B. This further reduces the demand for electricity (to just 40 MW average), forcing the network to further raise the price. Now the flat time-averaged price for electricity must be increased to $725/MWh.

At this price, it no longer makes sense for the customer to take electricity from the network at all. Now it is profitable for the customer to invest in 1400 MW of technology A, 300 MW of technology B and 800 MW of technology C. The customer then meets all of his/her own demand. The total cost of meeting demand is now $138 183/h (compared to the socially efficient outcome of $127 900/h). The network is now completely bypassed.

This example shows how the inefficient structure of charges can lead to inefficient customer investment decisions which can, in turn, lead to an inefficient reduction in demand for network services. This reduction in demand can lead to further price increases and the downward spiral as was saw in the example presented here.

20.5 Retail Tariff Structures

As noted earlier, in practice, no liberalised electricity market routinely exposes small customers (or their retailers) to the local nodal price. (In fact, as we have seen, we are not aware of instances where this price is even computed down into the distribution network).

It is important to emphasise that extending nodal pricing down into the distribution networks does not mean that customers will necessarily face a time-varying price. Retail customers have a varying degree of tolerance of exposure to a time-varying wholesale price. It is the job of the retailer to interface between the wholesale market conditions and the retail contract that customer's desire.

Many liberalised electricity markets already have a time-varying price for electricity at the wholesale level. Retailers routinely interface between this price and the simple tariffs that most smaller customers currently face. Extension of nodal pricing down into the distribution network would extend the current risk-management and pricing task of the retailers. However, customers would not necessarily be exposed to a time-varying wholesale price – that is, unless they chose to do so.

It might be asked: Why bother with nodal pricing for smaller customers if most of them choose to remain on time-averaged tariffs? The answer is that nodal pricing for smaller customers creates an environment in which retailers have an incentive to innovate to develop tariff structures that induce efficient behaviours by customers while simultaneously insulating them from the risks that they fear. Nodal pricing creates an environment in which retailers, in competing to meet the needs of customers, pursue ends consistent with overall economic efficiency.

Result: Nodal pricing of smaller customers does not imply that smaller customers will be exposed to a time-varying tariff. It is the task of retailers to interface between wholesale market conditions and the tariffs that customers desire. Nodal pricing of smaller customers exists to create an environment in which competition drives tariffs toward more efficient arrangements.

20.5.1 Retail Tariff Debates

There is a very long way to go from the simple time-averaged tariffs that most retail customers face today to some form of nodal pricing (as mediated by the retailer). Policy discussions have focused on several different types of tariffs for retail customers in the transition to fully efficient tariffs. It is worth discussing the pros and cons of these tariffs here.

Time-of-use (TOU) tariffs vary the prices faced by smaller customers in a fixed way over the course of the day or week. Time-of-use tariffs are an improvement on simple flat time-averaged tariffs, but still reflect a form of time-averaging (the difference is that the time-averaging is over shorter periods). Time-of-use tariffs are not dynamic in the sense that they do not and cannot reflect wholesale market conditions at a specific point in time. They therefore do not result in efficient usage decisions and they do not result in efficient investment decisions. The problem is that local network congestion does not occur precisely at certain times of the day in a predictable way. As a result, time-of-use tariffs will tend to create too strong incentives to curtail consumption (and increase local production) at times that are classified as peak times but for which the local network is not congested at those times. Similarly, time-of-use tariffs create too little incentive to curtail local consumption (and increase local production) at times when the local network is congested, especially at times that are not classified as peak times. In short, time-of-use tariffs are an improvement on tariffs that do not vary with time at all, but they are not sufficient to achieve an efficient outcome.

Critical-peak pricing (CPP) is a variant on time-of-use tariffs that incorporates a dynamic component. Under critical peak pricing, the system operator forecasts when network congestion is likely to arise and then charges a particularly high price for those periods. This is not true dynamic pricing – the episode of congestion must be forecasted in advance, the customer must be given notice (usually 24 h), and the price level during the period of congestion is predetermined. In addition, many implementations of CPP limit the number of times the system operator can announce a critical peak during a season.

Critical peak pricing is clearly an improvement on time-of-use pricing. The high price periods are much more likely to coincide with actual times of network congestion. However, CPP still has its drawbacks. The episodes of congestion must be forecast. CPP is therefore of no use for managing congestion that is not forecasted (e.g. due to an unplanned outage). In addition, there are limits on the number of such periods that can be called per year.

Another common problem in the implementation of CPP is that there is typically no attempt to make risk-management instruments available to allow retailers to hedge the risk of a CPP episode. An alternative approach would be to implement nodal pricing at the wholesale level and for the network business to offer a risk-management product that would allow retailers to insulate their customers from all but a fixed number of high-price periods per year (with 24 h notice).

Under *demand or capacity-based tariffs* the customer faces two charges: A fixed charge that is proportional to the maximum rate of consumption of the customer (typically measured in kW) and a variable charge (per kWh) per unit of consumption.

Capacity-based tariffs differ in how the maximum rate of consumption of the customer is determined. For example, the maximum rate of consumption could be simply an amount that is determined in advance through negotiation with the customer. Alternatively, the maximum rate of consumption could be determined on the basis of experience – for example, the maximum rate of consumption over a period of 12 months, or the maximum rate of consumption between the hours of 2:00 p.m. and 9:00 p.m.

Capacity-based tariffs coincide with nodal prices but only under certain strict conditions. Specifically, it must be that the peak net consumption of all customers occurs at the same time and this coincides with the period of maximum flows on the network. These conditions are highly unlikely to arise in practice.

For this reason, capacity-based tariffs give rise to inefficient operational and investment decisions by customers. The basic problem, as before, is that times of maximum consumption do not necessarily correspond to times of network congestion.

At times when the customer is approaching its maximum rate of consumption, it faces a very high effective price for additional consumption, even if the network is not congested at those times. It therefore faces too strong an incentive to lower consumption at those times. For example, a household that purchases a large air-conditioning unit and only uses it at times when the network is not congested may have a very high maximum rate of consumption but may not have any impact on consumption at times of local network congestion.

Similarly, a customer that is not consuming at its maximum rate of consumption faces little disincentive to consume even if the local network happens to be congested at such times.

The problem with demand or capacity-based tariffs is that the period of time over which the maximum rate of consumption is calculated must be specified in advance. In reality, a customer should be charged for his/her consumption during times of network congestion. Times of network congestion usually cannot be forecast with precision in advance.

Result: Contemporary debates regarding tariff choices for smaller customers focus on a choice between time-of-use tariffs, critical-peak-pricing and demand or capacity-based tariffs. None of these approaches achieves fully efficient operational and investment outcomes. It seems preferable to us to focus on implementing nodal pricing at the wholesale level combined with making available a range of hedging instruments to allow retailers to provide the arrangements that their customers desire.

20.6   Declining Demand for Network Services and Increasing Returns to Scale

Investment in embedded generation reduces demand for network services. We have noted earlier that existing tariff structures may be over-incentivising certain forms of investment in embedded generation. However, even if those tariff structures were made more efficient, there remains a chance that technological change will alter the long-term demand for network services. In particular, a technology might come along that makes it attractive for at least some customers to disconnect from the electricity network entirely. Does such a technological change give rise to issues for network pricing and regulation?

It turns out that a long-term decline in demand does have important implications in the presence of increasing returns to scale. In the presence of increasing returns to scale, (by definition) the sum of the incremental costs of each of the services provided (or the customers served) falls short of the total cost. Put another way, in the presence of increasing returns to scale, there are some *joint or common costs* that must be recovered, but that cannot be easily attributed to specific services. As a consequence, in the presence of increasing returns to scale, the average charge to a customer or service must be above the incremental cost of serving that customer or providing that service for at least some customers.

The fact that the revenue for a service or for a customer exceeds the incremental cost for that customer may give rise to inefficient incentives. Specifically, a customer may have an incentive switch to taking supply from an alternative technology even when it is efficient to take service from the network.

To see this, let us suppose that the incremental cost of serving a customer is $10, but the average cost is $18 (the difference is the contribution to the joint and common costs). The tariff charged to the customer is the average cost ($18) so that the network business breaks even. Now suppose that a new technology allows that customer to be served for $15. If the customer takes up the alternative technology, the customer is better off by $3, but the existing network loses $18 in revenue and only saves $10 in costs, so the network is worse off by $8. Overall, the social cost of providing services has increased by $5, so the overall economy is worse off by $5.

In this example, the total social cost increased when the customer chose the alternative technology. However, this is not always the case. If the new technology were cheap enough (say, $9), both the customer and the broader society would be better off if the customer chose the alternative technology.

How can we induce the customer to make the right social choice (that is, to choose the alternative technology if and only if it is socially beneficial)?

Let us suppose that the current charge for the service is R, the cost of the alternative service is C, and the incremental cost saving to the network if the customer chooses the alternative service is IC. Let us suppose that we can impose a penalty P on the customer if it chooses the alternative service, and/or we can offer a discount D, if it chooses the network service. The customer will choose the alternative service if and only if the cost of the alternative service (after paying any penalty) is less than the retail price for the current service (less any discount):

$$C + P < R - D$$

We would like the customer to choose the alternative service if and only if doing so involves a lower overall social cost, that is, if and only if

$$C < \text{IC}$$

We can therefore induce the customer to make an efficient decision if we set the size of the penalty or discount equal to the gap between the current charges and the incremental cost.

$$P + D = R - \text{IC}$$

In the case where we have a potential new customer that wants to connect to the network, we could and should allow that customer a discount on his/her tariffs when that potential customer has a bypass opportunity. In this case, there are no financial implications for existing customers from offering a discount, and doing so improves economic efficiency as we have noted. The formula shows that the discounted price charged to this customer should always be greater than or equal to the incremental cost of serving the customer.

But what about the case of existing customers? In this case the choice between the 'penalty' and the 'discount' affects the sharing of risks between the switching customer on the one hand, and the network and remaining customers on the other.

On the one hand, we could allow switching customers to leave the network at will. In this case the existing charge to the switching customer can be discounted down to (but not below) the incremental cost of providing the new service ($R - D = IC$). This ensures the switching customer has the right incentives to switch to the alternative technology or not. But whether the customer switches or not, either the remaining customers and/or the network will be left worse off.

On the other hand, we could adopt the rule that the switching customer remains liable for the penalty in the event it chooses to switch. In this case, by setting the penalty at the customer's existing contribution to the joint and common costs ($P = R - IC$), the switching customer will have the right incentive to switch to the alternative technology if and only if doing so lowers the overall social cost. Furthermore, whether or not the customer switches, the remaining customers and the network are not financially disadvantaged.

This can be made clearer using the earlier example. Let us suppose that there are five customers of the network. The incremental cost of serving each customer is $10, but the total cost is $90, so the joint and common cost is $40. Each customer is currently charged the same fee of $18. Due to a technological development, one customer has the opportunity to acquire the same services at a cost of either $15 or $5.

Let us suppose first that the customer has the right to switch. If the alternative technology costs $15, the customer could be offered a discount on its charges. This reduces its charge from $18 to $15. The customer is prevented from leaving the network, but the network revenue drops from $90 to $87. If the alternative technology costs $5, there is no discount (above incremental cost), which will prevent the customer from leaving, so the customer leaves and the network revenue drops to $72. The network can save on the cost of providing services to that customer, so the network costs drop from $90 to $80. The network is still left with a deficit of $8, which must be borne either by the network itself or by the remaining customers.

Now let us suppose that the customer cannot leave the network without paying a fee of $8. In this case, if the technology costs $15, the customer will not leave (since if it does so it will pay $15 + $8 = $23, whereas if it stays it pays only $18). The network revenue is left intact. Similarly, in the case where the technology costs $5, the customer will leave (since by leaving it pays $5 + $8 = $13, whereas if it stays, it pays $18). The network loses $18 in revenue but can save $10 in costs, so again the network revenue balance is left intact.

How should we choose between these two approaches? In our view, the charges for existing customers should not change as a result of another customer's decision. The basis for this is as follows: Existing customers have made a sunk investment in reliance on the network service. If the charges go up when other customers disconnect, the existing customers cannot be sure that their charges will not go up in the future for reasons outside their control. For this reason, they may be unwilling to invest or will distort their investments toward assets that are less reliant on the network service. This is an undesirable outcome. For this reason, it is preferable to insulate existing customers from other customer's switching decisions.

This leaves two possibilities: either the impact of switching should be borne by the network or the switching customer should remain liable for its contribution to the joint and common costs.

There are no clear grounds for choosing between these two possibilities. In fact, these issues should be resolved in the regulatory contract at the time both parties make their sunk investments. The regulatory contract should set out the obligations on the parties in the event of a substantial reduction in services or a disconnection from the network. Is the network responsible for the loss of the contribution to joint and common costs? This is a possible risk allocation, but the network would presumably be entitled to a degree of compensation for this risk. Or, are disconnecting customers responsible to make good their contribution to the joint and common costs? This liability could be in the form of on-going payments or in the form of one-off termination fees. Again, this is a possible risk allocation, but these obligations should be made clear in advance.

We should make the risk allocation clear in advance in the regulatory contract. Either customers have the right to disconnect with no on-going obligations to the electricity network, or they do not.

Result: In the presence of increasing returns to scale, the revenue received from a customer or service may exceed the incremental costs of providing that service or serving that customer. In this context, the customer may have too strong an incentive to choose a substitute technology (such as disconnection from the network). Furthermore, in the event the customer chooses a substitute technology, there can be financial implications for the network and the remaining customers of the network, due to the loss of the disconnecting customer's contribution to the joint and common cost of the network. The regulatory contract should make clear who bears this risk – either the network (facing the risk of stranded assets) or the disconnecting customer.

20.7 Summary

The vast majority of electricity customers are connected to lower-voltage distribution networks. These customers have their own production and consumption assets. Achieving efficient usage decisions of these production and consumption assets requires efficient price signals – the efficient nodal prices that we have studied throughout this text.

In practice, nodal pricing of distribution networks is still in its infancy. Most small customers face tariffs that are highly averaged – across time and across locations. These tariffs lead to inefficient operational and investment decisions. Customers have too little incentive to curtail consumption at times of local network congestion and too little incentive to draw on local production at such times. Customers have too much incentive to invest in consumption assets that are active at peak times and too little incentive to invest in local production assets that can be used at such times. As a consequence, if reliability is to be maintained, the network must be over-built.

Historically, the lack of efficient pricing at the distribution network level was arguably not too much of a problem. Customers did not have access to devices that could control their consumption or production in real time. However, the computer and communications revolutions have lowered the cost of such devices. It is likely that, in the future, many household appliances will be capable of responding to wholesale price signals. In addition, increasing concerns about climate change have increased investment in small-scale local generation, such as solar PV, and may drive further take-up of electric vehicles. Small customers are no longer merely passive consumers of electric power but have the potential to act dynamically, varying their net production or consumption in response to wholesale market conditions.

There are practical issues to be resolved in the computation of nodal prices for distribution networks. The computation of nodal prices requires information about supply and demand conditions at each node and the details of the physical network capability. At present, nodal prices are typically calculated for the high-voltage (transmission) networks. The representation of the high-voltage network in the dispatch process requires a very substantial amount of data. Extending this representation to the low-voltage networks would require at least an order of magnitude more information. It seems preferable to decentralise the task of computing the optimal dispatch. In principle, this might be achieved through a hierarchy of networks, with each network layer responsible for representing its own local network accurately. In principle, information about supply and demand conditions in each subnetwork could be communicated to the high-level network operator. If this approach proves feasible, it may facilitate the computation of nodal prices across the national, regional and local networks.

Local network operators often charge a time-averaged tariff for electricity consumption and also offer to pay a time-averaged tariff for local production. Differences in these prices give rise to incentives for customers with embedded generation to misrepresent the level of local production and consumption. Specifically, if the price for local production is above the price for local consumption, customers have an incentive to make both appear as high as possible. If the price for local consumption is above the price for local production, customers have an incentive to make both appear as low as possible through net-metering and by combining local loads.

A key question is the impact of time-averaged network charges on incentives for investment in local generation. From the earlier theory, we know that a customer faces the right incentives for investment in local generation when he/she faces the local nodal price. A time-averaged price reduces the incentive to invest in high-variable cost generation. In addition, if the time-averaged price is above the average of the local nodal price, the customer has too strong an incentive to invest in very low variable cost (baseload) local generation.

This analysis extends to the case of intermittent local generation, such as solar PV. Under nodal pricing, the average price received by a solar generator would likely be lower than the average local nodal price (since times of high solar output are likely correlated with times of lower local prices). This further reinforces the incentive to invest in solar PV generation. The time-averaged network charge induces over-investment in solar generation.

The over-incentive to invest in local generation may further reduce demand for network services, requiring further increases in prices and inducing further local generation investment, in a process known as a death spiral. In the presence of increasing returns to scale, the disconnection of customers results in the loss of the customers' contribution to the joint and common costs of the network. The regulatory contract should make clear who bears these risks – either the network (facing the risk of stranded assets) or the disconnecting customers (who may be made to pay some sort of termination fee).

Questions

20.1 What are the major pros and cons of nodal pricing at the distribution network level?

20.2 Does the presence of time-averaged distribution network charges result in inefficient decisions to invest in local (embedded) generation? Do small customers have too much incentive to invest in solar PV generation?

20.3 Is it possible that time-averaging of distribution network charges may threaten the viability of the distribution network itself? Why or why not?

20.4 What are the pros and cons of time-of-use pricing, critical-peak pricing and capacity-based charging?

20.5 Suppose that a new technology is developed that allows customers to obtain all their electricity needs without connection to the electricity grid. Should disconnection be allowed? Under what circumstances?

References

Baldick, R. (2009) Single Clearing Price in Electricity Markets, available at: http://www.competecoalition.com/files/Baldick%20study.pdf.

Baumol, William J. and Blinder, Alan S. (2008) *Economics: Principles and Policy*, South-Western Cengage Learning; 11th edition, 8 July, 2008.

Biggar, D.R. and Hesamzadeh, M.R. (November 2011) Towards a theory of optimal dispatch in the short run. In Innovative Smart Grid Technologies Asia (ISGT), 2011 IEEE PES (pp. 1–7), IEEE.

Bohn, R.E., Caramanis, M.C. and Schweppe, F.C. (1984) Optimal pricing in electricity networks over space and time. *Rand Journal of Economics*, **15** (3), 360–376.

Borenstein, S. and Bushnell, J. (2000) Electricity restructuring: Deregulation or Reregulation? Power Working Paper, PWP-074, University of California Energy Institute.

Borenstein, S., Bushnell, J. and Stoft, S. (2000) The competitive effects of transmission capacity in a deregulated electricity industry. *Rand Journal of Economics*, **31** (2), 294–325.

Borenstein, S., Bushnell, J. and Wolak, F. (2000) Diagnosing market power in California's restructured wholesale electricity market, Working Paper 7868. National Bureau of Economic Research.

Brennan, T.J. (2003) Mismeasuring electricity market power, *Regulation*, **26** (1), 60–65.

Brennan, T. (2006) Preventing monopoly or discouraging competition? The perils of price-cost tests for market power in electricity. *Electric Choices: Deregulation and the Future of Electric Power*, A. Kleit (ed.), Rowman and Littlefield. Available at SSRN: http://ssrn.com/abstract=868525

Brunekreeft, G., Neuhoff, K. and Newbery, D. (2005) Electricity transmission: an overview of the current debate. *Utilities Policy*, **13** (2), 73–93.

Bushnell, James B., Mansur, Erin T., and Saravia, C. (2008) Vertical arrangements, market structure, and competition: an analysis of restructured US electricity markets. *American Economic Review*, **98** (1), 237–266.

Chao, Hung-po and Wilson, R. (1999) *Design of Wholesale Electricity Markets*, Electric Power Research Institute.

Church, Jeffrey R. and Ware, R. (2000) *Industrial Organization: A Strategic Approach*, McGrawHill.

Counsell, K. and Evans, L. (2004) Day-Ahead Electricity Markets: Is There a Place for a Day-Ahead Market in the NZEM? New Zealand Institute for the Study of Competition and Regulation INC. (NZISCR).

Ding, F. and Fuller, J.D. (2005) Nodal, uniform, or zonal pricing: distribution of economic surplus. *IEEE Transactions on Power Systems*, **20** (2), 875–882.

Gilbert, R., Neuhoff, K. and Newbery, D. (2004) Allocating transmission to mitigate market power in electricity networks. *RAND Journal of Economics*, **35** (4), 691–709.

Glachant, J.M. and Finon, D. (2003) *Competition in European Electricity Markets – A Cross-country Comparison*, Edward Elgar, Northhampton, Massachusetts.

Grainger, J.J. (2003) *Power System Analysis*, Tata McGraw-Hill Education.

Harris, C. (2011) *Electricity Markets: Pricing, Structures and Economics*, Wiley.

Hesamzadeh, M.R., Biggar, D.R. (2012) Computation of extremal-nash equilibria in a wholesale power market using a single-stage MILP, Power Engineering Letters, *IEEE Transactions on Power Systems*, **27** (3), 1706–1707, August 2012.

The Economics of Electricity Markets, First Edition. Darryl R. Biggar and Mohammad Reza Hesamzadeh.
© 2014 John Wiley & Sons, Ltd. Published 2014 by John Wiley & Sons, Ltd.

Hesamzadeh, M.R., Galland, O. and Biggar, D.R. (2014) Short-run economic dispatch with mathematical modelling of the adjustment cost. *International Journal of Electrical Power and Energy Systems*, **58**, 9–18, June 2014.

Hesamzadeh, M.R., Biggar, D.R. and Hosseinzadeh, N. (2011) The TC-PSI indicator for forecasting the potential for market power in electricity networks. *Energy Policy*, **39** (10), 5988–5998.

Hogan, William W. (1992) Contract networks for electric power transmission. *Journal of Regulatory Economics*, **4** (3), 211–242.

Hogan, W., Harvey, S. and Schatzki, T. (2004) A Hazard Rate Analysis of Mirant's Generating Plant Outages in California, Center for Business and Government, Harvard University.

Ilic, M.D., Galiana, F. and Fink, L. (eds) (1998) *Power Systems Restructuring: Engineering and Economics*, Springer.

Irastorza, V. and Fraser, H. (2002) Are ITP-run day-ahead markets necessary? *The Electricity Journal*, **15** (9), 25–33.

Joskow, P. and Kahn, E. (2002) A quantitative analysis of pricing behaviour in California's wholesale electricity market during summer 2000. *The Energy Journal*, **23** (4), 1–35.

Joskow, Paul L. and Tirole, J. (2000) Transmission rights and market power on electric power networks. *RAND Journal of Economics*, **31** (3), 450–487, Autumn 2000.

Luenberger, D.G. (1995) *Microeconomic Theory*, McGraw-Hill, New York.

Mansur, E. (2007) Measuring welfare in restructured electricity markets. *Review of Economics and Statistics*, **90** (2), 369–386, May 2008.

Murray, B. (1998) *Electricity Markets: Investment, Performance, and Analysis*, Wiley.

Murray, B. (2009) *Power Markets and Economics: Energy Costs, Trading, Emissions*, Wiley.

Padhy, N.P. (2004) Unit commitment – a bibliographical survey. *IEEE Transactions on Power Systems*, **19** (2), 1196–1205.

Productivity Commission (2002) Price Regulation of Airport Services, Report No. 19, 23 January 2002.

Raniga, J. and Rayudu, R. (1999) *Stretching Transmission Line Capabilities – A Transpower Investigation*, (published by the) Institute of Professional Engineers of New Zealand.

Rassenti, S., Smith, V.L. and Bluffing, R.L. (1982) A combinatorial auction mechanism for airport slot allocation. *Bell Journal of Economics*, **13** (2), 402–417.

Rosellón, J. and Kristiansen, T. (2013) *Financial Transmission Rights*, Springer.

Saadat, H. (1999) *Power System Analysis*, WCB/McGraw-Hill.

Sauma, E.E. and Oren, S.S. (2006) Proactive planning and valuation of transmission investments in restructured electricity markets. *Journal of Regulatory Economics*, **30** (3), 261–290.

Sauma, E.E. and Oren, S.S. (2007) Economic criteria for planning transmission investment in restructured electricity markets. *IEEE Transactions on Power Systems*, **22** (4), 1394–1405.

Schweppe, F.C., Caramanis, M.C., Tabors, R.E. and Bohn, R.E. (1988) *Spot Pricing of Elecitricity*, Kluwer Academic Publishers.

Stoft, S. (2002) *Power System Economics: Designing Markets for Electricity*, Wiley-IEEE Press.

Stevenson, W.D. (1982) *Elements of Power System Analysis*, McGraw-Hill.

Twomey, P., Green, R., Neuhoff, K. and Newberry, D. (2005) A Review of the Monitoring of Market Power: The Possible Roles of TSOs in Monitoring for Market Power Issues in Congested Transmission Systems, Working Papers 0502, Massachusetts Institute of Technology, Center for Energy and Environmental Policy Research.

Twomey, P., Green, R., Neuhoff, K. and Newberry, D. (2005) A review of the monitoring of market power: the possible roles of TSOS in monitoring for market power issues in congested transmission systems. *Journal of Energy Literature*, **11** (2), 3–54.

Ventosa, M., Baıllo, A., Ramos, A. and Rivier, M. (2005) Electricity market modeling trends. *Energy Policy*, **33** (7), 897–913.

Wolak, F. (2005) Managing Unilateral Market Power in Electricity, *World Bank Policy Research, Working Paper 3691*, September 2005.

Wood, A.J. and Wollenberg, B.F. (2012) *Power Generation, Operation, and Control*, John Wiley & Sons.

Wu, F., Varaiya, P., Spiller, P. and Oren, S. (1996) Folk theorems on transmission access: proofs and counterexamples. *Journal of Regulatory Economics*, **10** (1), 5–23.

Index

The Economics of Electricity Markets, First Edition. Darryl R. Biggar and Mohammad Reza Hesamzadeh.
© 2014 John Wiley & Sons, Ltd. Published 2014 by John Wiley & Sons, Ltd.